Philosophy of Mathematics
and Natural Science

Philosophy
of Mathematics
and
Natural Science

BY

HERMANN WEYL
REVISED AND AUGMENTED
ENGLISH EDITION
BASED ON A TRANSLATION BY
OLAF HELMER

PRINCETON
PRINCETON UNIVERSITY PRESS
1949

Preface

Home is where one starts from.   As we grow older
The world becomes stranger, the pattern more complicated
Of dead and living.
  T. S. ELIOT, *Four Quartets*, East Coker, V.

A SCIENTIST who writes on philosophy faces conflicts of conscience from which he will seldom extricate himself whole and unscathed; the open horizon and depth of philosophical thoughts are not easily reconciled with that objective clarity and determinacy for which he has been trained in the school of science.

The main part of this book is a translation of the article, "Philosophie der Mathematik und Naturwissenschaft," that I contributed to R. Oldenbourg's *Handbuch der Philosophie* in 1926.   Writing it, I was bound by the general plan of the *Handbuch*, as formulated in broad outlines by the editors, that laid equal stress on both the systematic and historical aspects of philosophy.   I was also bound, though less consciously, by the German literary and philosophical tradition in which I had grown up, and by the limited circle of problems that had come to life for me in my own mental development.

Under the heading "Naturwissenschaft" my *Handbuch* article dealt almost exclusively with physics.   It is the only branch of the natural sciences with which I am familiar through my own work. There were additional reasons why biology was dismissed with a few general observations: the space allotted me was more than exhausted, and I could rely on the following article, "Metaphysik der Natur" by the biologist and philosopher Hans Driesch, to fill the gap.

Twenty odd years have since passed, a long and eventful time in the history of science.   But when (not of my own initiative) the plan arose to have the book translated into English I gave my consent, fully aware though I was of the accidental circumstances of its birth and the wrinkles of old age in its face.   For it seemed to me that its message of the interpenetration of scientific and philosophical thought is today as timely as ever.   But the events of the last two decades could not be ignored altogether.   For more than one reason the alternative of re-writing the book myself in English was out of the question; how could I hope to recapture the faith and spirit of that epoch of my life when I first composed it — after due literary preparations dashing off the manuscript in a few weeks? Thus a different course had to be followed.

In spite of numerous alterations in detail, I mention especially Sections 13–15 and the concluding Section 23, the substance of the old text has been preserved, the outlook still being that of a philosophically-minded mathematician at the time when the theory of relativity had reached completion and the new quantum mechanics was just about to rise. But the references are brought up to date and six essays have been appended for which the development of mathematics and physics in the intervening years, as well as biology, have provided the raw material. This arrangement, objectionable from the standpoint of esthetic unity, has a certain stimulating value. The appendices are more systematic-scientific and less historico-philosophical in character than the main text. With the years I have grown more hesitant about the metaphysical implications of science; "as we grow older, the world becomes stranger, the pattern more complicated." And yet science would perish without a supporting transcendental faith in truth and reality, and without the continuous interplay between its facts and constructions on the one hand and the imagery of ideas on the other.

One of the principal tasks of this book should be to serve as a critical guide to the literature listed in the references.

[Sections of historical and supplementary interest not necessary to the main course of development of the book, set off in the German edition by small print, are indicated in this volume by opening and closing brackets, such as these.]

Dr. Olaf Helmer, versed both in mathematics and philosophical logic, translated the whole *Handbuch* article, with the exception of sections 16 and 17 which were done by my son, Dr. Joachim Weyl. His and Dr. Helmer's manuscripts have been revised by the author. Unless an excessive amount of care and labor is bestowed on it, the translation of a work that depends to some degree on the suggestive power of language — and the communication of philosophical thoughts does — or that has any literary qualities is apt to be a compromise. I am afraid this book is no exception. But I can at least vouch for the absence of any gross errors or misunderstandings; that is more than can be said about the majority of translations.

HERMANN WEYL

Princeton, New Jersey
December 1947

BIBLIOGRAPHICAL NOTE

*Concerning editions and translations that have been used
for quotations*

R. DESCARTES, *Oeuvres*, ed. Victor Cousin, Paris 1824. The French versions of the *Meditationes de prima philosophia* = *Méditations* [*métaphysiques*] *touchant la première philosophie*, and of *Principia philosophiae* = *Les principes de la philosophie*, are contained in Vols. I and III respectively.

The monumental Swiss edition of L. EULER's *Opera omnia* is still far from complete and does not yet include the two works cited here (*Theoria motus*, and *Anleitungzur Naturlehre*).

G. GALILEI, *Opere*, Edizione nazionale, Florence 1890–1909, reprinted 1929—. "*Dialogo*" = "Dialogo sopra i dui massimi sistemi del mondo" is in Vol. VII; "*Discorsi*" = "Discorsi e dimostrazioni matematiche intorno à due nuoue scienze" in Vol. VIII, "Il saggiatore" in Vol. VI.

DAVID HUME's *Treatise of Human Nature* and JOHN LOCKE's *Enquiry concerning Human Understanding* are quoted by 'chapter and verse,' which makes the quotations independent of any special edition.

IMMANUEL KANT's *Critique of Pure Reason*, transl. by F. Max Müller, 2nd ed., New York 1905 [German original: *Kritik der reinen Vernunft*, first ed. 1781, second ed. 1787].

G. W. LEIBNIZ, *Mathematische Schriften*, ed. Gerhardt, Berlin 1849 seq., *Philosophische Schriften*, ed. Gerhardt, Berlin 1875 seq.

LEIBNIZ's letters to S. Clarke form part of the controversy Leibniz — Clarke that is to be found in G. W. LEIBNIZ, *Philosophische Schriften*, ed. Gerhardt, VII, pp. 352–440.

SIR ISAAC NEWTON's *Mathematical Principles of Natural Philosophy and his System of the World*, ed. F. Cajori, Berkeley, California, 1934, 2nd print 1946. [Original in Latin: *Philosophiae naturalis principia mathematica*, first ed. London 1687, second ed. 1713, third ed. 1726. The above translation is based on an old translation made after the third edition by Andrew Motte in 1729.]

NEWTON's *Opticks* was written in English. The 4th edition (London 1730) has been reprinted with an introduction by E. T. Whittaker: London 1931.

Contents

Part One. Mathematics

THE two parts of this book are intended to be a report on some of the more important philosophical results and viewpoints which have emerged primarily from research within the fields of mathematics and the exact empirical sciences. I shall point out connections with the great philosophical systems of the past wherever I have been aware of them. Illustrative examples will be chosen as simple as possible. In principle, however, knowledge of the sciences themselves must be upheld as a pre-requisite for anyone engaging in the philosophy of science.

The method of presenting the foundations of mathematics will lead from the surface into the depth; consideration of the more formal aspects will precede the study of problems connected with the infinite. Though these latter problems have stirred the imagination of all ages, their careful formal preparation and stringent treatment are recent achievements. Among the heroes of philosophy it was Leibniz above all who possessed a keen eye for the essential in mathematics, and mathematics constitutes an organic and significant component of his philosophical system.

CHAPTER I

Mathematical Logic, Axiomatics

To the Greeks we owe the insight that the structure of space, which manifests itself in the relations between spatial configurations and their mutual lawful dependences, is something entirely rational. Whereas in examining a real object we have to rely continually on our sense perception in order to bring to light ever new features, capable of description in concepts of vague extent only, the structure of space can be exhaustively characterized with the help of a few exact concepts and in a few statements, the axioms, in such a manner that all geometrical concepts can be defined in terms of those basic concepts and every true geometrical statement follows as a logical consequence from the axioms. Thereby geometry has become the prototype of a *deductive science*. And in view of this its character, mathematics is eminently interested in the methods by which concepts are *defined* in terms of others and statements are *inferred* from others. (Aristotelian logic, too, was essentially a product of abstraction from mathematics.) What is more, it does not seem possible to lay the foundations of mathematics itself without first giving a complete account of these methods.

1. RELATIONS AND THEIR COMBINATION, STRUCTURE OF PROPOSITIONS

In Euclidean geometry we are concerned with three categories of objects, *points*, *lines*, and *planes*, which are not defined but assumed to be intuitively given, and with the basic relations of *incidence* (a point lies on a line, a line lies in a plane, a point lies in a plane), *betweenness* (a point z lies between the points x and y), and *congruence* (congruence of line segments and of angles). Analogously, in the domain of natural numbers 1, 2, 3, . . . we have a single basic relation in terms of which all others are definable, namely that between a number n and the number n' immediately following upon n. Again, the kinship relations among people furnish an excellent illustration of the general theory of relations. In this case there are two basic categories, *males* and *females*, and two basic relations, *child* (x is child of y) and *spouse* (x is married to y).

The propositional scheme of a *relation*, e.g. 'x follows upon y,' contains one or more blanks x, y, . . . , each of which refers to a certain category of objects. From the propositional scheme a definite proposition is obtained, e.g. '5 follows upon 4,' when each blank is filled by (the name of) a certain object of the corresponding category. Language does not reflect the structure of such a relational proposition correctly; we have no subject, copula, and predicate, but a relation with two blanks, neither subordinate to the other, which are filled by objects. One might, in order to get rid of the grammatical accidents of language, represent the propositional schemata of relations by wooden boards provided with so many holes corresponding to the blanks, and the objects by little pegs which fit into the holes. In principle these would be symbols as suitable as words. Two propositions such as '5 follows upon 4' and '4 precedes 5' are expressions of one and the same relation between 4 and 5. It is unwarranted to speak here of two relations inverse to each other. The blanks in a relational proposition, though, do each have a specific position; and it is a particular property (commutativity) if the relation $R(xy)$ (e.g. x is a cousin of y) is equivalent (or coextensive) with $R(yx)$.

Properties will have to be counted among the relations, just as 1 is taken to be a natural number. Their propositional scheme possesses exactly one blank.

⟦In §47 of his fifth letter to Clarke, Leibniz speaks of a "relation between L and M, without consideration as to which member is preceding or succeeding, which is the subject or object." "One cannot say that both together, L and M, form the subject for such an *accidens;* for we would then have one *accidens* in two subjects, namely one which would stand, so to speak, with one foot in one subject and with the other in the other subject, and this is incompatible with the concept of an *accidens*. It must be said, therefore, that the relation . . . is something outside of the subjects; but since it is neither substance nor *accidens* it must be something purely ideal, which is nevertheless well worthy of examination." The (explicit or implicit) assumption that every relation must be based on properties has given rise to much confusion in philosophy. A statement asserting, say, that one rose is differently colored from a second is indeed founded on the fact that one is red, the other yellow. But the relation 'the point A lies on the left of B' is not based on a qualitatively describable position of A alone and of B alone. The same holds for kinship relations among people. The view here opposed evidently originates within the domain of sense data, which — it is true — can yield but quality and not relation. It is for this reason that Leibniz, in the above quotation, refers to the relation as something purely ideal. More than two-place relations are hardly ever mentioned in the logico-philosophical literature.

The introduction of propositional schemata with blanks represents

4

an important progress of mathematical beyond traditional logic. In analogy to mathematical functions, which yield a number when their arguments, or blanks, are filled by numbers, propositional schemata are often also referred to as "propositional functions."

Aside from relations, *operations* play a part in the axioms of arithmetic; e.g. the operation of addition which, when applied to two numbers, a and b, produces a third, $a + b$. This operation can be replaced, however, by the relation $a + b = c$ between the three numbers a, b, c; it is 'single-valued' with respect to the argument c, in the sense that for any two numbers a and b there exists one and only one number c which stands in the relation $a + b = c$ to them. Thus we are able to subordinate genetic construction to the static existence of relations. Later, however, we shall proceed conversely, inasmuch as we shall replace all relations by constructive processes. ⦄

The principles of the *combination of relations* are as follows:

1. In a relation scheme with several blanks it is possible to *identify* several of these blanks. For instance, from the scheme

$$N(xy): x \text{ is a nephew of } y$$

    we may obtain

$$N(xx): x \text{ is a nephew of himself.}$$

2. *Negation.* Symbol: $\sim$. $N(xy)$ becomes

$$\sim N(xy): x \text{ is not a nephew of } y.$$

3. *and.* Symbol: &. Thus $N(xy)$ and, say, $F(xy)$ — x is father of y — yield the relation with three blanks

$$F(xy) \ \& \ N(yz): x \text{ is father of } y \text{ and } y \text{ is nephew of } z.$$

    It must be stated which blanks of the combined schemata are to be identified. Symbolically this is indicated by choosing the same letter for the blanks.

4. *or.* Symbol: v. For instance,

$$F(xy) \ \text{v} \ N(yx): x \text{ is father of } y \text{ or } y \text{ is nephew of } x.$$

The combination by means of 'or' can also be expressed in terms of negation and the 'and' combination, and vice versa.[1]

---

[1] Leibniz employs the signs · and + for 'and' and 'or' respectively. We deviate from his notation in order to avoid confusion with the arithmetical operations of multiplication and addition. The formal analogy becomes apparent in J. H. Lambert's distributive law:

$$a \cdot (b + c) = (a \cdot b) + (a \cdot c)$$

(*Acta erudit.* 1765, p. 441). Our use of the product sign Π and summation sign Σ in 6 and 7 are in agreement with Leibniz's usage.

5. Filling a blank by an immediately given object of the corresponding category (substitution). $F(\text{I}, x)$ means: I am father of x. This is the scheme of that property with one blank x which appertains exclusively to my children.
6. *all.* Symbol: Π_x. For instance, $\Pi_x R(xy)$ means: all x (of the corresponding category) are in the relation $R(xy)$ to y.
7. *some.* Symbol: Σ_x. Thus $\Sigma_y R(xy)$ means: there exists a y to which x is in the relation $R(xy)$. Σ_x and Π_x are reducible to each other in the same way with the help of negation as ∨ and &. The presence of a prefixed symbol Π_x or Σ_x (with index x) deprives the blank x of its capability of substitution just as much as if it had been filled in according to 5. For the sake of these last two principles of construction, it will always be necessary to add the two-place relation of logical identity, $x = y$, to the immediately given relations of our domain of investigation.

{*Examples.* 1. Let (xl) mean: the point x lies on the line l. In plane geometry, according to Euclid, parallelism of two lines, $l \parallel l'$, consists in their having no point (x) in common:

$$\sim \Sigma_x \{(xl) \ \& \ (xl')\}$$

is therefore the definition of the relation $l \parallel l'$.

2. The statement that through two distinct points (x, y) there always exists a line (l) would have to be written thus:

$$\Pi_y \Pi_x ((x = y) \ \vee \ \Sigma_l \ \{(xl) \ \& \ (yl)\}).$$

3. In the domain of natural numbers, p is called a prime number if no numbers x and y, both different from 1, exist which stand to p in the relation $x \cdot y = p$. This property of p, of being a prime number, is to be defined as follows:

$$\Pi_y \Pi_x ((x = 1) \ \vee \ (y = 1) \ \vee \sim (x \cdot y = p)).\}$$

Starting with the immediately given basic relations of a field of objects we may by applying the above principles in arbitrary combination obtain an unlimited array of '*derived*' relations (among which the basic relations will of course be counted too). In particular we shall thus arrive at relations with only one blank, the 'derived properties.' How such a property $E(x)$ may serve as 'differentia specifica' in the sense of Aristotelian logic to demarcate a new concept within the 'genus proximum' of the object category to which its blank x refers, will be sufficiently clear from the definition of 'prime number' in Example 3. Among the derived propositional schemata we find, furthermore, those which no longer possess any blank at all, such as in Example 2; they are the *pertinent propositions* of our discipline. If we knew of each of these propositions whether or not it is true, then we should have complete knowledge of the objects of the basic categories as far as they are connected by the basic relations. The logical structure of a proposition of this kind can be described adequately only by stating the manner, order, and combination, in which our seven principles have contributed to its construction. This is a far cry from the old doctrine, according to which a proposition must always consist of subject, predicate, and copula. The syntax of relations, as indicated here, offers a firm starting point for a logical critique of language.

[Compare, for instance, Russell's remarks (*Introduction to Mathematical Philosophy*, Chap. 16) on the definite article in non-deictic application (such as in the proposition: *the* line through the distinct points A, B also passes through C).

A proposition is called *general* if it is constructed without recourse to the fifth principle, of substitution of an immediately given object ('this here'). A non-general proposition is called particular. (Here one might still distinguish between the singular case, in which Principle 5 only, and neither Π_x nor Σ_x, is used for elimination of a blank x, and the mixed general-singular case.) An object a shows itself to be an *individual* if it can be completely characterized by a pertinent general property; that is, if without recourse to Principle 5 a property can be constructed that applies to a but to no other object of the same category. Existence can be asserted only of something described by a property in this manner, not of something merely named, it being essential that Σ_x carries a blank x as an index. (This remark is of use in a critique of the ontological proof of the existence of God.) Within the domain of natural numbers, 1 is an individual, for it is the only such number which does not follow upon any other. Indeed, all natural numbers are individuals. The mystery that clings to numbers, the magic of numbers, may spring from this very fact, that the intellect, in the form of the number series, creates an infinite manifold of well distinguishable individuals. Even we enlightened scientists can still feel it e.g. in the impenetrable law of the distribution of prime numbers. On the other hand, it is the free constructibility and the individual character of the numbers that qualify them for the exact theoretical representation of reality. The very opposite holds for the points in space. Any property derived from the basic geometric relations without reference to individual points, lines, or planes that applies to any one point applies to every point. This conceptual homogeneity reflects the intuitive homogeneity of space. Leibniz has this in mind when he gives the following 'philosophical' definition of similar configurations in geometry, "Things are similar if they are indistinguishable when each is observed by itself." (*Math. Schriften,* V, p. 180.)}

2. THE CONSTRUCTIVE MATHEMATICAL DEFINITION

Aside from the combinatorial definition of derived relations, as discussed in Section 1, mathematics has a *creative definition* at its disposal, through which new ideal objects can be generated. Thus, in plane geometry, the concept of a circle is introduced with the help of the ternary point relation of congruence, $OA = OB$, which appears in the axioms, as follows, "A point O and a different point A determine a circle, the 'circle about O through A'; that a point P lies on this circle means that $OA = OP$." For the mathematician it is irrelevant what circles are. It is of importance only to know in what manner a circle may be given (namely by O and A) and what is meant by saying that a point P lies on the circle thus given. Only in statements of this latter form or in statements explicitly defined on their basis does the concept of a circle appear. Therefore the circle about O through A is identical with the circle about O' through A' if and only if all points lying on the first circle also lie on the second, and vice versa. The axioms of geometry show that this criterion, which refers to the infinite manifold of all points, may be replaced by a finite one: O' must coincide with O, and we must have $OA' = OA$.

{*Further examples.* 1. Nobody can explain what a *function* is, but this is what really matters in mathematics: "A function f is given whenever with every real number a there is associated a number b (as for example, by the formula $b = 2a + 1$). b is then said to be the *value* of the function f for the argument value a." Consequently, two functions, though defined differently, are considered the same if, for every possible argument value a, the two corresponding function values coincide.

2. In Euclidean geometry the "*points at infinity,*" in which parallel lines allegedly intersect, are such ideal elements added to the real points by a creative mathematical definition. By a suitable introduction of ideal points one can, more generally, extend a given limited portion S of space, the 'accessible' space, so as to comprise the whole space of projective geometry. The task is to decide through geometric constructions within S whether two real lines, i.e. lines passing through S, issue from the same ideal point. Such a point is defined most

simply as the vertex of a trilateral corner (formed by three real lines). Thus we arrive at the following definition: "Three non-coplanar lines a, b, c, any two of which are coplanar, determine an ideal point $[a, b, c]$. To say that a line l passes through this point means that l is coplanar with each of the lines a, b, c." Again this definition implies a criterion for the coincidence of two such ideal points. To every real point p there corresponds exactly one ideal point π such that every line through p passes, in the sense of our definition, through π. Thus a part of the ideal points may be identified with the real points. (Compare Pasch, *Vorlesungen über neuere Geometrie*, 2d ed., p. 40.) According to the same scheme, mathematics always accomplishes the extension of a given domain of operation through the introduction of ideal elements. Such an extension is made in order to enforce the validity of simple laws. For example, as a consequence of the addition of the points at infinity it is true not only that two distinct points can always be joined by a line, but also that two distinct coplanar lines always intersect at a point. The introduction of *imaginary* elements in geometry (in order to enforce simple and universally valid theorems on intersection of algebraic curves and surfaces) and the introduction by Kummer of *ideal numbers* in number theory (in order to restore the laws of divisibility, which at first were lost in the transition from rational to algebraic numbers) are among the most fruitful examples of this *method of ideal elements.* }

A special case is the process of *definition by abstraction*. A binary relation $a \approx b$ in a domain of objects is called an *equivalence* (a relation of the character of equality), if the following is universally true:

(i) $a \approx a$;

(ii) if $a \approx b$, then $b \approx a$ (commutativity);

(iii) if $a \approx b$ and $b \approx c$, then $a \approx c$ (transitivity).

By agreeing to consider two objects a and b as distinct if and only if they do not satisfy the equivalence relation $a \approx b$, a new object domain is derived *by abstraction* from the original one.

Examples and comments. 1. Similarity of geometrical figures is an equivalence. Every figure is attributed a certain *shape*, and two figures have the same shape if and only if they are similar. In a more philosophical mode of expression one is used to say that the concept of shape results from that of figure by abstracting from position and magnitude. In scientific practice the introduction of a concept thus abstracted expresses the intention of exclusively considering *invariant* properties and relations among the originally given objects. $R(xy)$ is invariant with respect to the equivalence $\approx$, if $R(ab)$ always entails $R(a'b')$, provided $a' \approx a$ and $b' \approx b$.

2. Two sets A and B of objects (say, the persons and chairs in a room) are said to be numerically equivalent, $A \approx B$, if it is possible to pair off the elements of A with those of B (if it is possible to assign one person to each chair, so that no chair remains vacant and no person remains unseated). Numerical equivalence obviously is an equivalence in the above sense. 'Every set determines a (*cardinal*) *number;* two sets determine the same number if and only if they are numerically equivalent.' (This explanation can already be found in Hume, *Treatise of Human Nature*, Book I, Part III, Section 1.)[2] In more careless formulation one would say that the concept of (cardinal) number results from that of set by abstracting from the nature of the elements of the set and merely considering their discernibility. The objection occasionally put forward that all elements, if degraded into mere Ones, collapse into one, is met by the above precise formulation.

[The example of number may serve to illustrate that the definition by abstraction is a special case of the creative definition. It is subordinated to the latter as follows: 'Every set A determines a number (A). To say that an arbitrary set M consists of (A) elements means that M and A are numerically equivalent.' Consequently the number (A) is the same as the number (B), if every set M that is $\approx A$ also is $\approx B$, and vice versa. But according to the rules (ii) and (iii) for equivalences, such is the case if and only if $A \approx B$. Finally, rule (i) guarantees that, in particular, A itself consists of (A) elements.

3. Two integers, according to Gauss, are congruent modulo 5 if their difference is divisible by 5. Congruence is a relation of the character of equality. Through the corresponding abstraction we obtain, from the integers, the congruence-integers modulo 5. Since the operations of addition and multiplication are invariant with respect to congruence, the result is a finite domain of only 5 elements, within which the usual algebra can be carried on just as well as in the infinite domain of the ordinary rational integers. We have here, for instance, $2 + 4 = 1$, $3 \cdot 4 = 2$ (modulo 5). Not only subtraction but even division can be carried out, by virtue of the fact that 5 is a prime number. This example is of fundamental importance for number theory.

4. The most significant physical concepts are likewise obtained in accordance with the scheme of mathematical abstraction. We shall return to this in Part II when the process of measurement is discussed.

---

[2] The passage is worth quoting. "When two numbers are so combined, as that the one has always an unit answering to every unit of the other, we pronounce them equal; and it is for want of such a standard of equality in extension, that geometry can scarce be esteemed a perfect and infallible science."

The principle of definition by abstraction I find alluded to by Leibniz in his fifth letter to Clarke, §47. He says there, "Incidentally, I have proceeded here by and large as Euclid did. The latter, since he found himself unable to define the concept of geometrical ratio absolutely, stipulated what was to be understood by equal ratios." And shortly before that, "The mind, however, is dissatisfied with this equality. It seeks an identity, a thing which would truly be the same, and it imagines it to be in a manner outside of the subjects." The principle has shown its full importance for mathematics only in the 19th century. It was consciously formulated in all generality by Pasch in his book quoted on p. 9 (1882), still more clearly by Frege (*Die Grundlagen der Arithmetik*, Breslau 1884, Sections 63–68). Compare also Helmholtz (Zählen und Messen, 1887, *Wissenschaftliche Abhandlungen*, III, p. 377).

Beside the above-mentioned mathematical form of abstraction one might be inclined to place another, the originary abstraction. In looking at a flower I can mentally isolate the abstract feature of color as such. This act of abstraction would here be primary while the statement that two flowers have the same color 'red' would be based on it; whereas in mathematical abstraction it is the equality which is primary, while the feature with regard to which there is equality comes second and is derived from the equality relation. But the integers of the same congruence class modulo 5 can also be characterized by the fact that upon division by 5 they all leave the same remainder; the similarity of two triangles by the fact that the angles in both have the same numerical values and corresponding sides have the same ratios. The general procedure of constructing these remainders and these numerical values of angles and ratios, respectively, takes the place of the feature 'color,' its identical result for two integers or triangles that of the identical 'red' of two flowers. Originary abstraction thus is subordinate to mathematical abstraction. But that which is common to all *congruent* triangles or to all bodies occupying the same spatial position, I find myself unable to represent by an objective feature (it is the latter example that Leibniz *loc. cit.* had in mind), but merely by the indication: congruent to *this* triangle, occupying *this* spatial position. Our question is connected with the problem of relativity (Section 13), with the difference between conceptual definition and intuitive exhibition. But in both cases alike, those of originary and of mathematical abstraction, the transformation of a common feature into an ideal object, e.g. of the property 'red' into an objectified 'red color,' of which the red things 'partake,' is an essential step (Plato's μέϑεξις).}

11

With every property $E(x)$ which is meaningful for the objects x of a given category we correlate a set, namely 'the set of objects x having the property E.' Thus we speak of the set of all even numbers, or of the set of all points on a given line. The conception that such a set be obtained by assembling its individual elements should by all means be rejected. To say that we know a set means only that we are given a property characteristic of its elements. Only in the case of a finite set do we have, in addition to such *general* description, the possibility of an *individual* description which would exhibit each one of its elements. [Formally, by the way, the latter mode of description is a special case of the former; e.g. the set consisting of three given objects a, b, c corresponds to the property of being either a or b or c: $(x = a) \lor (x = b) \lor (x = c)$.] It is possible that the same set is correlated with two properties E and E'. This happens when every object (of our category) having the property E also has the property E', and conversely. Hence, what is decisive for the identity of the two sets is not the manner of their definition (in terms of the principles enumerated in Section 1), but solely the question whether each element of one is an element of the other and vice versa, a question referring to a domain of *existing* objects and unanswerable by recourse to the meaning alone. If the concept of set is understood in this way, then the creative definition is seen to be nothing but the transition from a property to a set, so that the mathematical construction of new classes of ideal objects can quite generally be characterized as *set formation*. Now there is no longer anything objectionable in describing the circle about O through A as the set of all points P whose distance from O equals OA, or the color of an object as the set of objects having the same color, or the cardinal number 5 as the set of all those aggregates which are numerically equivalent to the exhibited aggregate of the fingers of my right hand. But it is an illusion — in which Dedekind, Frege, and Russell indulged for a time, because they apparently conceived of a 'set' after all as a collective — to think that thereby a concrete representation of the ideal objects has been achieved. On the contrary, it is through the principle of creative definition that the meaning of the general set concept is elucidated as well as safeguarded against false interpretations.

The properties employed in the creation of new abstracts Φ generally depend on one or more arguments u, v, . . . , which are allowed to vary freely within certain domains: Φ is a *function* of u, v, ... In the definition of a circle, for instance, the ternary point relation $OP = OA$ is interpreted as a property of P (relation with one blank P) depending on O and A; the 'circle about O through A' is a function of O and A. The criterion for the coincidence of two values of an abstract, $\Phi(u, v, \ldots)$ and $\Phi(u', v', \ldots)$, refers to a totality of existing objects. But of special importance are those cases where this transfinite criterion can be converted by virtue of certain universally valid facts into a finite criterion requiring recourse to the meaning of the defining relation only. Instances are our definition of the circle and the definitions by abstraction. Not only properties, but more generally *relations* may serve to define ideal elements. If we want to adhere to the set-theoretical terminology throughout, it will be necessary to have a 'binary set' (R) correspond to every binary relation R; such that (R) and (R') are identical if, for arbitrary elements a, b, it never happens that one of the statements $R(a, b)$, $R'(a, b)$ is true, the other false. The same for ternary, quaternary, . . . relations. We thus arrive at the following final version of the *principle of creative definition:* A relation $R(xy \ldots /uv \ldots)$, whose blanks are separated into two groups, $xy \ldots$ and $uv \ldots$, determines an abstract $\Phi(uv \ldots)$ depending on the arguments u, v, $\ldots$; equality $\Phi(uv \ldots) = \Phi(u'v' \ldots)$ for any two sets of values of the arguments, u, v, $\ldots$ and u', v', $\ldots$, holds if and only if any objects x, y, $\ldots$ of the appropriate categories which stand in the relation R to u, v, $\ldots$ also stand in the relation R to u', v', $\ldots$, and conversely.

3. LOGICAL INFERENCE

Having dealt with definitions we now come to *proofs.* If one turns a geometrical proposition into a hypothetical statement whose premiss consists of all geometrical axioms, replacing mentally at the same time any abbreviatory expressions by what they mean according to their definitions, one will arrive at a *'formally valid,' 'analytic' proposition,* the truth of which is in no way tied to the meanings of the concepts entering into it (point, line, plane, incidence, betweenness, congruence). The logic of inference has the task of characterizing those propositional structures which assure the formal validity of the proposition. *Barbara, Baralipton,* and so on, are of little help in this connection. Leibniz considered the doctrine of the *argumens en forme* "une espèce de Mathématique universelle, dont l'importance n'est pas assez connue" (*Nouveaux Essais,* Libre IV, Chap. 17, §4).

That part of logic which operates exclusively with the logical connectives 'not,' 'and,' 'or' will be referred to as *finite logic,* as opposed to *transfinite logic,* which in addition uses the propositional operators 'some' (or 'there is') and 'all'. The reason for this subdivision is as follows. Suppose several pieces of chalk are lying in front of me; then the statement 'all these pieces of chalk are white' is merely an abbreviation of the statement 'this piece is white & that piece is white & . . . '

13

(where each piece is being pointed at in turn). Similarly 'there is a red one among them' is an abbreviation of 'this is red ∨ that is red ∨ . . . .' But only for a finite set, whose elements can be exhibited individually, is such an interpretation feasible. In the case of infinite sets, the meaning of 'all' and 'some' involves a profound problem which touches upon the core of mathematics, the very secret of the infinite; it will unfold itself to us in the next chapter. The situation here may be compared to the transition from finite to infinite sums; the meaning of the latter is tied to special conditions of convergence, and one may not deal with them in every respect as with finite sums.

In the propositional calculus it is convenient to introduce, in addition to the symbols for 'not,' 'and,' 'or,' the symbol $a \rightarrow b$ (read: a implies b). It has the same meaning as $\sim a \vee b$ (a does not hold or b holds) and does not beyond that signify any deeper connection between the propositions a and b.

[Incidentally two of the four symbols $\sim$, &, ∨, $\rightarrow$ would suffice; for the propositional calculus it is convenient to choose $\rightarrow$ and $\sim$. Nay, it is even possible to get along with one symbol, a/b, denoting the incompatibility of the propositions a and b ($\sim a \vee \sim b$). For, in place of

$$\sim a, \quad a \rightarrow b, \quad a \ \& \ b, \quad a \vee b$$

we may write

$$a/a, \quad a/(b/b), \quad (a/b)/(a/b), \quad (a/a)/(b/b).$$

However, for the sake of greater lucidity, we shall here use all four symbols.]

In a finite-logical formula the letters (propositional variables) for which arbitrary propositions (without blanks) may be substituted are combined by those four symbols $\sim$, &, ∨, $\rightarrow$. For example:

$$b \rightarrow (a \rightarrow b).$$

There exists a general rule by which the formal validity of a formula of this kind can be recognized. In fact, assign to each letter occurring in the formula one of the values 'true' (T) or 'false' (F) in all possible combinations, and determine in each instance the value of the entire formula according to the following direction for evaluating compound propositions:

a	$\sim a$		a	b	$a \rightarrow b$	$a \,\&\, b$	$a \vee b$
T	F		T	T	T	T	T
F	T		T	F	F	F	T
			F	T	T	F	T
			F	F	T	F	F

(The number of combinations to be tested, for instance, when the formula contains 5 different propositional variables is 2^5.) If the resulting value of the formula is T in every case, then the formula is formally valid. This rule, which may be said to be based on the *law of contradiction* and the *law of the excluded middle* (*tertium non datur*), I shall call briefly *the finite rule*.

Example: $b \rightarrow (a \rightarrow b)$.

a	b	$a \rightarrow b$	$b \rightarrow (a \rightarrow b)$
T	T	T	T
T	F	F	T
F	T	T	T
F	F	T	T

On this level, then, it is possible to ascertain directly by a combinatorial procedure following a fixed scheme, whether a given assertion is a logical consequence of certain other propositions, provided premiss and conclusion are built up of propositions a, b, . . . (whose meanings do not matter) with the help of the four operations $\sim$, $\rightarrow$, &, $\vee$.

All this changes completely as soon as 'some' and 'all' (and their concomitants, the blanks) are admitted into our formulas. Σ_x and Π_x compel us to *construct* — we set up a number of formally valid basic formulas, the *logical axioms*, and state a rule by which further formally valid propositions may be obtained from formally valid propositions. The rule is none other than the one by which logic is applied to all theoretical sciences, namely the *syllogism*: if you have a proposition A, and a proposition $A \rightarrow B$ in which the first proposition A reappears on the left of $\rightarrow$, then set down the proposition B. All propositional structures obtained from the axioms by repeated application of this rule are of analytic character. It is impossible, however, to characterize descriptively the infinite manifold of these individual structures independently of the constructive manner in which they are generated. Hence the necessity of step by step demonstration. Using a phrase coined by J. Fries in a somewhat different sense, one

15

may speak of a 'primordial obscurity of reason.' We do not possess the truth, it wants to be attained by action.

{Galileo ("Dialogo," *Opere*, VII, p. 129) expresses a widely spread view when he interprets this as the difference between human and divine understanding. "We proceed in step-by-step discussion from inference to inference, whereas He conceives through mere intuition. Thus in order to gain insight into some of the properties of the circle, of which it possesses infinitely many, we begin with one of the simplest; we take it for a definition and proceed from it by means of inferences to a second property, from this to a third, hence a fourth, and so on. The divine intellect, on the other hand, grasps the essence of a circle *senza temporaneo discorso* and thus apprehends the infinite array of its properties." (But intensively, i.e. as regards the objective certainty of an individual mathematical truth, the human intellect does not fall short of the divine intellect).}

Concerning 'some' and 'all,' Σ_x and Π_x, we may lay down, first of all, the following two axioms, in which $a(x)$ stands for an arbitrary propositional scheme containing the one blank x, and c for any given object of the corresponding category:

$$\text{I. } \Pi_x a(x) \rightarrow a(c); \quad \text{II. } a(c) \rightarrow \Sigma_x a(x).$$

The first of these axioms tells us merely how to derive something from a universal proposition, but fails to show how other propositions can ever lead us to a universal proposition. The converse is true of the second axiom.

{Everybody knows the classical example of an inference: (α) all men are mortal, (β) Caius is a man; hence (γ) Caius is mortal. Our formalism decomposes it into several steps. Let H and M designate the properties of being (hu)man and mortal respectively, and let c designate Caius. Then

(α) $\qquad\qquad\qquad \Pi_x(H(x) \rightarrow M(x)),$

in connection with I,

$$\Pi_x(H(x) \rightarrow M(x)) \rightarrow (H(c) \rightarrow M(c)),$$

according to our syllogism rule of inference yields

$$H(c) \rightarrow M(c).$$

This, together with

(β) $\qquad\qquad\qquad H(c),$

16

again by the rule of inference, yields

(γ) $\qquad\qquad\qquad\qquad M(c).$

$\rightarrow$ does not, of itself, involve the idea of universality; but (α) illustrates how it may combine with 'all' to form a *universal hypothetical statement*. The grounds for the validity of a universal implication of the form

$$\Pi_x(a(x) \rightarrow b(x))$$

may of course be several. If they are solely to be found in the logical axioms, then the symbol $\rightarrow$ expresses purely logical consequence. But the grounds may well be of a factual nature, such as a causal relation or some other empirically observed regularity. This remark may suffice to clarify the question as to how the relation of cause and effect is connected with that of logical reason and consequence. The symbol $\rightarrow$ remains neutral with respect to all this.

The finite-logical axioms can be found listed in D. Hilbert and P. Bernays, *Grundlagen der Mathematik*, I, Berlin, 1934, p. 66. They are of course constructed in such a manner that their formal validity can be established by means of the 'finite rule.' Conversely it can be shown — but this already requires an essentially mathematical and not altogether trivial proof — that the list of axioms is complete, in the sense that all logical formulas containing only the symbols $\sim$, $\rightarrow$, &, v which are formally valid according to the 'finite rule' can be obtained from these few axioms by substitution and repeated application of the syllogism. The group of transfinite axioms, of which we know as yet only Axioms I and II, remains in need of supplementation.

From the syllogism rule of inference other, derived, rules of inference may be obtained by means of the logical axioms. Indeed every formally valid proposition of the form $A \rightarrow B$ (where A and B are built up from the propositional variables a, b, . . . with the help of the logical connectives), by virtue of the syllogism, leads to the following rule of inference: If you have a proposition of the form A, then you may set down the corresponding proposition of the form B. Conversely, the syllogism also has its representative among the logical formulas:

$$a \rightarrow ((a \rightarrow b) \rightarrow b).$$

And yet, since construction means action, one does not get along with formulas alone; some practical rule of inference that tells how to handle the formulas is needed. This is probably the truth behind the opinion as to the normative character of logic.

Kant's distinction between analytic and synthetic judgments (*Critique of Pure Reason,* Introduction) is so obscurely phrased as to render a comparison with the precise concept of formal validity in mathematical logic almost impossible. The latter concept is in agreement, however, with Husserl's definition (*Logische Untersuchungen,* II, 2d ed., p. 254): "Analytic laws are unconditionally universal propositions containing no concepts other than formal. As opposed to the analytic laws we have their particular instances, which arise through the introduction of material concepts or of ideas positing individual existence. And as particular cases of laws always yield necessities, so particular cases of analytic laws yield analytic necessities."}

REFERENCES

G. W. Leibniz, *Philosophische Schriften,* VII, pp. 43–247, 292–301.
—— *Mathematische Schriften,* VII, 49–76, 203–216.
L. Couturat, *La Logique de Leibniz,* Paris, 1901.
—— *Opuscules et fragments inédits de Leibniz,* Paris 1903.
G. Boole, *The Mathematical Analysis of Logic,* London and Cambridge, 1847.
—— *An Investigation of the Laws of Thought,* London, 1854; reprinted in: *Collected Logical Works,* Chicago and London, 1916.
E. Schröder, *Vorlesungen über die Algebra der Logik,* 3 Vols., Leipzig, 1890–95.
A. N. Whitehead and B. Russell, *Principia Mathematica,* 3 Vols., Cambridge, 1910–13; 2d ed., 1925–27.
B. Russell, *The Principles of Mathematics,* Cambridge, 1903; 2d ed., New York, 1938.
—— *Introduction to Mathematical Philosophy,* 2d ed., London, 1920.
L. Wittgenstein, *Tractatus Logico-Philosophicus,* New York and London, 1922.
C. I. Lewis and C. H. Langford, *Symbolic Logic,* New York, 1932.
A. Tarski, *Introduction to Logic* (translated by O. Helmer), New York 1941.

4. THE AXIOMATIC METHOD

The axiomatic method consists simply in making a complete collection of the basic concepts as well as the basic facts from which all concepts and theorems of a science can be derived by definition and deduction respectively. If this is possible, then the scientific theory in question is said to be *definite* according to Husserl. Such is the case for the theory of space. Of course, from the axioms of geometry I cannot possibly deduce the law of gravitation. Hence it was necessary to explain above what is to be considered a *pertinent* proposition of a given field of inquiry. Similarly the axioms of geometry fail to disclose whether Zurich is farther from Hamburg than Paris. Though this question deals with a geometrical relation, the relation is one between individually exhibited locations. Thus, precisely speaking, what is supposed to be deducible from the axioms are the *pertinent general true propositions*.

⟦"Such, then, is the whole art of convincing. It is contained in two principles: to define all notations used, and to prove everything by replacing mentally the defined terms by their definitions." Thus Pascal in a discourse *de l'esprit géométrique* (Oeuvres complètes, ed. F. Strowski, Paris (Librairie Ollendorff), I, p. 427). But this is more easily said than done. Euclid's *Elements* fail to afford a complete solution of the problem of axiomatizing geometry. He begins with ὅροι, definitions; but they are only in part definitions in our sense, the most important among them are descriptions, indications of what is intuitively given. Nothing else, in fact, is possible after all for the basic geometrical concepts such as 'point,' 'between,' etc.; but as far as the deductive construction of geometry is concerned, descriptions of this kind are evidently irrelevant. There follow, under the name of αἰτήματα, certain geometrical axioms, in particular the axiom of parallels: Given a plane P, a line l in P, and a point p in P not lying on l; all lines in P which pass through p, except one, intersect l. Finally a few general axioms of magnitude: κοιναὶ ἔννοιαι. They play their part in the development of geometry, inasmuch as certain geometrical relations such as congruence, or equality of areas, are tacitly assumed to satisfy these axioms. Behind them are concealed an indefinite number of proper axioms of geometry. In later books of the *Elements* the list of axioms is supplemented as the occasion demands. Because the geometrical postulates are intuitively self-evident and because a purely logico-deductive attitude is not natural to the human mind, it has required great pains to compile a complete list of geometrical axioms. 'Non-Euclidean geometry,' established by Bolyai and Lobatschewsky around 1830, becomes the driving force for axiomatic research in the second half of the 19th century. The most hidden axioms, those of order, are disclosed by Pasch around 1880. Finally, at the turn of the century, the goal is reached completely and finds its classical expression in Hilbert's *Grundlagen der Geometrie*. Hilbert arranges the axioms in five groups: the axioms of incidence, of order ('betweenness'), of congruence, of parallelity, and of continuity.

The axiomatic procedure of the ancients, which aside from Euclid was also handled by Archimedes with admirable facility, became exemplary for the foundation of modern mechanics. It dominates Galileo's doctrine of uniform and uniformly accelerated motion ("Discorsi," 3rd and 4th days), and even more so Huyghens' establishment of the laws of the pendulum in his *Horologium oscillatorium*. In more recent times the axiomatic program was carried out completely (outside of mathematics proper) for the statics of rigid bodies, the space-time theory of special relativity, and other parts of physics.

An axiom system is by no means uniquely determined by the discipline in question; rather, the choice of the basic concepts and basic facts is arbitrary to a considerable extent. The question as to whether it is possible to differentiate between essentially originary and essentially derived notions lies beyond the competence of the mathematician.[3] The definition of a geometrical relation concept that was originally chosen may with equal justification be replaced by any criterion which, in accordance with geometrical facts, is a necessary and sufficient condition under which the relation holds.]

An axiom system must under all circumstances be free from contradictions, in which case it is called *consistent;* that is to say, it must be certain that logical inference will never lead from the axioms to a proposition a while some other proof will yield the opposite proposition $\sim a$. If the axioms reflect the truth regarding some field of objects, then, indeed, there can be no doubt as to their consistency. But the facts do not always answer our questions as unmistakably as might be desirable; a scientific theory rarely provides a faithful rendition of the data but is almost invariably a bold construction. Therefore the testing for consistency is an important check; this task is laid into the mathematician's hands. Not indispensable but desirable is the *independence* of the individual axioms of an axiom system. It should contain no superfluous components, no statements which are already demonstrable on the basis of the other axioms. The question of independence is closely connected with that of consistency, for the proposition a is independent of a given set of axioms if and only if the proposition $\sim a$ is consistent with them.

The dependence of a proposition a on other propositions A (an axiom system) is established as soon as a concrete proof of a on the basis of A is given. In order to establish the independence, on the other hand, it is required to make sure that no combination of inferences, however intricate, is capable of yielding the proposition a. There are three methods at one's disposal of reaching this goal; by what has been said above, each of them qualifies also for proving the consistency of an axiom system.

(1) The first method is based on the following principle: if a con-

---

[3] Sometimes this is certainly the case; e.g. among the kinship relations, 'child' and 'spouse' are the essentially originary ones.

tains a new original concept, not defined in terms of those occurring in A, then a cannot be a consequence of A. For example: a ship is 250 feet long and 60 feet wide; how old is its captain? Only in the most trivial cases does this simple idea accomplish our objective.

(2) *The construction of a model.* Objects and relations are exhibited which, upon suitable naming, satisfy all of the propositions A, and yet fail to satisfy a. This method has been the most successful so far invented.

The most famous example is furnished by the axiom of parallels. From the beginning, even in antiquity, it was felt that it was not as intuitively evident as the remaining axioms of geometry. Attempts were made through the centuries to secure its standing by deducing it from the others. Thus doubt of its actual validity and the desire to overcome that doubt were the driving motives. The fact that all these efforts were in vain could be looked upon as a kind of inductive argument in favor of the independence of the axiom of parallels, just as the failure to construct a perpetuum mobile is an inductive argument for the validity of the energy principle. Negating the axiom of parallels amounts to the assumption that, given a point P and a line l not passing through P, there exist in the plane determined by P and l an infinitude of lines through P not intersecting l. Therefore this is what the constructors of non-Euclidean geometry did: they drew the consequences of that assumption, and in doing so they found, even though they made free use of the remaining axioms of Euclidean geometry, that no contradiction arose, *as far as they followed the matter up.* But they could not claim security for all future. Klein was the first to offer a Euclidean model for non-Euclidean geometry; the objects of Euclidean geometry itself, upon an assignment of names differing from the customary one, satisfy the non-Euclidean axioms. Let S be a sphere in Euclidean space. The dictionary which furnishes the translation into non-Euclidean language consists of only a few words (here characterized by quotation marks): by a *'point'* we understand any point in the interior of S. Several such 'points' are said to lie on a *'line'* or in a *'plane,'* and a 'point' is said to lie *'between'* two others, if they do so in the customary sense. A *'motion'* is any collinear transformation which transforms the sphere S into itself; two configurations are *'congruent'* if one results from the other by a 'motion.' For anyone who believes in the truth and thus in the consistency of Euclidean geometry, the consistency and thus conceivability of non-Euclidean geometry is thereby established.

The consistency of Euclidean geometry, on the other hand, can be demonstrated quite independently of the belief in its truth and of the intuitive content of its basic concepts. For analytic geometry, which can best be based on the concept of vector (see Section 12), has shown that Euclidean geometry is but a different expression of the facts of *linear algebra*, of the theory of linear equations and has thus provided us with a simple *arithmetical model* of Euclidean space. Linear algebra accounts for the affine concepts of geometry, while the adjunction of a positive definite quadratic form that serves as the 'metric ground form' leads to the metrical concepts. In algebra the number n of variables (or 'unknowns') may be left indeterminate. One has to choose $n = 3$ in order to get the geometry of the intuitive 3-dimensional space. Arithmetic and geometry, by virtue of this correspondence, are so closely interwoven that today even in pure analysis we constantly make use of geometrical terms. Any contradiction in geometry would at the same time show up as a contradiction in arithmetic. This may be looked upon as a reduction, since the numbers are to a far greater measure than the objects and relations of space a free product of the mind and therefore transparent to the mind.

These examples indicate that the method of models need not be restricted to those cases where the truth about the objects and relations employed in the construction of the model is known, but that it may serve to reduce the consistency of an axiom system A (e.g. that of geometry) to the consistency of another, B (e.g. that of arithmetic). This is achieved whenever the basic concepts of the system A are defined in terms of the basic concepts of the system B in such a manner that the axioms A become a logical consequence of the axioms B. No attention has to be paid for this purpose to the intuitive meaning of the basic concepts in A and in B; the assignment of the names given to the basic concepts of A to certain concepts derived from B is purely arbitrary.

More than anybody else has Hilbert, through the ingenious construction of suitable arithmetical models, contributed to the clarification of the logical relations that connect the various parts of the geometrical system of axioms.

If we are dealing with a finite number of objects only which are explicitly exhibited one by one and designated by symbols, we may be able to prove consistency by stating for each single instance in terms of the symbols whether or not the basic relations obtain. As an example we give a *combinatorial model* that ensures the consistency of the incidence axioms in plane projective geometry (which deal with the single relation 'point lies on line'). The model consists of seven symbols for points, 1, 2, 3, 4, 5, 6, 7, and seven symbols for lines, I, II, III, IV, V, VI, VII, and incidence is defined by the following table, in which a *, say, at the crossing of row 3 and column VI indicates that point 3 lies on line VI:

	I	II	III	IV	V	VI	VII
1		*	*	*			
2	*		*		*		
3	*	*				*	
4	*			*			*
5		*			*		*
6			*			*	*
7				*	*	*	

For example, verify from this table the axioms stating that through any two distinct points there goes exactly one line (i.e. any two rows contain exactly one pair of *'s in the same column) and that any two distinct lines intersect at exactly one point!

The case of a finite system of objects exhibited one by one is comparatively trivial. In all other cases the method of models is merely capable of reducing the consistency of one system to that of another. Ultimately it will become necessary to prove consistency in an absolute way for one basic system of axioms. For the larger part of mathematics and for the whole of physics this basic system deals with the concept of *real number*.

(3) For the purpose of an absolute proof of consistency we have none but the *direct method* at our disposal, which endeavors to show that by following the rules of deductive inference one will never arrive at two propositions of which one is the negation of the other. Complete enumeration of the logical rules of the game is here a necessary presupposition (compare Section 3); for only then can one apply the method to propositions, blind against their meaning, as one applies the rules of chess to chessmen. Only in recent years has Hilbert attacked the problem of securing the consistency of the arithmetical axioms in this manner. (Should a new and evidently stringent method of logical inference be discovered and thus the set of rules of the game be augmented, one would have to be prepared to see a consistency proof conducted by the direct method become obsolete. The method of models, on the other hand, is independent of the 'rules of the game.')

[The following might serve as an analogue in chess: it is required to see that a game of chess, no matter what the various moves, as long as it is played in accordance with the rules, can never lead to a position in which there are 10 queens of the same color on the board. Here the 'direct method' is applicable. For it can be gathered from the rules of the game that no move increases the sum of the numbers of queens and pawns of the same color. Hence, as this sum is 9 initially, it must remain $\leqq 9$. Incidentally, method (1) is a special case of the direct method, but it seemed to deserve special mention because of its simplicity.

In addition to consistency and independence, the *completeness* of the axioms which form the basis of a science will be required. What is meant by that? That for every pertinent general proposition a the question 'does a or $\sim a$ hold?' be decidable by logical inference on the basis of the axioms? Just as consistency guarantees that not *both* a and $\sim a$ can be obtained, completeness would then guarantee that always *one* of them can be obtained. Completeness in this sense would only be ensured by the establishment of such procedural rules of proof as would lead demonstrably to a solution for every pertinent problem. Mathematics would thereby be trivialized. But such a philosopher's stone has not been discovered and never will be discovered. Mathematics does not consist in developing the logical consequences of given assumptions omnilaterally, but intuition and the life of the scientific mind pose the problems, and these cannot be solved by mechanical rules like computing exercises. The deductive procedure that may lead to their solution is not predesigned but has to be discovered in each case. Analogy, experience, and an intuition capable of integrating multifarious connections are our principal resources in this task. As was already mentioned in Section 3, there is no descriptive characteristic of all propositions deducible from given premises; we have to rely on construction. It is not feasible in practice to proceed like Swift's scholar, whom Gulliver visits in Balnibarbi, namely, to develop in systematic order, say according to the required number of inferential steps, all consequences and discard the "uninteresting" ones; just as the great works of world literature have not come into being by taking the twenty-six letters of the alphabet, forming all 'combinations with repetition' up to the length 10^{10}, and selecting and preserving the most meaningful and beautiful among them.

Suppose we make a continuous deformation of space (as if it were filled out with plasticine), and suppose we understand now by lines, planes, and congruent figures such curves, surfaces, and figures as result from real lines, real planes, and really congruent figures by this deformation. Then evidently all the facts of geometry hold for these newly introduced concepts. It is therefore impossible to distinguish conceptually between the system of lines and the system of curves resulting from them by a spatial deformation. ⌉

24

This brings us to the idea of *isomorphism*, which is of fundamental importance for epistemology. Let us assume that we have a system Σ_1 of objects (such as the points, lines, and planes of geometry) and certain appertaining basic relations R, R', ... Let there be a second system Σ_2, with corresponding basic relations which (though they may have entirely different meanings) are correlated, say, by the use of the same names, to the relations R, R', . . . within the first domain of objects. Then, if it is possible to state a rule by which the elements of the system Σ_1 are paired in a mutually unique manner with the elements of the system Σ_2, so that elements in Σ_1 between which R (or R', . . .) holds correspond to elements in Σ_2 between which the relation with the same name R (or R', . . . respectively) holds, then the two domains are said to be *isomorphic*. The correlation in question is said to be an *isomorphic mapping* of Σ_1 into Σ_2. Isomorphic domains may be said to possess the same *structure*. For every pertinent true proposition about Σ_1 (whose sense can be understood by virtue of the meanings of R, R', . . . within Σ_1), there is a corresponding and identically phrased proposition about Σ_2, and conversely. Nothing can be asserted of the objects in Σ_1 that would not be equally valid in Σ_2. Thus, for example, Descartes' construction of coordinates maps the space isomorphically into the operational domain of linear algebra. These considerations induce us to conceive of an axiom system as a *logical mold* ('*Leerform*') *of possible sciences*. A concrete interpretation is given when designata have been exhibited for the names of the basic concepts, on the basis of which the axioms become true propositions. One might have thought of calling an axiom system complete if in order to fix the meanings of the basic concepts present in them it is sufficient to require that the axioms be valid. But this ideal of uniqueness cannot be realized, for the result of an isomorphic mapping of a concrete interpretation is surely again a concrete interpretation. Hence the final formulation has to be as follows: an axiom system is complete, or *categorical*, if any two concrete interpretations of it are necessarily isomorphic. In this sense the categoricity of Hilbert's axiom system of Euclidean geometry is guaranteed. Indeed it can easily be shown that a space satisfying these axioms is isomorphic to the algebraic model provided by Descartes' analytic geometry.

[A science can only determine its domain of investigation up to an isomorphic mapping. In particular it remains quite indifferent as to the 'essence' of its objects. That which distinguishes the real points in space from number triads or other interpretations of geometry one can only *know* (*kennen*) by immediate intuitive perception. But intuition is not blissful repose never to be broken, it is driven on toward the dialectic and adventure of cognition (Erkenntnis).[4] It would be folly to expect cognition to reveal to intuition some secret essence of things hidden behind what is manifestly given by intuition. The idea of isomorphism demarcates the self-evident insurmountable boundary of cognition. This reflection has enlightening value, too, for the metaphysical speculations about a world of things in themselves behind the phenomena. For it is clear that under such a hypothesis the absolute world must be isomorphic to the phenomenal one (where, however, the correlation needs to be unique only in the direction thing in itself → phenomenon); for "we are justified, when different perceptions offer themselves to us, to infer that the underlying real conditions are different" (Helmholtz, *Wissenschaftliche Abhandlungen*, II, p. 656). Thus even if we do not *know* the things in themselves, still we have just as much *cognition* about them as we do about the phenomena. The same idea of isomorphism clarifies the problem which Leibniz, stimulated by Hobbes' nominalistic theory of truth, treats in his dialogue on the connection between things and words; Leibniz evidently wrestles with giving expression to that idea (*Philosophische Schriften*, VII, pp. 190–193).

Through the disclosure of isomorphic relations it is possible to transfer any insights gained in one field to the isomorphic field. A service of this kind is rendered, for instance, by the principle of duality in plane projective geometry. Its only relational concept is the incidence of point and line (point lies on line, line passes through point). It is possible to pair off uniquely the points and the lines in the plane in such a manner that, whenever a point P lies on a line q, the line p paired with P passes through the point Q paired with q. Consequently any valid theorem of projective geometry (phrased in terms of the directionless relation of incidence) at once becomes another valid theorem if the words 'point' and 'line' are interchanged. S. Lie discovered that the lines of (complex) space may be uniquely correlated with the spheres in such a manner that intersecting lines correspond to tangent spheres. An important part of analytic function theory, the so-called theory of uniformization, may be treated most naturally in the language of Bolyai-Lobatschewskyan geometry. Let an electrical network be given which consists of individual homo-

---

[4] Unfortunately English uses the same word 'know' for the two meanings that the author's German distinguishes as *kennen* and *erkennen*, and that the Latin, French and Greek languages express by the pairs *cognoscere* vs. *scire*, *connaître* vs. *savoir*, γνῶναι vs. εἰδέναι. Our translation is inconsistent in so far as it uses the terms *cognize*, *cognition* in contrast to *know* and *knowledge* only in places where the distinction is essential. [Translator's note.]

geneous wires connecting at various branch points; if by a 'Point' we understand an arbitrary current distribution, which assigns to each (oriented) wire s the intensity I_s of the electric current in s, then these Points satisfy the laws of a Euclidean space with a center O and as many dimensions as there are branches in the network. Here the central point O is represented by the absence of current where every I_s vanishes, and the square of the distance of a 'Point' from O is defined as the Joule heat developed by the current distribution per unit of time. This isomorphism is of value since it correlates the simple and important notions of geometry and the simple and important physical notions concerning electrical circuits. For instance, the basic problem of finding the current distribution when the various electromotive forces in the wires s are given is identical with the geometrical problem of finding the perpendicular projection of a Point onto a plane. The existence of a unique solution is thereby at once mathematically established, and a method for computing the solution made available.⎫

Pure mathematics, in the modern view, amounts to a general hypothetico-deductive theory of relations; it develops the theory of logical 'molds' without binding itself to one or the other among the possible concrete interpretations. Concerning this *formalization*, as "a point of view, without which an understanding of mathematical methods is out of the question," compare Husserl, *Logische Untersuchungen*, I, Sections 67–72. "The presupposition for the erection of a general arithmetic," Hankel declares (*Theory of Complex Numbers*, 1867, p. 10) "is thus a purely intellectual mathematics, dissociated from all intuition, a pure theory of forms, which has as its object not the combination of quanta or their images, the numbers, but intellectual objects, to which there may (but need not) correspond actual objects or relations." The axioms become *implicit definitions* of the basic concepts occurring in them. The concepts, admittedly, retain a certain range of indeterminacy; but the logical consequences of the axioms are valid, no matter what concrete interpretation may be adopted within this range. Pure mathematics acknowledges but one condition for truth, and that an irremissible one, namely *consistency*.

⎧Perhaps there already is an inkling of this modern view in Euclid's term for axioms: αἰτήματα, postulates. Leibniz takes some decisive steps towards the realization of a *mathesis universalis* in the sense here indicated and clearly understood by him. The theory of groups above all, that shining example of "purely intellectual mathematics," belongs within the framework of his *ars combinatoria*. A finite *group*

G is a system of a finite number of objects within which, in some way, an operation is defined which generates, from two (equal or different) elements a, b (in this order), an object ab of the system. The only postulates, or axioms, are these:

the associative law $a(bc) = (ab)c$;
if $a \neq b$ (a different from b), then also $ac \neq bc$, $ca \neq cb$.

From these insignificant looking assumptions springs an abundance of profound relationships; and mathematics offers an astounding variety of different interpretations of this simple axiom system. The group is perhaps the most characteristic concept of the mathematics of the 19th century.

The method of implicit definition is of importance also within the sciences themselves, and not only in the laying of their foundations. The *area* of a *piece*, where by the latter I will understand a piece of the plane that is bounded by line segments, satisfies the following requirements:

(i) The area is a positive number.
(ii) If a piece is dissected into two parts by a sequence of line segments in its interior, then the area of the whole is equal to the sum of the areas of the parts.
(iii) Congruent pieces have the same area.

These are the really essential properties of the concept of area but they contain no explicit definition of it. It can be shown, however, that these requirements are consistent and that a procedure can be devised by which every piece γ is assigned a positive number $J(\gamma)$ as its area which satisfies requirements (ii) and (iii). The requirements fail to determine the concept unambiguously; they are also satisfied, apart from $J(\gamma)$, by $c \cdot J(\gamma)$, where c is any positive constant independent of γ. But beyond this there are no further possibilities. The remaining arbitrariness as expressed by the factor c, can only be eliminated by the exhibition of an individual piece, say, a square, and the stipulation that it be assigned the area 1 (relativity of size). The significance of the implicit definition within all sciences, not only mathematics, is expounded very aptly in Schlick's *Allgemeine Erkenntnistheorie* (Berlin, 1918, pp. 30–37). "From the viewpoint of exact science, which strings inference after inference, a concept is indeed nothing but that of which certain propositions may be asserted. Thereby it should consequently also be defined." A suitable field of application, aside from the exact sciences, might be jurisprudence.}

REFERENCES

M. Pasch, *Vorlesungen über neuere Geometrie*, 2d ed., Berlin, 1926.
D. Hilbert, *Grundlagen der Geometrie*, 1899; 7th ed. Leipzig, 1930.
—— *Axiomatisches Denken*, Mathematische Annalen 78, 1918.
M. Geiger, *Systematische Axiomatik der euklidischen Geometrie*, Augsburg, 1924.
F. Gonseth, *Les mathématiques et la réalité*, Paris, 1936.

CHAPTER II

Number and Continuum, the Infinite

5. RATIONAL NUMBERS AND COMPLEX NUMBERS

THE genetic construction of the mathematical realm of numbers takes as its point of departure the sequence of natural numbers 1, 2, 3, ... The first step to be made is the rise from the natural numbers to the fractions. Historically *fractions* owe their creation to the transition from counting to *measuring*. All measuring is based on a domain of magnitudes, such as the segments on a line. We have here (1) a relation of equality, $\mathbf{a} = \mathbf{b}$ (congruence), satisfying the axioms set up for such a relation (p. 9), and (2) an operation applicable to any two segments $\mathbf{a}$, $\mathbf{b}$ and producing a segment $\mathbf{a} + \mathbf{b}$. From the segment $\mathbf{a}$ we obtain, say, the segment 5$\mathbf{a}$ by forming the sum $\mathbf{a} + \mathbf{a} + \mathbf{a} + \mathbf{a} + \mathbf{a}$ with 5 terms $\mathbf{a}$. This brings out the connection between counting and measuring. This process of iteration which leads from $\mathbf{a}$ to 1$\mathbf{a}$, 2$\mathbf{a}$, 3$\mathbf{a}$, . . . can be exactly explained as follows:

α) $1\mathbf{a} = \mathbf{a}$;

β) if n is a natural number, then $(n + 1)\mathbf{a}$ results from $n\mathbf{a}$ in accordance with the formula
$$(n + 1)\mathbf{a} = (n\mathbf{a}) + \mathbf{a}.$$

Within the domain of line segments, the operation of iteration admits of a unique inversion, partition: given a segment $\mathbf{a}$ and a natural number n, there exists one and (in the sense of equality) only one segment $\mathbf{x}$ such that $n\mathbf{x} = \mathbf{a}$; it is denoted by $\mathbf{a}/n$. The operation of partition may be combined with that of iteration. Thus e.g. we get 5$\mathbf{a}$/3, called '5/3 times' $\mathbf{a}$. The fractional symbol m/n serves as the symbol of the composite operation, so that two fractions are equal if the two operations denoted by them lead to the same result, no matter to what segment $\mathbf{a}$ they are being applied. *Multiplication* of fractions is performed by carrying out one after another the operations denoted by them. The possibility of *adding* fractions is due to the fact that the operation (applied to an arbitrary segment $\mathbf{x}$) that is expressed by

$$(m\mathbf{x}/n) + (m^*\mathbf{x}/n^*)$$

can be represented by a single fraction.

It is unnecessary to introduce special fractions for each domain of magnitudes. Since their laws are independent of the nature of these magnitudes, it is more expedient to define them in purely arithmetical terms.[5] This can be achieved by simply choosing as domain of magnitudes in the above considerations the natural numbers themselves. The fact that in this domain a relation between x and y, such $5x = 3y$, cannot always for a given x be solved with respect to y, does not impede the development of the theory. We thus arrive at the following formulation: "Two natural numbers m, n determine a fraction m/n. The statement that, of any two natural numbers x and y, the second is m/n times the first is merely another form of expressing the equation $mx = ny$." This is a creative definition in the sense of Section 2. Two fractions m/n, m^*/n^* are equal provided any numbers x, y which stand in the relation $mx = ny$ also stand in the relation $m^*x = n^*y$, and conversely. The operational rules for natural numbers permit one to replace this transfinite criterion, whose phrasing seems to require a checking through of all possible numbers x, y, by the following finite one:

(C) $$m \cdot n^* = n \cdot m^*.$$

Hence we deal with a special case of the definition by abstraction: the equality of the fractions m/n, m^*/n^* may be explained directly by (C), after one has convinced oneself that this relation is an equivalence. The introduction of fractions as 'ideal elements' can also be motivated purely arithmetically without reference to applications. Indeed after the numerical operations have been suitably extended to fractions it is found that all the important arithmetical axioms remain in force. Moreover division, the inverse operation of multiplication, can now always be carried out while this was only exceptionally so in the arithmetic of natural numbers.

⟦If the same idea is applied once again for the purpose of ensuring the invertibility of addition, then we get from the fractions to the *rational numbers* (which include 0 and the negative). (This, though, calls for one rather serious sacrifice — the possibility of division has to be abandoned for the divisor 0.) There are nowhere in this pro-

---

[5] This is in line with the oldest mathematical tradition, that of the Sumerians. Only after the discovery of the irrational did the Greeks abandon the algebraic road and find themselves compelled to couch algebraic facts in geometric terms. The post-classical Occident, partly stimulated by the algebraic achievements of the Arabs, reversed this development. There was little justification, however, for the modern viewpoint subsuming all quantities under a universal concept of number, before Dedekind gave Eudoxus' analysis of the irrational its constructive twist (cf. Section 7).

cedure any logical obscurities or philosophical difficulties. A much more serious matter is the starting point, the system of natural numbers, and then the irrational, the transition from the rational numbers to the continuum of real numbers. But once we have climbed to this level, the further advance toward the complex and hypercomplex numbers no longer leads past any abysses. In order to introduce the *complex numbers* it is only necessary to describe how any such number is given and how one is to operate with them. A complex number is given by its two components; thus we might as well say that we understand by a complex number any pair (α, β) of real numbers (Hamilton, 1837). We shall not set down the rules of operation here explicitly. According to them, $e = (1, 0)$ plays the part of unity in the complex domain, since its multiplication by any complex number (α, β) reproduces (α, β). And $(0, 1)$ is that imaginary unit i which satisfies the equation $i \cdot i = -e$. The inner reason for the stipulations is again to be seen in their extending the formal rules of computation from real to complex numbers. Nothing remains of the mystic flavor that was so long attached to the imaginary quantities.[6] From the complex it is possible to ascend to the *hypercomplex numbers* with 3 or more components. But it could be shown quite generally that, no matter how addition and multiplication be defined in their domain, the continued validity of all rules of operation of arithmetic is unattainable. In this respect the complex numbers denote a natural boundary for the extension of the number concept. Yet also hypercomplex number systems play their role in mathematics; thus the 4-component *quaternions*, which satisfy all rules of operation except the commutative law of multiplication, are a useful tool in dealing with the rotations of a rigid body in space.

Instead of constructing the realm of numbers genetically, arithmetic may also be based on an axiom system. From this viewpoint the genesis merely serves to reduce the consistency of that system to the consistency of the axioms governing the natural numbers. The

[6] For instance, Huyghens declares in 1674 (see Leibniz, *Mathematische Schriften*, II, p. 15) with reference to a complex formula: "Il y a quelque chose de caché là-dedans, qui nous est incompréhensible." Even Cauchy, in 1821, still has a somewhat obscure idea as to the manipulation of complex quantities. But negative quantities had produced almost as many headaches at an earlier time. Referring to the rule "minus times minus is plus," Clavius says in 1612: "debilitas humani ingenii accusanda (videtur), quod capere non potest, quo pacto id verum esse possit." Descartes, in accordance with contemporary usage, still designates the negative roots of an algebraic equation as false roots. The explanation, surviving in some textbooks, of i as that number which, when multiplied by itself, yields -1 is of course pure nonsense as long as only the real numbers are at one's disposal; it merely contains the demand that the number concept be so expanded and the sense of multiplication be so extended to the expanded number domain as to produce the desired equation.

axioms of arithmetic fall into two groups, the algebraic axioms and the axioms of magnitude.   The algebraic group deals with the operations of addition and multiplication.   It contains the formal rules of operation (such as $a + b = b + a$), requires the existence of a 0 and a 1 with the properties

$$a + 0 = 0 + a = a, \quad 1 \cdot a = a \cdot 1 = a$$

and the invertibility of addition and multiplication (with the exception of division by 0).   The axioms of magnitude (which do not carry over to the domain of complex numbers) deal with the relation $a > b$ (*a* greater than *b*).   Compare the table in Hilbert's *Grundlagen der Geometrie.*}

REFERENCES

W. R. HAMILTON, *Lectures on Quaternions*, Dublin, 1853.
H. HANKEL, *Theorie der komplexen Zahlen*, Leipzig, 1867.
D. HILBERT, *Grundlagen der Geometrie*.
O. HÖLDER, *Die Arithmetik in strenger Begründung*, 2d ed., Leipzig, 1928.

6. *THE NATURAL NUMBERS*

"The integers were created by God; all else is man-made," is a frequently quoted statement of Kronecker's.   In the natural numbers, the problem of cognition presents itself to us in its simplest form. Let us once more begin with the purely mathematical aspect.

The sequence of natural numbers commences with 1 and is generated by a process which yields from a number already obtained the next following number; never does an earlier number recur in this progression.   A concept (a characteristic or an operation) referring to arbitrary numbers can therefore be introduced only by *complete induction* (also called mathematical induction), namely by stating (α) what the concept means for the first number 1, and (β) how it carries over from any number n to the next following n' ($= n + 1$). Examples: The definition of na in the preceding section.   The concepts even and odd: (α) 1 is odd; (β) n' is even or odd according as n is odd or even.   The general notion of addition $a + n$ of two natural numbers a and n:

$$(\alpha) \quad a + 1 = a'; \quad (\beta) \quad a + n' = (a + n)'.$$

What is true of the concepts similarly holds for the proofs.   To prove that a certain theorem holds for every number one shows (α) that it holds for 1, and (β) that it holds for n' if n is a number for which it holds.   With the help of this method of definition and of proof by complete induction, of inference from n to $n + 1$, the theory of natural numbers can be completely built up step by step. That inference introduces an entirely new and peculiar feature unknown to Aristotelian logic into the mathematical method; it is the very essence of the art of mathematical demonstration. The first explicit mention of the principle of complete induction seems to be with B. Pascal (1654) and Jacob Bernoulli (1686).

In the building up of number theory by complete induction, the *succession* of numbers appears as their constitutive characteristic. They occur primarily as *ordinal numbers* and are distinguished only by their position in the sequence. Justly Schopenhauer (*Vierfache Wurzel vom zureichenden Grunde*, Section 38) says of this conception of number, "Every number presupposes the preceding ones as reasons for its being: I can get to ten only through all the preceding numbers. . . . " The well-known method of counting, applied to a given aggregate of objects, produces a certain natural number as the *number* (*Anzahl*) *of elements* in the aggregate. By virtue of the counting process the elements of the aggregate are themselves arranged in a sequence (first, second, third, . . .); and a special consideration is required to ensure the fundamental fact that the result of counting is independent of the order. Only thus is the concept of cardinal number put on a safe basis. Compare, for instance, the treatment by Helmholtz (Zählen and Messen, *Wissenschaftliche Abhandlungen*, III, p. 356); further L. Kronecker (*Werke*, III, 1, p. 249).

{The question has been argued extensively whether the concept of cardinal, rather than ordinal, number is not the primary one. The former, if it is to be introduced independently of an ordinal arrangement, has to be defined by abstraction (as on p. 10). This definition is not even restricted to finite sets; a theory of infinite cardinal numbers based thereon was developed by G. Cantor within the framework of his general set theory. But the criterion of numerical equivalence makes use of the possibility of pairing, which can only be ascertained if the acts of correlation are carried out one after another in temporal succession and the elements of the sets themselves are thereby arranged in order. Even if one follows the road of abstraction and splits up the act of numerical comparison of two sets by first ascribing a number to each set and then comparing these numbers, it remains indispensable to order each individual set itself by exhibiting its elements one by one in temporal succession. (Such a one-by-one exhibition is necessary anyhow if an aggregate is to be considered as concretely given; and the numbers employed by us in everyday life concern only such aggregates.) For this reason it seems to me unquestionable that the concept of ordinal number is the primary one. Modern research in the foundations of mathematics, which has destroyed dogmatic set theory, confirms this view.

Another point of debate is the question whether the numbers are independent ideal objects or whether arithmetic is concerned merely with the concrete *numerical symbols* "whose shape is recognizable by us with certainty independently of place and time, of the particular conditions of their manufacture, and of trifling differences in their execution" (Hilbert). Thus e.g. Helmholtz (Zählen und Messen, *loc. cit.*, p. 359): "I consider arithmetic, or the theory of pure numbers, as a method built upon purely psychological facts, by which the consistent application of a system of symbols of unlimited extent and unlimited possibility of refinement is taught. In particular, arithmetic investigates what different modes of combination of these symbols (numerical operations) lead to the same result." Only recently Hilbert carried this point of view consistently into effect (compare Section 10), in a manner unassailable even by the criticism directed against it by Frege (*Grundgesetze der Arithmetik*, 1893). A succession of strokes ('ones') offers itself as a suitable symbol. If I hear a sequence of tones, I put down a stroke upon hearing each one, placing one stroke after another: ////. A second time I proceed similarly, again obtaining a symbol consisting of a succession of strokes. If I were immediately able to judge the equality or disparity of the 'shape' of the two symbols, a numerical comparison would be accomplished. Here the representation of the data by strokes has the function of putting these data into a 'normal form' of such a kind that a difference in shape at once indicates a difference in number. (For a directly given whole, number is meant to describe a relation between the whole and such parts of it as are considered as units. A difference in the shape of two wholes does not necessarily imply a difference in the number of units; e.g. ∶∶ and ∴. An act of assembling is said to be the basis for determining the number of elements. It seems to me that the application of the symbolic method of counting to a structural whole of units does not require that a mere 'aggregate' be abstracted by dissolving the structural tie; nor need individually given elements, such as successive tones, be assembled to form an aggregate. The statement 'there were this many tones: ////' is quite intelligible in itself, and it is unnecessary to search for an 'aggregate of the tones heard.') The immediate recognition of equality or disparity of two symbols consisting of successions of strokes is possible, however, for the lowest numbers only. In general one has to proceed by using the strokes recorded during the first sequence over again, say, by crossing them out one by one; for this purpose it is required that the first sequence stays put (and does not disappear like the tones themselves). In principle, symbols can be dispensed with for the verification of a statement such as 'this time there were more tones than the first time,' provided the tones of the first sequence (which may have been falling in pitch) can be reproduced in their temporal succession while the second sequence is being listened to. Symbols become indispensable only when the comparison is torn up into two number determinations ('the first time there were 4, now there are 5 tones; 5 is greater than 4'); for then part of the mental operation ('5 is greater than 4') is shifted onto the permanent symbols, which are at the same time expedient for preservation and communication. Thus it is not the comparison of numbers but the determination of numbers which is of an essentially symbolic character. 'There were 4 tones' is unintelligible without reference to a symbol.

If one wants to speak, all the same, of numbers as concepts or ideal objects, one must at any rate refrain from giving them independent existence; their being exhausts itself in the functional role which they play and their relations of more or less. (They certainly are not concepts in the sense of Aristotle's theory of abstraction.)

The employment of several digits and the *positional system* (developed in Mesopotamia and later consistently by the Indians for written numbers) permit a quick decision about greater and smaller for much larger numbers than the simple numerical symbols composed of successive Ones; this considerable practical advantage is not one of principle however. The basis of the number system, which in our system is ten, is different with different cultures. The Indian, and particularly the Buddhistic, literature revels in the possibilities of producing and designating prodigious numbers by means of the positional system, that is, by combination of addition, multiplication, and exponentiation. In spite of their fantastic aspect there is something truly great in these efforts; the human mind for the first time senses its full power to fly, through the use of the symbol, beyond the boundaries of what is attainable by intuition. Something akin we find among the Greeks only in the latest epoch, namely, in Archimedes' paper addressed to Gelon "The Sand-reckoner"; and here is manifested the delight, not in the step by step opening-up of the infinite, but in the rational subjugation of the unbounded.

Regarding the *relation of number to space and time* we may say that time, as the form of pure consciousness, is an essential, not an accidental, presupposition for the mental operations on which the sense of a numerical statement is founded. Contrary to the opinion of some philosophers (e.g. Hobbes), this does not apply to space, although permanent symbols having a spatial configuration are the

36

most convenient means of putting down a result of counting, of storing and communicating it, and of safeguarding the manipulation of numbers. Kant above all has emphasized the bond between the number concept and time, but it would be going too far if one were to claim arithmetic as the science of time in the same sense that geometry is the science of space.

With reference to two concretely given numerical symbols, m and n, the sense of the proposition $m + n = n + m$ can be described without having to 'generate' any other numbers. It is also possible to see that this proposition holds in *any* concrete situation. Something new happens, however, when I imbed the actually occurring numerical symbols in the *sequence of all possible numbers.* That sequence is produced through a generating process according to the principle that any given number gives rise to a new, the next following, number by the addition of One. Here the being is projected upon the background of the *possible*, of a manifold of possibilities which is produced by a fixed process and yet is open towards infinity. This is the standpoint held by us at the beginning of the present section when arithmetic was founded on the principle of complete induction. We rely on it when we speak of a trillion ($= 10^{12}$) dollars. By repeated application of definitions by complete induction we obtain from the prime arithmetical process of changing n into $n + 1$ the operation of multiplication by 10, and by performing this operation 12 times (beginning with 1), we arrive at the desired number 10^{12}. The numbers 10 and 12 can be written out in strokes; as for 10^{12}, this has never been done, and yet we maintain the 'fiction' of such a number.}

Thus it is already in the field of numbers that we encounter the following basic features of *constructive cognition:*

1. We ascribe to that which is given certain characters which are not manifest in the phenomena but are arrived at as the result of certain mental operations. It is essential that the performance of these operations is held universally possible and that their result is held to be uniquely determined by the given. But it is not essential that the operations which define the character be actually carried out.

2. By the introduction of symbols the assertions are split so that one part of the operations is shifted to the symbols and thereby made independent of the given and its continued existence. Thereby the free manipulation of concepts is contrasted with their application, ideas become detached from reality and acquire a relative independence.

3. Characters are not individually exhibited as they actually occur, but their symbols are projected on the background of an ordered mani-

fold of possibilities which can be generated by a fixed process and is open into infinity.

Cognition has not stopped here. The leap into the beyond occurs when the sequence of numbers that is never complete but remains open toward the infinite is made into a closed aggregate of objects existing in themselves. Giving the numbers the status of ideal objects becomes dangerous only when this is done. The belief in the absolute is deeply implanted in our breast; no wonder, then, that mathematics was bold and naive enough to perform that leap. Whoever accepts as meaningful the definition 'n is an even or odd number according as a number x does or does not *exist* such that $n = 2x$,' which refers to the infinite totality of all numbers (the definition of even and odd by complete induction, as mentioned earlier, is a different matter), already stands on the other shore; for him the system of numbers has become a realm of absolute existences which is 'not of this world' and from which only gleams here and there are caught and reflected in our consciousness. The vindication of this transcendental point of view forms the central issue of the violent dispute which has flamed up again today over the foundations of mathematics. The issue is symptomatic for all knowledge and may, in the field of mathematics sooner than elsewhere, lead to a clear decision.

REFERENCES

R. DEDEKIND, *Was sind und was sollen die Zahlen?*, Braunschweig, 1888; 3rd ed., 1911.

G. FREGE, *Die Grundlagen der Arithmetik*, Breslau, 1884; reprinted 1934. *Grundgesetze der Arithmetik*, 2 vols., Jena, 1893–1903.

E. HUSSERL, *Philosophie der Arithmetik*, Halle, 1891.

7. THE IRRATIONAL AND THE INFINITELY SMALL

In a different form than in the sequence of integers we encounter the infinite in the *continuum*, which is capable of infinite division. Cases of special importance are the continua of time and of space. Here we find the second open place in the above described construction of the mathematical realm of numbers. Antiquity has bequeathed to us two important contributions to the problem of the continuum: (a) a far-reaching analysis of the mathematical question of how to fix a single position in the continuum, and (b) the discovery of the philosophical paradoxes which have their origin in the intuitively manifest nature of the continuum.

[(a) The pure geometry of the Greeks, in elevating itself above the inexactitude of the sense data, applies the idea of existence (not only to the natural numbers but also) to the points in space. The discovery of the irrationality of the ratio $\sqrt{2}$ of the diagonal and side of a square made it clear that the fractions are not the only possible quantities measuring ratios of line segments, and thus not the only 'real numbers.' In the Platonic dialogues the deep impression can be sensed which this mathematical discovery made upon the rising scientific consciousness of his time. Independently of the particular geometrical constructions which led to individual irrationalities such as $\sqrt{2}$, Eudoxus recognized the general foundations of this phenomenon.

1. In place of the untenable commensurability he sets down the axiom: if **a** and **b** are any two segments, then **a** can always be added to itself so often that the sum $n\mathbf{a}$ exceeds **b**. This means that all segments are of a comparable order of magnitude, or that there exists neither an actually infinitely small nor an actually infinitely large in the continuum.

2. And what is it that characterizes the individual segment ratio? Eudoxus replies: two segment ratios, $\mathbf{a}:\mathbf{b}$ and $\mathbf{a}':\mathbf{b}'$, are equal to each other if, for arbitrary natural numbers m and n, the fulfillment of the condition in the first line below invariably entails the validity of the corresponding condition in the second line:

$$(I) \begin{cases} n\mathbf{a} > m\mathbf{b} \\ n\mathbf{a}' > m\mathbf{b}' \end{cases} \quad (II) \begin{cases} n\mathbf{a} = m\mathbf{b} \\ n\mathbf{a}' = m\mathbf{b}' \end{cases} \quad (III) \begin{cases} n\mathbf{a} < m\mathbf{b} \\ n\mathbf{a}' < m\mathbf{b}' \end{cases}$$

Hence what is characteristic of the individual real number α is the cut which it creates in the domain of rational numbers by dividing all fractions m/n into three classes, those which are (I) less than α, (II) equal to α, and (III) greater than α. The second class is either empty or contains only a single fraction. The first axiom guarantees that no two different segments can have the same ratio to the fixed unit segment. Euclid's theory of proportions is likewise erected on this foundation, while Archimedes bases on it his general method of exhaustion.

Only in the 19th century did mathematics go beyond Eudoxus, and settled the problem in a more definite fashion. For Eudoxus the real number is given as the ratio of two given segments, and thus it is up to the axioms of geometry to tell us what segment ratios exist. But in Euclidean geometry it is not possible to construct (by means of ruler and compass) from a given segment 1, the segment $\sqrt[3]{2}$, which would solve the Delian problem of duplicating the cube, or of the segment π, which equals the circumference of a circle of diameter 1. Yet we are convinced of their existence on the basis of continuity considerations: if the edge of a cube increases from 1 to twice that size, the

39

volume of the cube rises continuously from 1 to 8, hence must pass the value 2 at some time. As for the segment π, we can approximate it from below and from above with any degree of accuracy by the Euclideanly constructible perimeters of regular 6-, 12-, 24-, . . . sided polygons inscribed to and circumscribed about the circle. Thus we are turning the tables: *any arbitrarily given cut* in the domain of rational numbers, that is to say, any division of all rational numbers into three classes I, II, III, no matter in what way effected, determines a real number. (The only requirements to be satisfied are the following: neither I nor III must be empty; II contains at most one fraction; I contains no largest, III no smallest fraction; any number in I is smaller than any number in II or III; any number in III is greater than any number in I or II.) According to Dedekind (*Stetigkeit und Irrationalzahlen*, 1872), we have no reason to admit only part of these cuts as real numbers. And in geometry we then postulate (Dedekind's axiom) the existence of that segment which stands to the given unit segment in the ratio determined arithmetically by the cut. Since conversely, according to Eudoxus, the ratio of any segment **a** to the unit segment determines a cut, the axiom of Dedekind guarantees the *completeness* of the geometrical elements: the system of points is incapable of extension, provided all axioms (including that of Eudoxus) are maintained (Hilbert). This logical completeness (absence of gaps) reflects the intuitive continuity among the points in space. With Dedekind's number concept, analysis makes itself independent of geometry. Thereby, at last, it is in a position to analyze continuity and to provide geometry with the means of proving continuity theorems of the following kind: a continuous curve joining the center of a circle to a point outside the circle meets the circumference. In Euclid, the proofs of such theorems are incomplete, as was already pointed out by Leibniz with reference to the first construction occurring in Euclid, namely that of the equilateral triangle ABC from the points A and B; Euclid fails to show that the circle about A through B and the circle about B through A have a point in common.

Another means of characterizing a real number, equivalent to that of the cut, is the infinite sequence of 'nested' rational intervals $a_n b_n$ ($n = 1, 2, 3, . . .$), each of which lies within the preceding one, and the length $b_n - a_n$ of which converges to 0 as the index n increases indefinitely (compare the example of π). Since the fraction is logically no more complicated than the natural number — it is determined by two natural numbers, its numerator and denominator — we may sum up the result of the historical development of Problem (a) as follows:}

The individual natural numbers form the subject of number theory, the possible sets (or the infinite sequences) of natural numbers are the subject of the theory of the continuum.

(b) The essential character of the continuum is clearly described in this fragment due to Anaxagoras: "Among the small there is no smallest, but always something smaller. For what is cannot cease to be no matter how far it is being subdivided." The continuum is not composed of discrete elements which are "separated from one another as though chopped off by a hatchet." Space is infinite not only in the sense that it never comes to an end; but at every place it is, so to speak, inwardly infinite, inasmuch as a point can only be fixed step-by-step by a process of subdivision which progresses *ad infinitum*. This is in contrast with the resting and complete existence that intuition ascribes to space. The 'open' character is communicated by the continuous space and the continuously graded qualities to the things of the external world. A real thing can never be given adequately, its 'inner horizon' is unfolded by an infinitely continued process of ever new and more exact experiences; it is, as emphasized by Husserl, a limiting idea in the Kantian sense. For this reason it is impossible to posit the real thing as existing, closed and complete in itself. The continuum problem thus drives one toward epistemological idealism. Leibniz, among others, testifies that it was the search for a way out of the "labyrinth of the continuum" which first suggested to him the conception of space and time as orders of the phenomena. "From the fact that a mathematical solid cannot be resolved into primal elements it follows immediately that it is nothing real but merely an ideal construct designating only a possibility of parts" (correspondence Leibniz-De Volder, Leibniz, *Philosophische Schriften*, II, p. 268).

[In contrast to this nature of the continuum, Leibniz conceives the idea of the *monads*, since — differently from Kant — he feels compelled to give the phenomena metaphysically a foundation in a world of absolute substances. "Within the ideal or the continuum the whole precedes the parts. . . . The parts are here only potential; among the real [i.e. substantial] things, however, the simple precedes the aggregates, and the parts are given actually and prior to the whole. These considerations dispel the difficulties regarding the continuum — difficulties which arise only when the continuum is looked upon as something real, which posesses real parts before any such division as we may devise, and when matter is regarded as a substance" (letter to Remond, *Philosophische Schriften*, III, p. 622).

The impossibility of conceiving the continuum as rigid being cannot be formulated more concisely than by Zeno's well-known paradox of the race between Achilles and the tortoise. The remark that the successive partial sums $1 - \dfrac{1}{2^n}$ $(n = 1, 2, 3, \cdots)$ of the series

$$\frac{1}{2} + \frac{1}{2^2} + \frac{1}{2^3} + \cdots$$

do not increase beyond all bounds but converge to 1, by which one nowadays thinks to dispose of the paradox, is certainly relevant and elucidating. Yet, if the segment of length 1 really consists of infinitely many subsegments of lengths ½, ¼, ⅛, . . . , as of 'chopped-off' wholes, then it is incompatible with the character of the infinite as the 'incompletable' that Achilles should have been able to traverse them all. If one admits this possibility, then there is no reason why a machine should not be capable of completing an infinite sequence of distinct acts of decision within a finite amount of time; say, by supplying the first result after ½ minute, the second after another ¼ minute, the third ⅛ minute later than the second, etc. In this way it would be possible, provided the receptive power of the brain would function similarly, to achieve a traversal of all natural numbers and thereby a sure yes-or-no decision regarding any existential question about natural numbers!

Descartes struggles with the idea that the material corpuscles of a liquid in motion have to divide *in infinitum,* "or at least *in indefinitum,* and that into so many parts that it is impossible to imagine one, however small, of which one would not know that it was actually subdivided into still smaller parts." To him this remains a mystery, confronted with which he takes recourse to the incomprehensibility of the divine omnipotence. Euler, in his "Anleitung zur Naturlehre" (*Opera postuma,* II, 1862, pp. 449–560), which in magnificent clarity summarizes the foundations of the philosophy of nature of his time, declares that although the bodies are infinitely divisible the statement that every body consists of infinitely many ('ultimate') parts is entirely false and is even obviously incompatible with the infinite divisibility (Euler, *op. cit.,* Chap. II, §12). In the Kantian system, the first two antinomies of pure reason refer to the continuum.[7]}

Three attempts have been made in the history of thought to conceive of the continuum as Being in itself. According to the first

[7] The first of these, however, is formulated misleadingly. According to the argument presented, it is not a question of whether the world does or does not have a temporal beginning, but whether the number of temporal moments up to the present time is finite or infinite. In a continuously filled time, the latter will be the case, no matter whether (by virtue of an intrinsic or extrinsic measuring principle) it be of finite or infinite length.

and most radical the continuum consists of countable discrete elements, atoms. With regard to *matter*, this path, initiated by Democritus in antiquity, has been followed with brilliant success in modern physics. Plato, clearly conscious of the goal of 'saving' the phenomenon by means of the idea, was the first to design a consistent atomism with respect to *space*. In Islamic philosophy the atomistic theory of space was renewed by the Mutakallimûn (see Lasswitz, *Geschichte der Atomistik*, I, 1890, pp. 139–150), and in the Occident by Giordano Bruno's doctrine of the minimum. Hume, too, in his space-time theory (*Treatise of Human Nature*, Book I, Part II, Section 4) transforms the vagueness of the sense data, at which he aims, into a composition out of indivisible elements. Stimulated by quantum theory the idea again arises today in discussions about the foundations of physics. But so far it has always remained mere speculation and has never achieved sufficient contact with reality. How should one understand the metric relations in space on the basis of this idea? If a square is built up of miniature tiles, then there are as many tiles along the diagonal as there are along the side; thus the diagonal should be equal in length to the side. Hume, consequently, is forced to admit that the "just as well as obvious" principle of comparing the measures of curves and surfaces by means of the number of component elements is, in fact, *useless*. B. Riemann, in his inaugural lecture *Über die Hypothesen, welche der Geometrie zugrunde liegen* (1854), states the alternative "that for a discrete manifold the principle of measurement is already contained in the concept of this manifold, but that for a continuous one it must come from elsewhere."

The second attempt is that of the infinitely small. This is discussed ingeniously and in detail on the first day of Galileo's "Discorsi." Just as I can bend a straight line segment into an octagon or a thousand-sided polygon, so, according to Galileo, I may also transform it into a polygon with infinitely many infinitely small sides by simply winding it around a circle, and thus do not have to rely on a limiting process which never reaches the goal.[8]

[If a wheel is rolled off along a horizontal line, then every one of the smaller concentric circles appears to be stretched out in the form

---

[8] Hankel says (*Zur Geschichte der Mathematik im Altertum und Mittelalter*, Leipzig, 1874): "The idea of never reaching the area of the circle, no matter how far one might go in the sequence of polygons, although one approaches it arbitrarily closely, strains the power of imagination to such a degree that it will tend, at all cost, to bridge this gap extending, as it were, between reality and the ideal. Under this psychological pressure the — infinitely small or infinitely large? — step is taken that leads to the assertion: the circle is a polygon with infinitely many infinitely small sides. The Ancients, however, have refrained from this step; as

of a line h of equal length (*rota Aristotelis*). However, if the circular wheel is replaced by a many-sided regular polygon, then the 'covered' segments along h, into which the sides of the polygon fall successively, form a disrupted line. Thus, in the case of the circular wheel, one must assume that h consists of an infinitely dense succession of covered and uncovered segments. "This method," says Galileo in the "Discorsi" (*Opere* VIII, p. 93), "perhaps better than any other, enables us to avoid many intricate labyrinths such as are encountered in the question of cohesion in solids, mentioned before, and that of rarefaction and contraction, without forcing upon us the objectionable admission of empty spaces and thereby of the penetrability of bodies. We escape all these difficulties, so it seems to me, by assuming a composition out of indivisibles." If a curve consists of infinitely many straight 'line elements,' then a tangent can simply be conceived as indicating the direction of the individual line element; it joins two 'consecutive' points on the curve. However, he who rejects Galileo's hypothesis has to define the tangent at the point P of a curve as the limiting line approached indefinitely by the secant line PP' as the second moving point P' on the curve converges toward P. The discussion between Johann Bernoulli and Leibniz on this question is very instructive. Leibniz says (*Mathematische Schriften*, III, p. 536), "For if we suppose that there actually exist the segments on the line that are to be designated by $\frac{1}{2}, \frac{1}{4}, \frac{1}{8}, \ldots$, and that *all* members of this sequence actually exist, you conclude from this that an infinitely small member must also exist. In my opinion, however, the assumption implies nothing but the existence of any *finite* fraction of arbitrary smallness." But Bernoulli replies (*op. cit.*, p. 563). "If 10 members are present the 10th necessarily exists, if 100 then necessarily the 100th, $\ldots$, if therefore their number is ∞ then the ∞^{th} [infinitesimal] member must exist."}

The limiting process was victorious. For the *limit* is an indispensable concept, whose importance is not affected by the acceptance or rejection of the infinitely small. But once the limit concept has been grasped, it is seen to render the infinitely small superfluous. Infinitesimal analysis proposes to draw conclusions by integration from the behavior in the infinitely small, which is governed by elementary laws, to the behavior in the large; for instance, from the universal law of attraction for two material 'volume elements' to the magnitude of attraction between two arbitrarily shaped bodies with homogeneous or non-homogeneous mass distribution. If the infinitely

---

long as there were Greek geometers, they have always halted in front of the precipice of the infinite. . . . "

44

small is not interpreted 'potentially' here, in the sense of the limiting process, then the one has nothing to do with the other, the processes in infinitesimal and in finite dimensions become independent of each other, the tie which binds them together is cut. Here Eudoxus undoubtedly saw right.

〔Incidentally, as far as I can see, the 18th century remained far behind the Greeks with regard to the clarity of its conception of the infinitely small. More than one writer of this enlightened era complains of the 'incomprehensibilities of mathematics,' and vague and incomprehensible indeed is their notion of the infinitesimal. As a matter of fact, it is not impossible to build up a consistent 'non-Archimedean' theory of quantities[9] in which the axiom of Eudoxus (usually named after Archimedes) does not hold. But as was just pointed out, such a theory fails to accomplish anything for analysis. Newton and Leibniz seemed to have the correct view, which they formulated more or less clearly, that the infinitesimal calculus is concerned with the approach to zero by a limiting process. But they lack the ultimate insight that the limiting process serves not only to determine the value of the limit but also to establish its existence. For that reason Leibniz is still quite unclear as to the summation of infinite series. Only slowly does the theory of limits gain a foothold. In 1784 D'Alembert declares emphatically in the *Encyclopédie*, "La théorie de la limite est la base de la vraie métaphysique du calcul différentiel. Il ne s'agit point, comme on le dit ordinairement, des quantités infiniment petites; il s'agit uniquement des limites des quantités finies." It was left to Cauchy, at the beginning of the 19th century, to carry these ideas out consistently. In particular he discovers the correct criterion for the convergence of infinite series, the condition under which a number is generated as limiting value through an infinite process. The proof of the criterion, however, requires that fixation of the number concept which was later accomplished by the principle of the Dedekind cut.〕

The third attempt to 'save' the continuum in the Platonic sense may be seen in the modern set-theoretic foundations of analysis.

REFERENCES

R. Dedekind, *Stetigkeit und irrationale Zahlen*, Braunschweig, 1872; third edition 1905.
　　Concerning the ancient history of the problem of the continuum and the irrational cf.:

---

[9] Compare, for instance, Hilbert, *Grundlagen der Geometrie*, Chapter II, §12. An example of infinitely small magnitudes, discussed already by Leibniz and Wallis, are the *anguli contactus* (between, say, a circle and its tangent) as opposed to the angles formed by straight lines.

P. TANNERY, *Pour l'histoire de la science hellène*, Paris, 1887.

E. FRANK, *Platon und die sogenannten Pythagoreer*, Halle, 1923.

H. HASSE and H. SCHOLZ, *Die Grundlagenkrise der griechischen Mathematik*, Berlin, 1928.

B. L. VAN DER WAERDEN, *Mathematische Annalen*, 117 (1940), pp. 141–161.

K. VON FRITZ, *Annals of Mathematics*, 46 (1945), pp. 242–264.

8. SET THEORY

At a first glance it might seem as though with the limiting process the rigid *Being* is definitely resolved into *Becoming;* as though, thereby alone, Aristotle's doctrine is mathematically realized which taught that the infinite is forever being on the way and therefore exists only δυνάμει not ἐνεργείᾳ (potentially, not actually). This appearance is deceptive. For the individual convergent sequence, such as the sequence of partial sums of the Leibniz series

$$ \tfrac{1}{1} - \tfrac{1}{3} + \tfrac{1}{5} - \tfrac{1}{7} + \ldots , $$

which converges to $\pi/4$, does not unfold itself according to a lawless process which we have to accept blindly in order to find out what it produces step by step; but it is fixed once and for all by a definite *law*, which correlates with every natural number n the corresponding approximate value (the nth partial sum). A classification of the infinitely many rational numbers into the three classes I, II, III of a Dedekind cut is not made by taking one fraction after another and assigning it to its class, but rather according to a law, namely, by stating that all rational numbers with such and such a property are to belong to class I. (It suffices to define class I, since the other two classes are defined automatically along with it.) This law, or this property, fixes the intended real number exactly.

It is said that a function $f(x)$ is continuous at the place $x = a$ if $f(x)$ converges to $f(a)$ when the variable x approaches a. But how is this notion of convergence defined? "For *every* positive ϵ there should *exist* a positive δ of such a kind that, for *all* real numbers x which satisfy the condition $a - \delta < x < a + \delta$, we also have $f(a) - \epsilon < f(x) < f(a) + \epsilon$." Our attitude thus remains static. It is characterized by the unlimited application of the terms 'there exists' and 'all' not only to natural numbers but also to the places in the continuum, i.e. to the possible sequences or sets of natural numbers. This is the essence of set theory: It considers not only the sequence of numbers but also the totality of its subsets as a closed aggregate of objects existing in themselves. In this sense it is based on the actually infinite. But once this has been accepted, the vast structure of analysis has an unshakeable firmness; it is securely founded, in all its parts based on sound argument, exact in its concepts, without gaps

46

in its proofs. It has thus gained a foundation which guarantees the unconditional intersubjective agreement of all workers in its field.

〔To be sure, considerable mathematical acumen was required thus to establish such general facts concerning continuity as are suggested by intuition; for instance, that a continuous function assumes all intermediate values, that a closed planar curve without multiple points divides its plane into two domains, or that a two-dimensional domain cannot be mapped continuously and in a one-to-one fashion into a three-dimensional domain. We experience again and again with our students what assiduous training is necessary in order to acquire that freedom from prejudice which is indispensable for a proper understanding of these proofs and their stringency. Besides such theorems confirming our intuition, analysis also reveals numerous occurrences which appear to run counter to it, such as continuous curves being everywhere without a tangent or filling out an entire square. It was the work of the 19th century from Cauchy and Gauss to Weierstrass to test all unproved suppositions of analysis on the above foundation.〕

The set-theoretical method has permeated not only analysis but also the first beginning of mathematics, the theory of the natural numbers. From the point of view of set theory, the number sequence is a completed set N, within which a mapping $n \to n'$ is defined that uniquely correlates an element n' with every element n of the set. This very fact, the existence of a one-to-one mapping of N onto a subset of N that is not identical with the entire N (the correlations $n \to 2n$ or $n \to n^2$ have the same effect), shows N to be an *infinite* set. The finiteness of a set is established only when the impossibility of such a mapping has been demonstrated.

〔Thus, for set theory, there is no difference in principle between the finite and the infinite. The infinite even appears to it as the simpler of the two (in agreement with Descartes, who maintained that the infinite is prior to the finite [letter to Clerselier, *Corr.*, ed. Adam and Tannery, V, p. 356, "Or je dis que la notion que j'ai de l'infini est en moi avant celle du fini"; also Méditations métaphysiques, third meditation, Oeuvres de Descartes, I, pp. 280–281]). The fact that, in the definite sense stated, Euclid's axiom of magnitude καί τὸ ὅλον μέρους μεῖζον ("the whole is greater than the part") fails to hold for an infinite set was pointed out already by Galileo (Discorsi, *Opere*, VIII, p. 79). From this, Leibniz concludes (letter to Bernoulli, *Math. Schriften*, III, p. 536) that "the number, or set, of all numbers entails a contradiction if one conceives of it as a completed whole." Bolzano sees in it a "paradox of the infinite" (*Paradoxien des Unendlichen*, 1851, §20). Dedekind, finally, elevates this fact to the status of a definition of the infinite (*Was sind und was sollen die Zahlen?*, 1887).⎦

Following Dedekind, a set C of natural numbers is said to be a *chain* if, for every number x contained in C, its 'image' $x' = x + 1$ likewise belongs to C. The fact that every natural number can be reached by starting with 1, going on to its image $1'$ ($= 2$), obtaining $2'$ ($= 3$) by repeating the mapping, *and so on*, — the idea of this 'and so on,' that seems logically irreducible, but constitutes the essence of the natural number sequence, is then expressed in the form of the following principle: *Every chain which contains 1 as an element is identical with the whole of N.* Complete induction can therefore be based on the transfinite use of the concepts 'all' and 'there is'; in this way set-theory abolishes the partition between mathematics and logic. The investigations of Dedekind, Frege, and Russell aim at logicizing mathematics completely.

The question as to when a natural number n is less than a given number m, which common sense answers by the finite specifically arithmetical criterion: 'if the enumeration of the numbers from 1 to m leads to n before m is reached,' is decided in set theory by the following transfinite purely logical criterion: 'if there exists a chain containing m but not n.' But such a thing is possible only after one has climbed to that level of application of 'there exists' where this term refers to the *sets* of natural numbers.

And it is for this purpose alone that we require that objectification of sets which everyday language, strangely enough, has carried out all along. A proposition such as 'the rose is red' is no longer subordinated to the scheme 'x is red,' having one blank, x, but to the more general one 'x has the property X,' from which the proposition results by the substitution $x =$ rose, $X =$ red. The words 'has the property' denote a certain relation ε, which may hold between the arbitrary object x and the arbitrary property X. Only in this connection do we encounter the *copula* ε; it changes the originally bipartite proposition into a tripartite one, $x \, \varepsilon \, X$. (The grotesque confusion of the copula with existence and with equality is one of the saddest indications of the dependence of philosophical speculation on accidental linguistic forms.) The way is now open for a formal application of the definitional principles 6. and 7. of Section 1 to the blank X. The introduction of the general set concept thus consists of two essentially different steps; the first is the objectification just described, the second is the stipulation that two properties X and Y, or the corresponding sets, be considered equal if all elements of X also belong to Y, and vice versa.

From an aggregate of individually exhibited objects we may by selection produce all possible subsets and thus make a survey of them one after another. But when one deals with an infinite set like N, then the existential absolutism for the subsets becomes still more objectionable than for the elements. Since one can lay hands on such subsets only as are determined by a characteristic property of their elements, it is difficult to rid oneself of the feeling that a chaotic abundance of possibilities, of sets put together haphazardly and without rule or law, goes by the board. But the paradoxical character of the elusive 'aggregate of all possible properties of natural numbers' can be laid bare even more precisely. Suppose we had somehow succeeded in the demarcation of an 'extensionally definite' aggregate of such properties (I shall call them properties of the first level), so that we have the right to believe that the question 'is there a property of the first level of such and such a well-described kind A?' is answered by the facts with a clear-cut yes or no. We may then speak of the property P_A which applies to a number x if and only if there *exists* any property at all of the first level which appertains to x. This property P_A, however, according to its meaning certainly lies outside the circle of properties of the first level; it belongs to a higher, the second level, since it has been defined in terms of the totality of properties of the first level. Russell formulates this insight somewhat vaguely by his "vicious circle principle": "No totality can contain members defined in terms of itself." Similarly, the third level is constructed above the second, and so forth. Correspondingly, sets of natural numbers — and hence real numbers — of the first, second, third, . . . levels should be distinguished. The mode of construction of the property P_A occurs in analysis, for instance, in determining the least upper bound of a point set on a line. The obliteration by the existential absolutism of these differences in level, which were first brought out in Russell's theory of types, constitutes an unquestionable vicious circle.

[One could escape this dilemma only if, for every property of the second level, there existed a property of the first level equal to it (not in meaning but) in extension. As long as the sequence of natural numbers is accepted as an extensionally definite aggregate, one might consider as the properties of the first level those which are generated by the definitional principles of Section 1 from the one basic relation 'n follows upon m.' In this case, our wish will hardly be fulfilled. We would have the task of extending the principles of construction for the properties of the first level in such a manner that every set of the second level demonstrably coincides with one of the first. But there is not the slighest indication that this is possible. Russell, in order to extricate himself from the affair, causes reason to commit hara-kiri, by postulating the above assertion in spite of its lack of support by any evidence ('axiom of reducibility'). In a little book *Das Kontinuum*, published in 1918, I have tried to draw the honest consequence and constructed a field of real numbers of the first level, within which the most important operations of analysis can be carried out.]

In spite of its paradoxical character, the idea of absolute existence in the domain of natural numbers and sets of natural numbers has so far not ~~yet~~ led to any contradiction. G. Cantor, however, freed himself of all fetters and manipulated the set concept without any restriction, in particular permitting the formation of the set of all subsets of any given set. He developed a general theory of cardinal and ordinal numbers of infinite sets. Here, at the farthest frontiers of set theory, actual contradictions did show up. But their root can only be seen in the boldness perpetrated from beginning in mathematics, namely, of treating a field of constructive possibilities as a closed aggregate of objects existing in themselves.

REFERENCES

B. BOLZANO, *Paradoxien des Unendlichen*, posthumous edition Přihouský, Leipzig 1851.
R. DEDEKIND, *Was sind und was sollen die Zahlen?*
G. FREGE, *Die Grundlagen der Arithmetik.*
G. CANTOR, *Gesammelte Abhandlungen*, Berlin, 1932; in particular Sections III and IV.
B. RUSSELL, *The Principles of Mathematics.*
H. WEYL, *Das Kontinuum*, Berlin, 1918.
A. FRAENKEL, *Einleitung in die Mengenlehre*, 3d. ed., Berlin, 1928.

9. INTUITIVE MATHEMATICS

This situation was first clearly recognized by L. E. J. Brouwer (since 1907). He designed a system of mathematics which does not make that leap into the beyond of which we spoke at the end of Section 6. An existential statement, such as 'there exists an even number,' is not considered a proposition in the proper sense that asserts a fact. An 'infinite logical summation' such as is called for by a statement of this kind (1 is even or 2 is even or 3 is even or . . . *ad infinitum*) is evidently incapable of execution. '2 is an even number,' this is a real proposition (provided 'even' has been defined recursively as on p. 33); 'there exists an even number' is nothing but a *propositional abstract* derived from that proposition. If I consider an insight a valuable treasure, then the propositional abstract is merely a document indicating the presence of a treasure without disclosing its location. Its only value may lie in the fact that it causes me to look for the treasure. It is a worthless piece of paper as long as it is not endorsed by a real proposition such as '2 is an even number.' Whenever nothing but the *possibility* of a construction is being asserted, we have no meaningful proposition; only by virtue of an effective construction, an executed proof, does an existential statement acquire meaning. In any of the numerous existential theorems in mathematics, what is valuable in each case is not the theorem as such but the construction carried out in its proof; without it the theorem is an empty shadow.

The question, put in Section 3, as to how conclusions may be drawn from existential statements, must here be answered by denying that possibility in principle. It can be done only after the existential statement has been replaced by the meaningful whole from which it was isolated as a propositional abstract. All proofs that depend on the construction of auxiliary elements fall under this remark. On the other hand, how do we obtain universal theorems on natural numbers? In order to explain this by means of a very simple example, let the number-theoretical function $\varphi(n)$ be defined by complete induction as follows:

$$(\alpha) \; \varphi(1) = 1; \quad (\beta) \; \varphi(n') = (\varphi(n))'.$$

Here, (β) represents a universal proposition, from which, in connection with (α), we may infer by complete induction that generally $\varphi(n) = n$. Thus the definition itself is seen to be the root of universality, and from there it spreads by complete induction. The principle of complete induction (as an instrument of definition or inference), not pressed into a formula but concretely applied at every step, is the true and only power of mathematics, the mathematical prime intuition. In this point Brouwer is in agreement with Poincaré (*"Science et hypothèse"*). The negation of a universal proposition about numbers would be an existential proposition; since this is void in itself, universal propositions are incapable of negation. Even a universal statement does not refer to a fact, it is not to be interpreted as the logical product of infinitely many singular propositions but as a *hypothetical* statement: if applied to a single definite given number it yields a definite proposition. There is no occasion here for the application of a principle of *tertium non datur* (either all numbers have the property A, or else there exists a number with the property $\sim A$). The belief in it, according to Brouwer (*Jahresberichte der Deutschen Mathematiker-Vereinigung*, 28, 1920) "was caused historically by the fact that, firstly, classical logic was abstracted from the mathematics of the subsets of a definite finite set [i.e. a set given by exhibition of its elements], that, secondly, an *a priori* existence independent of mathematics was ascribed to this logic, and that, finally, on the basis of this supposititious apriority it was unjustifiably applied to the mathematics of infinite sets."}

In Brouwer's analysis, the individual place in the *continuum*, the real number, is to be defined not by a set but by a *sequence* of natural numbers, namely, by a law which correlates with every natural number n a natural number $\varphi(n)$. (The two definitions cease to be equivalent, as soon as the natural numbers may no longer be treated as an extensionally definite aggregate.) How then do assertions arise which concern, not all natural, but all real numbers, i.e. all values of a real variable? Brouwer shows that frequently statements of this form in traditional analysis, when correctly interpreted, simply concern the totality of natural numbers. In cases where they do not, the notion of sequence changes its meaning: it no longer signifies a sequence determined by some law or other, but rather one that is created *step by step by free acts of choice*, and thus necessarily remains in statu nascendi. This 'becoming' *selective sequence* (*werdende Wahlfolge*) represents the continuum, or the variable, while the sequence determined *ad infinitum* by a law represents the individual real number falling into the continuum. The continuum no longer appears, to use Leibniz's language, as an aggregate of fixed elements but as a medium of free 'becoming.' Of a selective sequence *in statu nascendi*, naturally only those properties can be meaningfully asserted which already admit of a yes-or-no decision (as to whether or not the property applies to the sequence) when the sequence has been carried to a certain point; while the continuation of the sequence beyond this point, no matter how it turns out, is incapable of overthrowing that decision.

{In accordance with intuition, Brouwer sees the essential character of the continuum, not in the relation between element and set, but in that between part and whole. The continuum falls under the notion of the 'extensive whole,' which Husserl characterizes as that "which permits a dismemberment of such a kind that the pieces are by their very nature of the same lowest species as is determined by the undivided whole" (*Logische Untersuchungen*, second edition, II, p. 267). The division scheme of the one-dimensional continuum is best illustrated by the example of a finite line segment. By halving it, one decomposes it into two parts, a left (10) and a right one (11); each of the latter, by again halving them, decomposes into a left and right one, 100, 101 and 110, 111 respectively, and so on. This process may be described purely combinatorially and thus furnishes the arithmetical blank-form of the open one-dimensional continuum. This must

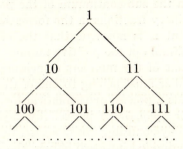

be distinguished from the realization of the process for a concretely exhibited continuum, such as the segment in space. In carrying out its continued subdivision according to the arithmetical scheme it is clearly irrelevant whether the two parts are always of the same length, as long as only the fineness of the parts eventually drops below any possible threshold of exactness. (It may even be that comparison of length has no foundation in the nature of the given continuum.) The process of subdivision, which *in concreto* can never have been carried out except to a certain point, determines a coordinate system within the continuum and thus makes it possible to designate the individual points in arithmetical terms by binary fractions. Since in a concrete continuum no exact boundaries can be set, one must imagine that the division framework is at no stage of the procedure fixed with complete accuracy, but that, as the subdivision continues, the earlier points of division steadily increase in precision. Any two adjacent parts of the ith division step may be joined into a 'division interval of the ith level.' The division intervals of the ith level overlap in such a manner that for any approximately given number, as soon as the approximation is sufficiently accurate, a division interval of the ith level can be found into which that number falls. Thus the individual real number will have to be defined as an *infinite sequence of nested division intervals of increasing level*.

Two real numbers α, β coincide if, for every value of n, the nth interval of the sequence α and the nth interval of the sequence β partially or wholly overlap; they are distinct if a number n exists for which these two intervals are disjoint. Because of the inapplicability of the *tertium non datur* to statements of this kind, Brouwer does not recognize this as a clear-cut alternative. This view fits in well with the character of the intuitive continuum. For there the separateness of two places, upon moving them toward each other, slowly and in vague gradations passes over into indiscernibility. In a continuum, according to Brouwer, there can be only continuous functions. *The continuum is not composed of parts*. Thus I may well distinguish within the continuum the sub-continuum of the positive numbers by using only positive binary fractions in the formation of intervals and interval sequences; but it is not true that the entire continuum is composed of the continuum of the positive numbers, that of the negative numbers, and that of the numbers coinciding with zero, in the sense that every number must belong to one of these three continua. An old truth thus finds a precise mathematical formulation; one that Aristotle (περὶ ἀτόμων γραμμῶν) expressed by saying, "That which moves does not move by counting," or (*Physics*, Bk. VIII, Ch. 8), "If the continuous line is divided into two halves, the one dividing point is taken for two; it is both beginning and end. But as one divides in this manner, neither the line nor the motion are any longer continuous . . . In the continuous there is indeed an unlimited number of halves, but only in possibility, not in reality." Compare in this connection the passages quoted earlier from Leibniz's letters. The principle comes into its own again that "nothing is separable which is not already separate" (Gassendi).

Mathematics with Brouwer gains its highest intuitive clarity. He succeeds in developing the beginnings of analysis in a natural manner, all the time preserving the contact with intuition much more closely than had been done before. It cannot be denied, however, that in advancing to higher and more general theories the inapplicability of the simple laws of classical logic eventually results in an almost unbearable awkwardness. And the mathematician watches with pain the larger part of his towering edifice which he believed to be built of concrete blocks dissolve into mist before his eyes.⌐

REFERENCES

L. E. J. Brouwer, *Over de grondslagen der wiskunde*, Dissertation, Amsterdam and Leipzig, 1907.
—— *Intuitionisme en Formalisme*, Groningen, 1912 (English translation in *Bull. Am. Math. Soc.*, 20 (1913–14));
—— Zur Begründung der intuitionistischen Mathematik, *Mathematische Annalen*, 93, 95, 96 (1924–27).
H. Weyl, Über die neue Grundlagenkrise der Mathematik, *Mathematische Zeitschrift*, 10 (1921).
O. Becker, Beiträge zur phänomenologischen Begründung der Geometrie und ihrer physikalischen Anwendungen, *Husserls Jahrbuch für Philosophie*, 6; especially pp. 398–436, 459–477.

10. SYMBOLIC MATHEMATICS

Is there no way to escape such radical consequences? The resolution to make this sacrifice is doubly hard in view of the historical fact that in set-theoretical analysis we find, in spite of the boldest and most elaborate combinations, complete certainty of deduction and an obvious accord among all the results. Hilbert has set himself the goal of saving mathematics in its entirety through the axiomatic method. He, too, admittedly is convinced that the power of intuitive thought does not reach farther than was asserted by Brouwer, that it is incapable of supporting the transfinite modes of deduction in mathematics, and that none of the transfinite statements of mathematics can be justified as being evident material truths (*einsichtige inhaltliche Wahrheiten*). What Hilbert proposes to secure is not the truth but the *consistency* of traditional analysis.

For this purpose he has to formalize mathematics, including logic, so that it becomes a game with symbols played according to fixed rules. (The symbols are not meant to be symbols *for* something.) The mathematical formulas which are made up of these symbols do not throughout admit of a material interpretation. Along with the meaningful propositions, 'ideal propositions' had to be introduced in order to reestablish artificially the validity of the simple logical rules that, as Brouwer had shown us, were lost in the transition to the infinite — just as in algebraic number theory ideal numbers were introduced in order to enforce the validity of the simple divisibility theorems. There are four different kinds of symbols,[10] which are distinguished, like the pawns and knights in chess, by the different rules of the game that apply to them: *constants* (such as 1), *variables* (symbols for blanks, x, y, . . .), one-place and many-place *operations*, and *integrations*. The most important one-place operations are $\sim$ (negation), σ (transition from a natural number to the next following one), and N (Na, to be read: a is a natural number). The most important two-place operations are $\rightarrow$, $=$, and ε. We construe all these as operations; in particular, N is the operation which, when applied to a, produces the proposition: a is a number; $=$ is the operation which, when applied to a and b, produces the proposition: a equals b. In order to arrive at a convenient general formulation of the rules of the game, these operational symbols may consistently be written in front of the terms (formulas) to which they apply, e.g. $\varepsilon \big\langle {}^{a}_{b}$ instead of $a\,\varepsilon\,b$. Among the integrations (which are always followed by one formula only) we have, above all, the quantifiers Σ_x, Π_x and the symbol ϵ_x to be introduced presently; they carry one (or several) arbitrary variables as subscripts. A prefixed integration symbol with the

[10] Deviating somewhat from Hilbert's original version, I here follow von Neumann's simplified formalism (Zur Hilbertschen Beweistheorie, *Mathematische Zeitschrift*, 1926).

subscript x has the effect that the variable x becomes 'bound' at all places in the formula following the symbol, thus losing its capacity of being substituted for. In the course of the development of mathematics it is always possible to introduce new symbols. What a *formula* is, is defined recursively: "(α) every constant or variable by itself is a formula; (β) from one or two (or several) formulas already constructed a new formula is obtained by writing down respectively a one- or two- (or several-) place symbol o of operation or a (one-place) symbol of integration, and having it followed by the formula(s) in question in their proper order, each written on a separate line and its initial symbol joined to o by a dash." The complete formula then looks like a (parthenogenetic) genealogical tree of symbols, from which the "grammatical structure" of the formula, i.e. the manner of its recursive construction can be read off unambiguously. One also can decide in this way whether a given tree-like arrangement of symbols is or is not a formula.

⌈The linear arrangement, which is more convenient to print, has to make use of parentheses if the recursive construction is to remain uniquely recognizable. We return to the usual symbolism, which is less systematic, whenever it is a question of merely outlining the procedure in its essential features.

It is unnecessary to worry about the fact that in the formal construction the operations are applied indiscriminately to all kinds of things. Who is afraid of such generosity may prefer to discriminate between 'numerical' and 'factual' formulas, in accordance with the following recursive stipulations: "(α) A constant or variable by itself, as well as any formula beginning with σ or ϵ_x, is a *numerical* formula; formulas beginning with $\sim$, $\rightarrow$, $\vee$, &, N, $=$, ϵ, Σ_x, Π_x, on the other hand, are *factual*. (β) The symbols σ and N must be followed by one, $=$ ϵ by two *numerical* formulas, while $\sim$, ϵ_x, Σ_x, Π_x must be followed by one, $\rightarrow$, &, $\vee$ by two *factual* formulas." Similar restrictions will then have to accompany the axiomatic rules and the syllogistic rule of inference below.

If $A(x)$ (as always in what follows) is an arbitrary formula containing only the one 'free' variable x (free in the sense that it is not bound at every place where it occurs), and if b is a 'closed formula' (i.e. one containing no free variables), then b may be substituted for x in A wherever x occurs free (i.e. is not bound). The result of this process of *substitution*, which thus has been described intuitively, is again a formula; it is denoted by the abbreviating sign $A(b)$.[11]

---

[11] Here the letters A, b are clearly not symbols of the game, but are used as signs of communication that enable us to speak of formulas etc. in general. Hil-

Formulas serving as *axioms* form the starting point of any proof. Instead of individual axioms, however, we formulate general rules for the formation of axioms. First come the axiomatic rules of finite logic, such as

$$c \to (b \to c).$$

It says: take any two formulas b and c without free variables and construct out of them the formula $c \to (b \to c)$; the result you may use as an axiom. Secondly, there are the two axiom rules of equality; they establish the connection between logic and arithmetic:

$$b = b.$$
$$(b = c) \to (A(b) \to A(c)).$$

Thirdly, we have specifically arithmetical rules of a finite character. In them the constant 1 appears, which is the material starting point of all construction:

$$N1.$$
$$Nb \to N(\sigma b).$$
$$(\sigma b = \sigma c) \to (b = c).$$
$$\sim (\sigma b = 1). \}$$

Next we come to the *transfinite* part. Taking for granted the alternative, denied by Brouwer, that either an honest man exists or all men are dishonest, one is sure to find an Aristides of whom it can be said: if any man be honest then Aristides is. For, in the first case, we may choose as Aristides one of the honest, and, in the second case, any man at all. In order to be able to construct such an Aristides, not just for the property of honesty, but for every property, i.e. for every formula A containing one free variable x, we invent a fictitious divine automaton which produces, whenever an arbitrary property A is fed into it, that individual $\epsilon_x A$ which certainly possesses the property A provided such an individual exists at all. ϵ_x is an integration symbol. (Indulgently following the fatal custom of employing the word 'is' to denote both the copula and existence we too use the same letter ϵ for both; but the confusion is avoided by the variable attached as subscript to the existential ϵ.) If such an automaton were at our disposal, then we would be rid of all the trouble caused by 'some' and 'all.' But the belief in its existence is, of course, sheer nonsense. Mathematics, however, proceeds as if it existed. This can be expressed in the form of an axiom rule, and if the application of this rule does not lead to

bert employs the Gothic alphabet to distinguish them from the symbols proper. Because of the aversion of the English-speaking reader to Gothic type, this practice has not been followed in our translation, although it is undoubtedly a valuable help in keeping the issue clear.

contradictions, then its use is legitimate in formalized mathematics. Thus we have the following transfinite logical axiom rules:

$$A(b) \to \Sigma_x A(x); \qquad \Pi_x A(x) \to A(b);$$
$$\Sigma_x A(x) \to A(\epsilon_x A); \quad A(\epsilon_x(\sim A)) \to \Pi_x A(x).$$

Those stated in the second line were still omitted in Section 3; they permit us to infer something from Σ_x and to infer Π_x from other formulas. Of course, they do not offer the same service as the fictitious automaton; for, given a formula A, they fail to reveal the identity of $\epsilon_x A$. Only in special circumstances may a formula such as $\epsilon_x A = 1$ appear as the terminal formula of a proof starting with the axioms.

{Among the arithmetical axioms, the principle of complete induction is still absent. It may be interpreted as a transfinite arithmetical axiom rule, expressing the fact that a property appertaining to 1 and 'handed on' from x to σx is a property of every arbitrary number. But, as we know, this rule becomes superfluous if it is admissible to introduce for every property A a new object y, namely, the corresponding set, such that the proposition 'x is an element of y' is equivalent with the subsistence of $A(x)$. If this hypothesis is formulated as an axiom rule, it turns out that its application leads inescapably to a contradiction — a fact tantamount to a *forfeiture of the unlimited right of objectification*. For the purposes of analysis, however, it is sufficient to restrict the argument x to the range of natural numbers, so that we may lay down the following narrower transfinite set-theoretical rule:

(**I**) $$\Sigma_y \Pi_x \{Nx \to ((x \; \varepsilon \; y) \rightleftarrows A(x))\},$$

where $B \rightleftarrows C$ serves as an abbreviation of $(B \to C) \& (C \to B)$. It seems to be desirable, though not indispensable, for the construction of analysis to add the *axiom of definiteness*, according to which two sets of numbers are equal if they contain the same elements:

$$\Pi_x \{Nx \to ((x \; \varepsilon \; b) \rightleftarrows (x \; \varepsilon \; c))\} \to (b = c).\}$$

A mathematical *proof* consists in manufacturing axioms by means of the given rules — these axioms never contain free variables — and in progressing to ever new formulas by applying the syllogistic rule of inference to such axioms or to formulas already obtained. We repeat the rule (cf. Sect. 3): Given two formulas b and $b \to c$ in the second of which the first reappears at the left of $\to$ one may pass on to the formula c. To survey in advance what *demonstrable formulas* will be obtained as the result of this game is impossible, mainly because the syllogism leads from two formulas b and $b \to c$ to the new formula c, which is shorter than the second of the premises, so that in the proof game shrinkage interchanges with growth.

Up to this point all is game, not knowledge. But now the game is made the subject of investigation in what Hilbert calls *metamathematics*, the aim being to make certain that the game will never lead to a contradiction. Such a contradiction would arise if the actual play of two proof games would terminate, the one with a formula b, the other with the opposite formula $\sim b$. Only in order to arrive at this one insight does Hilbert require the finite, material, meaningful mode of thought, which cannot be pressed into any 'axioms.' In particular, this material thinking makes use of an intuitive inference by complete induction, such as we drew when we came to the conclusion (Section 4) that a correctly played game of chess can never produce 10 queens of the same color.

⎰One of the rules of the elementary propositional calculus that either figures among the axiomatic rules or is readily deduced from them is

(1) $$\sim b \to (b \to c),$$

where b and c are any closed formulas. Let c be an arbitrary formula of this kind, and suppose that a certain formula b and its negation $\sim b$ have been demonstrated. Under these circumstances, two syllogistic steps lead from (1) first to $b \to c$ and then to c. Hence in case the formalism is known to be inconsistent, *any* closed formula c may be demonstrated, and thus the proof game loses all interest. Consistency may also be defined by saying that the formula $\sim (1 = 1)$ is not demonstrable.⎱

The axiom system may be continually expanded, but it must be shown that the consistency is not overthrown by the expansion. In particular, definitions may be introduced in the form of new axiom rules; e.g.

$$\sigma 1 = 2, \quad \sigma(\sigma b) = \sigma_2 b.$$

This applies especially to the recursive definitions of $b + c$, $b \cdot c$ and other arithmetical operations. It can be shown once and for all that consistency, if it prevailed before, is preserved by the addition of axioms of this kind, that stand for simple or recursive definitions.[12]

[12] It can also be shown that once the definitional axioms for $b + c$, $b \cdot c$ and the corresponding operational symbols $+$, $\cdot$ have been introduced, all other recursively definable arithmetical operations are expressible in the formalism. Compare e.g. Hilbert and Bernays, *Grundlagen der Mathematik*, vol. I, pp. 412–422.

Regarding the natural numbers, Hilbert's construction, in contrast to Brouwer's, gets along without that 'possibility *ad infinitum*' which was described in Section 6 as the third step of constructive cognition. For Hilbert, 10^{12} is a transfinite symbol, which does not denote a number of the form $\sigma\sigma \ldots \sigma 1$. Geometry and physics may be adjoined, as soon and insofar as they have been strictly axiomatized. Hilbert even believes (*Axiomatisches Denken*, 1917), "Every potential subject of scientific thought, as soon as it is ripe for the formation of a theory, is bound to fall under the axiomatic method and, therefore, indirectly to the lot of mathematics."[13]

[As long as the transfinite components are left out of consideration, the consistency proof can easily be carried out by means of a 'valuation' of formulas. By a precisely described recursive procedure, every formula, according to its origin, is ascribed one of the values T or F (true or false) in such a manner that the finite axioms obviously get the value T and that the rules of evaluation given in Section 3 hold for the logical combinations. *Hence, as long as the transfinite is excluded, the syllogism and thus the deductive method, remains impotent;* for a decision as to the truth or falsehood of the premiss $b \rightarrow c$ is made only after the conclusion c has been evaluated.

The consistency proof can no longer be carried out along those lines if the transfinite axiom rules are taken into consideration. This brings out the fact that, with them, the insight into true and false ceases. After Hilbert and P. Bernays had developed more indirect methods, W. Ackermann and J. von Neumann in 1926 seemed to have succeeded in establishing the consistency of 'arithmetic,' i.e. of an axiomatic system including the transfinite logical axioms and the principle of complete induction, excluding however the dangerous axiom (**I**) about the conversion of predicates into sets. This result would vindicate the standpoint taken by the author in *Das Kontinuum*, that one may safely treat the sequence of natural numbers as a closed aggregate of existing objects. Justification of the same standpoint with respect to the 'aggregate of all possible sets of natural numbers' would depend on extending the consistency proof to the set-theoretical axiom rule (**I**); at the moment we do not see how that could be done.

[13] Out of an entirely different conception of mathematics, Kant (*Metaphysische Anfangsgründe der Naturwissenschaft*, Preface) comes to the conclusion "that in every specific natural science there can be found only so much science proper as there is mathematics present in it." In the same sense as Hilbert, on the other hand, Husserl (*Logische Untersuchungen*, I, §71) declares with particular reference to mathematical logic that "the mathematical form of treatment . . . is for all strictly developed theories (this word taken in its true sense) the only scientific one, the only one that affords systematic completeness and perfection and gives insight into all possible questions and their possible forms of solution."

Even in the consistency proof for arithmetic just referred to a serious gap was later discovered. Concerning this development after 1926 and the catastrophe precipitated by an important discovery by K. Gödel in 1931, see Appendix A. But whatever the ultimate value of Hilbert's program, his bold enterprise can claim one merit: it has disclosed to us the highly complicated and ticklish logical structure of mathematics, its maze of back-connections, which result in circles of which it cannot be gathered at a first glance whether they might not lead to blatant contradictions.

The described symbolism evidently attacks again, in a refined form, the task which Leibniz had set himself with his "general characteristic" and *ars combinatoria*. But is it really more than a bloodless ghost of the old analysis that confronts us here? Hilbert's mathematics may be a pretty game with formulas, more amusing even than chess; but what bearing does it have on cognition, since its formulas admittedly have no material meaning by virtue of which they could express intuitive truths? The subject of mathematical investigation, according to Hilbert, is the concrete symbols themselves. It is without irony, therefore, when Brouwer says (*Intuitionisme en formalisme*, p. 7), "Op de vraag, waar de wiskundige exactheid dan wel bestaat, antwoorden beide partijen verschillend; de intuitionist zegt: in het menschelijk intellect, de formalist: op het papier." The question why he sets up just these rules must remain unanswered by the consistent formalist. He will have to refer us to philosophy, psychology, or anthropology, so Brouwer thinks, in order to justify his "lustgevoel van echtheitsovertuiging" and his belief that the chosen axiom system is more suitable than any other to be projected onto the world of experience.]

This last remark reminds us that it is the function of mathematics to be at the service of the natural sciences. The propositions of theoretical physics, however, certainly lack that feature which Brouwer demands of the propositions of mathematics, namely, that each should carry within itself its own intuitively comprehensible meaning. Rather, what is tested by confronting theoretical physics with experience is the system as a whole. It seems that we have to differentiate carefully between phenomenal knowledge or insight — such as is expressed in the statement: 'This leaf (given to me in a present act of perception) has this green color (given to me in that same perception)' — and theoretical construction. Knowledge furnishes truth, its organ is 'seeing' in the widest sense. Though subject to error, it is essentially definitive and unalterable. Theoretical construction seems to be bound only to one strictly formulable rational principle, that of concordance (compare Section 17, p. 121), which in mathematics, where the domain of sense data remains untouched, reduces to consistency; its organ is creative imagination. In connection with physics we shall have to discuss in greater detail the question what its determining factors, besides concordance, are. Intuitive truth, though not the ultimate criterion, will certainly not be irrelevant here. Hilbert himself expresses the following opinion (Über das Unendliche, *Mathematische Annalen*, 95, p. 190), "The function left to the infinite . . . is merely that of an idea — if, with Kant, one understands by an idea a concept of reason (*Vernunftbegriff*) transcending all experience and supplementing the concrete in the sense of totality." But perhaps this question can be answered only by pointing toward the essentially historical nature of that life of the mind of which my own existence is an integral but not autonomous part. It is light and darkness, contingency and necessity, bondage and freedom, and it cannot be expected that a symbolic construction of the world in some final form can ever be detached from it.

REFERENCES

Concerning LEIBNIZ, see references at the end of Section 3.
D. HILBERT and P. BERNAYS, *Grundlagen der Mathematik*, 2 Vols., Berlin, 1934–39.

11. ON THE CHARACTER
OF MATHEMATICAL COGNITION

[From time immemorial mathematics has been looked upon as the science of quantity, or of space and number. (Though we also find this definition with Leibniz, the *mathesis* thus delineated is to him but a part of the more comprehensive *ars combinatoria*.) Today this view appears much too narrow in consideration of such fields as projective geometry or group theory. Consequently we need not worry particularly over an exact determination of what is meant by quantitative. In fact, the development of mathematics itself raises doubts as to whether quantity is a well-determined and philosophically important category. Geometry, inasmuch as it is concerned with real space, is no longer considered a part of pure mathematics; like mechanics and physics, it belongs among the applications of mathematics. Under the influence of the general arithmetic of hypercomplex numbers and later of the axiomatic investigations, of set theory and symbolic logic, the distinction between mathematics and logic is gradually obliterated. "Mathematics is the science which draws necessary conclusions," B. Peirce declares in 1870. The definition of 'mathematics or logic' is discussed in detail in Chapter XI of Husserl's

Logische Untersuchungen (Vol. I, *Die Idee der reinen Logik*) and in the last chapter of Russell's *Introduction to Mathematical Philosophy*.

The crisis brought on by the set-theoretical antinomies — no matter if one follows Brouwer's radical intuitionism or Hilbert's symbolism — again throws into sharper relief the peculiar character of mathematics. Like Plato, Brouwer looks upon the two-oneness as the root of mathematical thinking. "Dit neo-intuitionisme zieht het uiteenvallen van levensmomenten in qualitatief verschillende deelen, die alleen gescheiden door den tijd zich weer kunnen vereenigen, als oergebeuren in het menschelijk intellect, en het abstraheeren van dit uiteenvallen van elken gevoelsinhoud tot de intuitie van twee-eenigheid zonder meer, als oergebeuren van het wiskundig denken." We have seen how the division scheme of the one-dimensional continuum results from "one becoming two"[14] again and again (compare the diagram on p. 53). The integers when written in the binary system are obtained in the same manner. Stenzel (*Zahl und Gestalt bei Plato und Aristoteles*, 1924) makes it appear probable that Plato thought of his numbers as being arranged according to this scheme; but since the splitting of one into two here leads to larger and larger numbers, while in the continuum we descend to smaller and smaller parts, he refers to that two-ness as the "great-and-small." (See, however, for a different interpretation: H. Cherniss, *The Riddle of the Early Academy*, Univ. of Calif. Press, 1945.) More appropriate for the integers is their natural order, which Aristotle (*Metaphysics* A6 and M6) sets up in opposition to Plato's number concept. But it, too, can be generated out of the two-oneness; starting with an undivided whole, we separate it into an element (the 1), to be preserved as a unit, and an undivided remainder, the latter we then separate again into an element (2) and an undivided remainder, and so forth. (This can be visualized as the continued chopping-off of a segment from a half-line; time is open toward the future, but whenever we stop we find that another segment of time has been lived through.) In this scheme, not every part but only the last remaining part is subject to further bipartition. }

Independently of the value attached to this last reduction of the mathematical thought process to the two-oneness, *complete induction* appears, from the intuitionist point of view, as that which prevents mathematics from becoming one huge tautology and which confers upon its assertions a synthetic non-analytic character. The pro-

[14] Allusion to the phrase "Da wurde Eins zu Zwei" by which Nietzsche described his Zarathustra experience in several of his poems; e.g. in "Sils Maria":

"Da, plötzlich, Freundin, wurde Eins zu Zwei —
— und Zarathustra ging an mir vorbei . . . "

cedure of complete induction is, indeed, a decisive feature throughout. If at first it does not appear to play any part in elementary geometry (especially in elementary projective geometry), the reason is to be seen in the naive application of 'some' and 'all' to the points. In the intuitionist view, this is inadmissible; the field of construction of geometry is a continuum and hence capable of exact mathematical treatment only after it has been spun over with a division net as described above (compare also Section 15).

From the formalist standpoint, the transfinite component of the axioms takes the place of complete induction and imprints its stamp upon mathematics. The latter does not consist here of evident truths but is *bold theoretical construction*, and as such the very opposite of analytical self-evidence. The material reasoning of metamathematics, on the other hand, in running over the steps of a proof, operates by means of an intuitive inference from n to $n + 1$ and concerns itself with "extra-logical, concrete objects, which can be overlooked completely in all their parts and whose exhibition, differentiation, and succession or coordination are intuitively given along with the objects as something neither capable nor in need of reduction to anything else" (Hilbert). Thus Hilbert agrees with Kant — who, incidentally, likewise emphasized the symbolic construction with concrete tokens in algebra (*Critique of Pure Reason*, ed. Max Müller, p. 576, = p. 717 of the first edition, 1781) — that "mathematics possesses a content that is secure independently of all logic and therefore can never be based upon logic alone" (*Über das Unendliche*, p. 171).

{It should be recognized, however, that according to the Kantian usage of the words 'analytic' and 'synthetic' at least an individual equation such as $3 + 2 = 5$ ought to be called analytic; for, as Leibniz explained, it follows logically from the definitions

$$3 + 1 = 4, \quad 4 + 1 = 5, \quad (a + 1) + 1 = a + 2,$$

and thus "lies in the concepts" of the numbers 3, 5 and of the operation $+2$. Or else what meaning did Kant connect with these symbols?

Mathematics undoubtedly is *a priori*. It is not, as J. S. Mill wants to make us believe, founded on experience, in the sense that only repeated observations of numerical examples confer an increasing measure of verisimilitude upon such arithmetical theorems as

$$m + n = n + m$$

that are pretended to hold for arbitrary numbers.}

A conspicuous feature of all mathematics, which makes it so inaccessible to the layman, is the abundant use of *symbols*. The intuitionist does not consider this an essential characteristic, he sees in them, as he does in all spoken or written language, merely a tool of communication and of support for the memory by fixation. Not so the formalist. He thinks of mathematics as consisting wholly of symbols, which have no meaning verifiable in sensual or mental intuition and which are manipulated according to fixed rules. Language, on the other hand — for instance in the description of substitution or of the practical rule of inference, as well as in metamathematical reasoning — serves him as a means for communicating modes of procedure and acts of meaningful thought. (Communication remains forever exposed to the risk of misunderstanding.) "In the geometrical figure and, later, in the mathematical formula," A. Speiser says (*Klassische Stücke der Mathematik*, 1925, p. 148), "mathematics has liberated itself from language; and one who knows the tremendous labor put into this process and its ever-recurring surprising success, cannot help feeling that mathematics nowadays is more efficient in its particular sphere of the intellectual world than, say, the modern languages in their deplorable condition of decay or even music are on their fronts." In his transcendental methodology (*Critique of Pure Reason*, Part II), Kant sees the essence of mathematics in the construction, "Philosophical knowledge is that which reason gains from concepts, mathematical that which it gains from the construction of concepts" (ed. Müller, p. 572, = p. 713 of the first edition, 1781). Using the theorem of the sum of the angles in a triangle as an example, he illustrates how geometrical theorems are found, not by conceptual analysis, but by construction of suitable auxiliary points and lines. The details of his description of the constructive procedure can no longer be considered satisfactory today. This much is true, however, that in the proof of a mathematical theorem it is almost always necessary to go far beyond its immediate content. The reason is to be seen in the fact emphasized before that a proof proceeding according to the syllogistic rule of inference is not a monotonically progressing construction — in contrast to a formula, whose manufacture always advances in the same direction and whose constructive parts are therefore preserved in the final form — but a constant change of adding on and removing. This circumstance, together with the points 1, 2, and 3 enumerated in Section 6 (p. 37), seem to me to give a fairly adequate characterization of construction as opposed to pure reflection.

The stages through which research in the foundations of mathematics has passed in recent times correspond to the three basic possibilities of epistemological attitude. The set-theoretical approach is the stage of *naive realism* which is unaware of the transition from the given to the transcendent. Brouwer represents *idealism*, by demanding the reduction of all truth to the intuitively given. In axiomatic formalism, finally, consciousness makes the attempt to 'jump over its own shadow,' to leave behind the stuff of the given, to represent the *transcendent* — but, how could it be otherwise?, only through the *symbol*. Basically, the idealist viewpoint in epistemology has been adhered to by occidental philosophy since Descartes; nevertheless, it has searched again and again in metaphysics for an access to the realm of the absolute, and Kant, who meant to shoot the bolt once and for all, was yet followed by Fichte, Schelling, and Hegel. It cannot be denied that a theoretical desire, incomprehensible from the merely phenomenal point of view, is alive in us which urges toward totality. Mathematics shows that with particular clarity; but it also teaches us that that desire can be fulfilled on one condition only, namely, that we are satisfied with the symbol and renounce the mystical error of expecting the transcendent ever to fall within the lighted circle of our intuition. So far, only in mathematics and physics has symbolical-theoretical construction gained that solidity which makes it compelling for everyone whose mind is open to these sciences. Their philosophical interest is primarily based on this fact.

{If in summing up a brief phrase is called for that characterizes the life center of mathematics, one might well say: mathematics is the *science of the infinite*. It was the great achievement of the Greeks to have made the tension between the finite and the infinite fruitful for the analysis of reality. It has been attempted here to bring out the past and present importance of this tension — and of the attempts to overcome it — for the history of theoretical knowledge. "The infinite, like no other problem, has always deeply moved the soul of men. The infinite, like no other idea, has had a stimulating and fertile influence upon the mind. But the infinite is also more than any other concept, in need of clarification" (Hilbert, *Über das Unendliche*).

For a survey of the various issues and problems in which mathematical research is interested today, the reader may be referred to Courant and Robbins, *What is Mathematics?*}

REFERENCES

I. Kant, *Kritik der reinen Vernunft*.
  A thorough discussion of Kant's philosophy of mathematics is given by L. Couturat in *Revue de métaph. et mor.*, May 1904.
H. Poincaré, *La science et l'hypothèse*, Paris; many editions.
R. Courant and H. Robbins, *What is Mathematics?*, New York, 1941.
L. E. J. Brouwer, Mathematik, Wissenschaft und Sprache, *Monatshefte der Mathematik und Physik* 36, 1929, pp. 153–164.
H. Weyl, The Mathematical Way of Thinking, Univ. Penn. Bicenten. Conf. 1940; also *Science* 92, 1940, pp. 437–446.

CHAPTER III

Geometry

Nowhere do mathematics, natural sciences, and philosophy permeate one another so intimately as in the problem of space. The presuppositions for the discussion of this problem, inasmuch as they have emerged from mathematical investigation, are to be briefly outlined in this chapter.

12. NON-EUCLIDEAN, ANALYTIC, MULTI-DIMENSIONAL, AFFINE, PROJECTIVE GEOMETRY; THE COLOR SPACE

[Little has to be added concerning the topic of non-Euclidean geometry to what has been said in Section 4 in connection with axiomatics. If all remaining axioms are maintained, then there are these three possibilities: given a point P and a line l in a plane, with P not on l, there are either infinitely many lines in that plane which pass through P but do not intersect l, or just one such line, or none (known since Klein as the 'hyperbolic,' 'parabolic,' and 'elliptic' cases respectively). The sum of the angles in a triangle in these cases is respectively less than, equal to, and greater than, 180°. The last-named possibility, pointed out only toward the middle of the 19th century by Riemann, exists only if the axioms of order are modified to the effect that the line appears no longer as an open but as a closed curve. Plane elliptic geometry is none other than that which holds on a sphere in Euclidean space, except that diametrically opposite points have to be identified. Or, in other words, while all other terms referring to the geometry of the plane p retain their ordinary 'Euclidean' meaning, the meaning of the notion of congruence is to be modified to the effect that two configurations in p are considered 'congruent' if their projections from a central point O, not in p, onto a sphere about O are congruent in the ordinary sense. The plane, in this case, has to be enriched by the inclusion of the 'points at infinity,' whose rays of projection are the lines through O parallel to p. The mappings of p which map 'congruent' configurations into each other can be characterized, without reference to the space, as collinear transformations that have an invariance property similar to that prevailing in the Klein model of Bolyai-Lobatschewskyan geometry. Thus the way is open to the development of an elliptic geometry not only in the plane but also in space. The true relation between the three kinds of geometry is brought out best if the non-metrical projective space is taken as the starting point and a 'Cayley metric' built into it. According to the type of absolute conic on which this metric is based, one or another of the three metric geometries is obtained. Klein himself interpreted his construction in this sense, namely, as endowment of projective space with a Lobatschewskyan metric, not as construction of a model by means of metric Euclidean space.}

Analytic geometry reduces every geometrical problem to an algebraic one. This presupposes that the number concept, by the inclusion of fractions and irrational numbers, has acquired that width which makes it suitable, not only for counting, but also for measuring. The Greeks had been deterred from this step because they took the discovery of the irrational seriously.[15] The post-classical Western civilization, less scrupulous than they, resumed the old algebraic traditions of the Sumerians, Indians, and Arabs. It attained to independent achievements in geometry only after the science of space, through Descartes' *Géométrie* (1637), became subjected to algebraic calculus.

{Today probably the best approach to analytic geometry is by means of the vector concept, following the procedure of Grassmann's *Ausdehnungslehre*. The vector calculus is a computational device whose objects are not numbers but simple geometrical entities. A treatment of geometry along these lines was demanded and even partially executed by Leibniz in his work *De analisi situs* and his design for a geometrical characteristic (*Mathematische Schriften*, **V,** p. 178, and II, p. 20), which belong within the framework of his "universal characteristic." The translations, or parallel displacements, of space are called *vectors*. A point A is mapped by a translation $\vec{a}$ into a point $A\vec{a} = B$, the 'endpoint of the vector $\vec{a}$ laid off from A.' Conversely, if A, B are any two points in space, there exists one and only one translation $\vec{a}$ which carries A into B. Among the translations we have the 'identity,' under which all points remain fixed; this is the vector 0. Translations can be combined, they form a *group;* the effect of carrying out first one translation, $\vec{a}$, then another, $\vec{b}$, is the same as that of a single translation, the resultant $\vec{a} + \vec{b}$. The number concept enters geometry through the process of iteration of a

[15] Descartes speaks of the "misgivings of the Ancients regarding the use of terms of arithmetic in geometry, which can only have had their origin in a lack of understanding concerning the connection between these two disciplines."

translation $\vec{a}$ (consisting in adding $\vec{a}$ arbitrarily often to itself; compare the beginning of Section 5). Starting with a point A and repeating the same step $\vec{a}$ again and again, one obtains the skeleton of a line, namely, a sequence of equidistant points beginning with A. The line itself results, so to speak, by continuous iteration of the same infinitely small translation. By partition (as in Section 5) we contrive to apply not only integral but also fractional multipliers λ to the vector $\vec{a}$, and the continuity requirement finally removes the restriction to rational numbers. Thus arises an axiomatic construction of geometry (strictly speaking, of affine geometry, in which only parallel line segments can be measured against one another) that presupposes the fully formed concept of real number — into which the entire analysis of continuity is thrown — and uses as the only basic geometric concepts 'point' and 'vector.' Three basic operations connect these objects: (1) two vectors $\vec{a}, \vec{b}$ generate a third vector, $\vec{a} + \vec{b}$; (2) a number λ and a vector $\vec{a}$ generate the vector $\lambda\vec{a}$; (3) a point A and a vector $\vec{a}$ generate a point $A\vec{a}$. The axioms referring to these operations form a system that, also in logical respect, is of a much more transparent and homogeneous structure than the purely geometrical axioms of Euclid or Hilbert. Indeed, they determine, as has already been pointed out in Section 4, nothing other than the operational field of linear algebra. They reveal a wonderful harmony between the given on one hand and reason on the other. Moreover the simplest derived geometrical concepts, to which here belong especially the line and the plane, correspond to those which suggest themselves most naturally from the logical standpoint. All vectors $\vec{x}$ which are obtained from two given ones, $\vec{e_1}$ and $\vec{e_2}$, through the formula

(1) $$\vec{x} = x_1\vec{e_1} + x_2\vec{e_2},$$

with arbitrary numerical coefficient x_1, x_2, form a 'linear vector manifold of dimension 2.' For the sake of the uniqueness of the representation (1) it is assumed here that $\vec{e_1}, \vec{e_2}$ are linearly independent, i.e. that the expression on the right furnishes the vector 0 only if x_1 and x_2 are both equal to 0. If all these vectors $\vec{x}$ are laid off from a fixed initial point O, then the endpoints $O\vec{x} = P$ form a 'linear point manifold of dimension 2,' or a plane. The coordinate system here consists of the point O and the two linearly independent vectors $\vec{e_1}, \vec{e_2}$. Relative to these, the point P is characterized by its 'coordinates' x_1, x_2. Similarly

69

linear vector manifolds and linear point manifolds of dimensions 1, 2, 3, . . . (line, plane, . . .) may be introduced.}

Only here do we meet the concept of *dimension*. In real space we cannot go beyond the third dimension; there exist 3, but no more, linearly independent vectors. Measured against the transparent lawfulness that finds its expression in our axiom system, this dimension number 3 appears as a *contingent* feature. We might just as well replace the number 3 by any number n of dimensions, by postulating that there be n, but no more, linearly independent vectors. A coordinate system for the space then consists of an initial point O and n such vectors. For $n = 1, 2, 3$ we thus obtain respectively the geometry of the line, of the plane, and of space. Only on the basis of the notion of an n-dimensional geometry to which this formalization leads in a cogent manner does the *problem of the number of dimensions* become meaningful: What inner peculiarities distinguish the case $n = 3$ among all others? If God, in creating the world, chose to make space 3-dimensional, can a 'reasonable' explanation of this fact be given by disclosing such peculiarities?

{If all vectors are laid off from a fixed initial point, it is seen that the geometry of vectors is identical with the (affine) geometry of a point space provided with an absolute center O. If one identifies any two non-vanishing vectors resulting from one another through multiplication by a number, i.e. if one considers as elements the *rays* through O, the n-dimensional affine vector geometry becomes the $(n - 1)$-dimensional *projective geometry* (of the family of rays through O).

The projective geometry holds in the space of the perceptively given *color qualities* of colored light. (The manifold of the objective physical colors has infinitely many dimensions; of these, the normal non-colorblind eye produces a 2-dimensional 'projection,' a huge manifold of physically different colors giving the same color impression.) If two colors of definite intensity are composed (mixed), the result is a new definite color of a definite intensity. The various intensities of one color may be compared with one another, so that, after a unit intensity has been chosen, every intensity can be measured by a number (iterated composition of a color of unit intensity with itself producing a scale of intensities without change of the color quality). The intensities of two different color qualities, on the other hand, are incommensurable. Thus the colors with their various qualities and intensities fulfill the axioms of vector geometry if addition is interpreted as mixing; consequently, projective geometry applies to the color qualities. All colors resulting from the mixing of three basic colors A, B, C form the 'triangle' ABC. The color space turns out to be 2-dimensional, by virtue of the fact that three basic colors suffice to produce all colors by mixing, or at least that the entire color field can be composed of such color triangles. For the real colors fill out only a restricted section of the entire projective plane. But it can, by the procedure described in Section 2, be extended ideally into a full projective plane; *ideal* colors must be chosen as the basic colors A, B, C if the field of real colors is to fall entirely within the triangle ABC. In the projective color plane, the pure spectral colors lie on a curve whose extremities come very close together and are connected by purple. Epistemologically it is not without interest that in addition to ordinary space there exists quite another domain of intuitively given entities, namely the colors, which forms a continuum capable of geometric treatment. }

REFERENCES

F. KLEIN, Über die sogennante nicht-euklidische Geometrie, *Gesammelte mathematische Abhandlungen*, I, Berlin, 1921, pp. 254–305, 311–350.

O. VEBLEN and J. W. YOUNG, *Projective Geometry*, 2 Vols., New York, 1910–1918.

H. WEYL, *Raum Zeit Materie*, fifth edition, Berlin, 1923; Sections 1–4.

H. VON HELMHOLTZ, *Handbuch der physiologischen Optik*, 3 Vols., third edition, Hamburg and Leipzig, 1909–11; Section 20.

13. THE PROBLEM OF RELATIVITY

Our knowledge stands under the norm of *objectivity*. He who believes in Euclidean geometry will say that all points in space are objectively alike, and that so are all possible directions. However, Newton seems to have thought that space has an absolute center. Epicurus certainly thought that the vertical is objectively distinguishable from all other directions. He gives as his reason that all bodies when left to themselves move in one and the same direction. Hence the statement that a line is vertical is elliptic or incomplete, the complete statement behind it being something like this: the line has the direction of gravity at the point P. Thus the gravitational field, which we know to depend on the material content of the world, enters into the complete proposition as a contingent factor, and also an individually exhibited point P on which we lay our finger by a demonstrative act such as is expressed in words like 'I,' 'here,' 'now,' 'this.' Only if we are sure that the truth of the complete statement is not affected by free variation of the contingent factors and of those that are individually exhibited (here the gravitational field and the point P) have we a right to omit these factors from the statement and still to claim objective significance for it. Epicurus's belief is shattered as soon as it is realized that the direction of gravity is different in Princeton and in Calcutta, and that it can also be changed by a redistribution of matter. Without claiming to give a mechanically applicable criterion, our description bears out the essential fact that objectivity is an issue decidable on the ground of experience only. It also accounts for the two main sources of the error so often committed in the history of knowledge, that of mistaking a statement for objective that is not: (1) one overlooked certain relevant circumstantial factors on which the meaning of the statement depends although they are not mentioned explicitly in its elliptic form, (2) though these factors were recognized, one did not investigate carefully enough whether or not the truth of the statement is affected by their variation. It is no wonder then that at several phases in the course of the history of science the realm of that which is considered objective has shrunk.

Whereas the philosophical question of objectivity is not easy to answer in a clear and definite fashion, we know exactly what the adequate mathematical concepts are for the formulation of this idea. Let us start with a completely axiomatized science like Euclidean geometry. For simplicity's sake we assume only one fundamental category, the points of space. According to Hilbert the fundamental relations that enter into the axioms would then be (1) the ternary relation: three points lie on a straight line, (2) the relation: (three distinct points A, B, C lie on a straight line and) B lies between A and C; (3) the relation: four points lie in a plane; (4) the relation of congruence $AB \equiv CD$ between two pairs of points AB and CD. What we are going to say applies to any domain of objects the axioms of which deal with a few basic relations. Without prejudicing what the objects are we may call them points and thus speak of the domain as the point-field.

In Section 4 the notion of isomorphic mapping was introduced. We now consider the special case when our domain of objects is mapped not upon another domain but upon itself, and thus arrive at the notion of *automorphism:* an automorphism is a one-to-one mapping $p \to p'$ of the point-field into itself which leaves the basic relations undisturbed; i.e. whenever points a, b, . . . satisfy the basic relation $R(ab \ . . .)$ then the points a', b', . . . into which a, b, . . . are carried over by the mapping satisfy the same relation, and vice versa. In other words $R(ab \ . . .)$ implies $R(a'b' \ . . .)$, and $R(a'b' \ . . .)$ implies $R(ab \ . . .)$. A mapping σ carries every point of the point-field into a point $p' = p\sigma$. The simplest mapping is the identity ι carrying every point p into p itself. Two mappings σ: $p \to p'$ and τ: $p' \to p''$ may be carried out one after the other and then give rise to a new mapping $\sigma\tau$: $p \to p''$. A mapping σ: $p \to p'$ is one-to-one if it has an inverse σ^{-1} that carries p' back into p: $\sigma\sigma^{-1} = \sigma^{-1}\sigma = \iota$. Then σ^{-1} is also one-to-one. The identity is a one-to-one mapping; and if σ and τ are, so is $\sigma\tau$, its inverse being $\tau^{-1}\sigma^{-1}$. The word transformation will be used as a synonym for one-to-one mapping. The fundamental fact about the automorphisms is that they form a *group*. This means the following three things: (1) the identity is an automorphism; (2) if σ is an automorphism, then σ^{-1} is; (3) if σ and τ are automorphisms, then $\sigma\tau$ is. These three facts are an immediate consequence of the definition.

A figure F in its widest sense, or a configuration of points, is nothing but a point-set; F is given if for every point p it is determined whether or not it belongs to F. A ternary relation $R(xyz)$ between points is *invariant* with respect to a given transformation σ: $p \to p'$ and its inverse $p' \to p$ if $R(abc)$ always implies $R(a'b'c')$ and vice versa. We can now say in precise terms what is meant by the objective equality or 'indiscernibility' of all points in Euclidean space. It means that, given any two points p_1 and p_0, there is always an automorphism carrying p_0 into p_1. Two figures F and F' are *similar* if one can be carried into the other by an automorphism. That is now our interpretation of Leibniz's definition of similar figures as figures that are indiscernible if each is considered by itself. The three postulates for a group simply state that each figure is similar to itself and that similarity is symmetric and transitive (see the axioms for equivalence on p. 9). A point relation is said to be objective if it is invariant with respect to every automorphism. In this sense the basic relations are objective, and so is any relation logically defined in terms of them by means of the principles enumerated in Section 2, provided no use is made of Principle 5 permitting a blank to be filled out by an individually exhibited point. (Whether every objective relation may be so defined raises a question of logical completeness which is as unlikely to be answerable as the corresponding question of completeness for the axioms in the form whether every true universal statement about points can be deduced from the axioms.)

When our task is to investigate the real space, neither the axioms nor the basic relations are given to us. On the contrary: in our attempt to axiomatize geometry we select as our basic relations some of the point relations of which we are convinced that they have an objective significance (for instance Epicurus would have included the basic relation: A, B lie on a vertical; Euclid did not). Hence in order to do justice to the real state of affairs we shall have to invert the order in the development of our ideas. We start with a group Γ of transformations. It describes, as it were, to what degree our point field is homogeneous. Once the group is given we know what likeness or similarity means — namely two figures are similar (or alike, or equivalent) that arise from each other by a transformation of Γ —, and also under what condition a relation is objective, namely if it is invariant with respect to all transformations of Γ. It is in this sense that Felix Klein in his famous Erlanger Program (1872) promulgated the conception that a geometry is determined by a group of transformations. The question of axiomatizing this geometry is now relegated to the background. (As a first step it would require the finding of a few objective relations R_1, R_2, . . . such that the group of all transformations leaving R_1, R_2, . . . invariant is not larger than Γ but coincides with Γ.) While we need not close our eyes to the fact that objective relations can be logically constructed from other such relations, we refrain from making a distinction between basic and derived. We are equally interested in all invariant relations.

[If Newton were right in ascribing to space an absolute center O, the true group Γ_0 of automorphisms would consist of those transformations of the Euclidean group Γ of automorphisms which leave O fixed; Newton's Γ_0 is a subgroup of Euclid's Γ. On the other hand, in studying Euclidean geometry we may be primarily interested in such properties as are invariant with respect to all affine or all projective transformations. (The affine and the projective transformations of a plane are those that result from carrying out one after the other any number of parallel projections or central projections respectively.) The groups Γ' and Γ'' of these transformations are wider than Γ; more precisely, Γ is part of Γ', and Γ' part of Γ''. The importance of affine and projective geometry for the theory of perspective is obvious. One sees how helpful Klein's point of view proves in surveying, and bringing to light the mutual relationship of, various kinds of geometries such as are either suggested by the nature of things or spring from arbitrary but logically useful abstraction. (Klein had a predecessor in Möbius who stressed the group-theoretical viewpoint for a number of special types of geometries.) The widest group of automorphisms one can possibly envisage for a continuum consists of all continuous transformations; the corresponding geometry is called topology. It was a lucky chance for the development of mathematics that the relativity problem was first tackled, not for the continuous point space, but for a system consisting of a finite number of distinct objects, namely the system of the roots of an algebraic equation with rational coefficients (Galois theory). This circumstance has greatly benefited the exactness of the relevant concepts. The objective relations are here those which can be constructed by means of the four basic operations of algebra (addition, subtraction, multiplication, division), in other words the algebraic relations with rational coefficients. This sort of problems gave rise to a general theory, not only of transformation groups, but also of abstract groups.}

Having explained automorphism we now come to a second phase of the relativity problem. How is it possible to assign to the points of a point-field marks or labels which could serve for their identification or distinction? The labels are supposed to be self-created, distinctive and always reproducible symbols, such as names, numbers (or number triples x, y, z, etc.). Only after this has been accomplished can one think of representing the spectacle of the actually given world by construction in a field of symbols. All knowledge, while it starts with intuitive description, tends toward symbolic construction. No serious difficulty is encountered as long as one deals with a domain consisting of a finite number of points only, which can be 'called up' one after the other. The problem becomes a serious one when the point-field is infinite, in particular when it is a continuum. A conceptual fixation of points by labels of the above-described nature that would enable one to reconstruct any point when it has been lost, is here possible only in relation to a *coordinate system*, or frame of reference, that has to be exhibited by an individual demonstrative act. The objectification, by elimination of the ego and its immediate life of intuition, does not fully succeed, and the coordinate system remains as the necessary residue of the ego-extinction.[16] It is good to remember here that in practice two- or three-dimensional point-sets are usually given by actually putting a body or a figure drawn with pencil on paper before our eyes, and not by a logico-arithmetical construction of set-defining properties. It took a long time for mathematics before it had acquired the constructive tools to cope with the complexity and variety of such intuitively given figures. But once it had reached that stage the superiority of its symbolic methods became evident.

{Take as an example the points on a line. The coordinate system consists here of a point O and a unit segment OE, or of two distinct points O, E. When this frame of reference is given, any point P can be characterized by its abscissa x, the number measuring the length OP with OE as the unit yardstick (x is positive for points lying on the same side of O as E, negative for points on the opposite side). Any

[16] Against the establishment of an essential difference between conceptual determination and intuitive exhibition, the objection might be raised that even the objective geometrical relations upon which the conceptual determination is based require intuitive exhibition. But these are a few isolated relational concepts, while the points themselves form a continuum. I am inclined to admit that this fact alone constitutes the essential difference.

two frames of reference, OE and $O'E'$, are objectively alike, for there is exactly one automorphism (similarity) that maps O into O' and E into E'. Hence by exhibiting an individual coordinate system no more is exhibited than is absolutely necessary. The field for the symbol x consists of all real numbers. Relative to a given coordinate system the correspondence $P \leftrightarrows x$ is a one-to-one mapping of the point-field onto the variability range of the symbol. The coordinates x and x' of the same arbitrary point in two coordinate systems are connected by a relation $x = ax' + b$ where $a \neq 0$ and b are two constants characteristic of the relative position of the two coordinate systems. }

With this example in mind, one will be able to understand the following general description. A class Σ of frames of reference $\mathfrak{f}$ is supposed to be given. The class as such should be objectively distinguished; i.e. if $\mathfrak{f}$ belongs to it, so does any similar frame $\mathfrak{f}\sigma = \mathfrak{f}'$ arising from $\mathfrak{f}$ by an automorphism σ. But the class is supposed to contain no more elements than this requirement makes absolutely necessary, i.e. any two frames $\mathfrak{f}$, $\mathfrak{f}'$ of the class are similar. Moreover an objective rule A is supposed to be given by which each point p with respect to any frame $\mathfrak{f}$ of the class Σ determines a definite (reproducible) symbol $x = A(p; \mathfrak{f})$ as its coordinate. For a given $\mathfrak{f}$ the correspondence $p \rightleftarrows x$ between points p and symbols x is one-to-one. That x is objectively determined by p and $\mathfrak{f}$ means that

$$(1) \qquad\qquad A(p; \mathfrak{f}) = A(p\sigma; \mathfrak{f}\sigma)$$

for any automorphism σ.

From these conditions there flow the following consequences. Let σ be an automorphism $p \rightarrow p'$ of the point-field and $\mathfrak{f}$ be a fixed frame of class Σ. The coordinates x of p and x' of p' in this frame are connected by a transformation S, $x' = xS$, which represents the automorphism σ in terms of $\mathfrak{f}$. To the identity $\sigma = \iota$ there corresponds the identity $S = I$; σ^{-1} and $\sigma\tau$ are represented by S^{-1} and ST if S and T represent σ and τ. In this sense the transformations S corresponding to the several σ of Γ form a group $\mathbf{G}$ that is isomorphic with Γ. $\mathbf{G}$ is nothing but the representation of Γ in terms of $\mathfrak{f}$. Take, on the other hand, a fixed frame $\mathfrak{f}$ of our class Σ and an arbitrary frame $\mathfrak{f}' = \mathfrak{f}\sigma$ that arises from $\mathfrak{f}$ by the automorphism σ. I maintain that the coordinates x, x' of the same arbitrary point with respect to $\mathfrak{f}$ and $\mathfrak{f}'$ are connected by the equation $x = x'S$. Indeed, denote the arbitrary point by $p\sigma$ instead of p; we then have $x = A(p\sigma; \mathfrak{f})$ and, because of (1),

$$x' = A(p\sigma; \mathfrak{f}\sigma) = A(p; \mathfrak{f}),$$

76

and thus our assertion follows. The group **G** that represents Γ in terms of f must be independent of f. Indeed, representation of Γ by two different groups **G, G*** in terms of the two similar frames f and f* would constitute an objective difference between f and f*, which is impossible. It is easy to verify this explicitly. Let f* = fγ, where γ is an automorphism. Moreover, let x and x' be the coordinates of an arbitrary point p and its image $p' = p\sigma$ with respect to f, and y and y' with respect to f*. The transformations representing γ and σ in terms of f may be called C and S. Write the equation $x' = xS$ in the more suggestive form $x \xrightarrow{(S)} x'$. After what has been said, we then have the following diagram

$$
\begin{array}{ccc}
 & (S) & \\
x & \longrightarrow & x' \\
(C)\uparrow & & \downarrow(C^{-1}). \\
y & & y'
\end{array}
$$

Hence the transformation that leads from y to y' and thus represents σ in terms of the frame f* is $S^* = CSC^{-1}$. With S also $CSC^{-1} = S^*$ is in the group **G**, and vice versa: $S = C^{-1}S^*C$.

As long as the points could not be characterized conceptually, the transformations of the point field could not be either, and it was thus perhaps not perfectly clear what was meant by saying that the group of automorphisms is known or given. A stage has now been reached where this last shadow of obscurity disappears. Every point is replaced by its coordinate x (with respect to a fixed frame), and thus the group Γ of automorphisms σ appears as a group **G** of transformations S. The individual transformation S carrying x into $x' = xS$, is a reproducible symbol like any individual value of x. But while the coordinate x is not only dependent on p but also on f, the group **G** is independent of f and hence free from anything in need of individual exhibition. To fulfill the demand of objectivity we construct an image of the world in symbols. The pure mathematician will say: Given a group **G** of transformations in a field of symbols, a geometry is established by agreeing to study, and consider as objective, only such relations in that field as are invariant under the transformations of **G.**

{A last remark of a purely logical nature concerns the frames. It is quite legitimate to regard as the frame of reference f the coordinate assignment $p \rightarrow x = f(p)$ itself established by f. This seems even preferable if one has to be prepared for a group of automorphisms so wide as to comprise all continuous transformations. The symbol f is then simply a token for the function f whose argument ranges over the points p and whose value is an element x in the field of symbols. If $\sigma: p \rightarrow p'$ is any transformation, then the transformed function $f' = f\sigma$ will be defined by the equation $f'(p') = f(p)$ for $p' = p\sigma$, or $f'(p) = f(p\sigma^{-1})$. When we write $x = A(p; \mathfrak{f})$ for $x = f(p)$ then A stands for the universal logical operator 'value of'; $x = A(p; \mathfrak{f})$ means: x is the value of the function $\mathfrak{f}$ for the argument p.]

REFERENCES

F. KLEIN, *Vergleichende Betrachtungen über neuere geometrische Forschungen*, Erlangen, 1872 (*Gesammelte mathematische Abhandlungen*, I, 460–497).

14. CONGRUENCE AND SIMILARITY.
LEFT AND RIGHT

There is no doubt that the conviction which Euclidean geometry carries for us is essentially due to our familiarity with the handling of that sort of bodies which we call rigid and of which it can be said that they remain the same under varying conditions. The portions of space which such a solid fills in two of its positions are called *congruent*. Measurement depends on rigid bodies to the same degree as counting does on the use of concrete number symbols. (About the physical foundation of geometry cf. also Sections 16 and 18.) Once geometry has been abstracted from the behavior of actual bodies that are approximately rigid it provides a standard for the physical investigation of all bodies, and we can judge how far a given body realizes the ideal of rigidity. This process is not essentially different from the one by which a scale of temperature is first based on the behavior of actual gases and then reduced to the 'ideal gas scale' by postulating the exact validity of such laws as are approximately satisfied by the existing gases. Since places on a rigid body can be tagged, congruence is a point-by-point mapping of the two congruent volumes V and V'. The notion of congruence at first is relative to a given rigid body b. Its factual independence of b is one of our most fundamental experiences. Indeed, let V, V' be two portions of space filled by the solid b in two of its positions. Let b^* be another solid that fits into V; then it may be so moved as to fill V'. Since one may extend a rigid body so as to cover any given point, the mapping $V \rightarrow V'$ can be extended to the whole space. The congruent mappings of space form a group Δ^+ of transformations which we call the *group of Euclidean motions*. Once this group is known, congruent volumes may be defined as portions of space that can be carried into each other by a transformation S of Δ^+. The facts suggest an interpretation according to which the

78

group Δ^+ of congruent mappings expresses an intrinsic structure of space itself; a structure stamped by space on all spatial objects.

If this view is correct, congruence should be made the one and only basic concept of geometry. Let us first investigate what the consequences of this conception of geometry are for the automorphisms of space (*similarities*). We know quite generally that once the basic relational concepts of a geometry are fixed the group Γ of automorphisms is also fixed. In our case the criterion for an automorphism C is this: C as well as C^{-1} must transform any pair of congruent portions of space v_1, v_2 into a congruent pair. Consider the pair v_1^*, v_2^* arising from v_1, v_2 by the transformation C. Let S be the motion that

$$
\begin{array}{c}
(S) \\
v_1 \longrightarrow v_2 \\
(C^{-1}) \uparrow \quad \downarrow (C) \\
v_1^* \quad v_2^*
\end{array}
$$

carries v_1 into v_2. As the above diagram indicates, v_1^* goes into v_2^* by the mapping $C^{-1}SC$. Hence the criterion demands that the transformations $C^{-1}SC$ and CSC^{-1} should belong to Δ^+ whenever S does. A transformation C is said to commute with a given group Δ of transformations if $C^{-1}SC$ and CSC^{-1} are in Δ whenever S is. The transformations commuting with Δ form a group called the normalizer of Δ. This group necessarily contains Δ as a subgroup, be it that Δ is identical with its normalizer or a proper part of it. Our analysis can now be summarized thus: *The group Γ of similarities is the normalizer of the group Δ^+ of motions.* Hence congruent figures are necessarily similar. The converse need not be true. Indeed, since Δ^+ happens to be a proper subgroup of its normalizer there exist similar figures in Euclidean space which are not congruent; as for instance a body and its mirror image, or a building and a small scale model of it.

Let us now invert the procedure and follow Klein by starting with a given group Γ of automorphisms. Take a subgroup Δ of Γ and declare two figures to be Δ-equivalent if one is carried into the other by a transformation of Δ. Under what circumstances has this relation of Δ-equivalence objective significance? If and only if Δ-equivalent figures are carried into Δ-equivalent figures by every transformation C of Γ, or in other words, if every element C of Γ commutes with Δ. In that case the mathematician says that Δ is an invariant subgroup of Γ. Hence Δ-equivalence is an objective relation provided Δ is an invariant subgroup of Γ. For instance, the parallel displacements form an invariant subgroup of the group of Euclidean similarities; and indeed the relation $\parallel$ between two figures arising from each other by parallel displacement is clearly of objective geometric significance — although our language lacks a suggestive word for it. The normalizer of the group of parallel displacements consists of all affine transformations; hence affine geometry may be based on the one relation ‖ between figures. Or, still more simple, the subgroup consisting of the identity only is an invariant subgroup, and indeed the relation of identity between two figures is of objective significance. (There is none that has a better claim to objectivity, owing to the fact that the identity is contained in every possible group Γ of transformations.) The smaller the group Δ the larger its normalizer, and thus the wider the gap between congruence and similarity; or more precisely, if Δ' is a subgroup of Δ then the normalizer Γ' of Δ' contains the normalizer Γ of Δ. The normalizer of an invariant subgroup Δ of Γ always comprises Γ. A geometry whose group of automorphisms is Γ can be based on the objective relation of Δ-equivalence *alone*, provided the normalizer of Δ is not larger than Γ but coincides with Γ.

〔A last remark will conclude this analysis. Space is a continuum, and when we speak of any transformation in space it is reasonable to interpret this as meaning any continuous transformation. We indicate by Ω the group of those transformations that are taken into account at all; in the case of a continuum this would be the group of all continuous transformations. By putting this explicitly in evidence our definition of normalizer may be repeated as follows. Given a subgroup Δ of the group Ω; those elements of Ω that commute with Δ constitute the normalizer Γ of Δ. In this form the notion of normalizer makes sense even for abstract groups Ω and Δ.

Kant speaks about the divergence between congruent and similar in *Prolegomena*, §13, and claims that "by no single concept, but only by pointing to our left and right hand, and thus depending directly on intuition [Anschauung] can we make comprehensible the difference between similar yet incongruent objects (such as oppositely wound snails)"; and in his opinion only transcendental idealism offers a solution for this riddle. No doubt the meaning of congruence is based on spatial intuition, but so is similarity. Kant seems to aim at some subtler point, but just this point is one which can be completely clarified by an analysis in terms of a group Γ and its invariant subgroups Δ, or of a group Δ and its normalizer Γ. Whenever Δ is a proper invariant subgroup of Γ, the notions of congruence $=$ Δ-equivalence and similarity $=$ Γ-equivalence do not coincide although the former is of objective significance ($=$ Γ-invariant). The phenomenon about which Kant wonders can thus be most satisfactorily subsumed under general and abstract 'concepts.'〕

Whoever raises congruence to the rank of the only basic relation of geometry is obliged to develop geometry from this one notion. Several ways are open to accomplish this. A deeper insight than by the elementary approach in the style of Euclid's axioms would be gained if one succeeded in formulating the fundamental facts of geometry as simple axioms concerning the group Δ^+ of Euclidean motions. Following Ueberweg, Helmholtz first carried out this program with surprising success in his essay "Ueber die Tatsachen, die der Geometrie zugrunde liegen." Later S. Lie, who established a general theory of transformation groups, resumed the problem with his more powerful mathematical tools and generalized it from 3 to n dimensions. The Euclidean group of motions Δ^+ turns out to be almost completely characterized by the fact that it permits the rigid body that measure of free mobility with which we are familiar by experience. In more exact terms: it is possible by suitable congruent mappings to carry any point into any other, and, if a point is kept fixed, to carry any line direction at that point into any other at the same point; furthermore, if a point and line direction are kept fixed, it is possible to carry by congruent mapping any surface direction through them into any other such direction, and so forth, up to the $(n - 1)$-dimensional direction elements. If, on the other hand, a point and a line direction through it, and a surface direction through the latter, and so forth, up to an $(n - 1)$-dimensional direction element, are given, then there exists no congruent mapping besides the identity under which this system of incident elements remains fixed. We just said that this axiom *almost* completely characterizes the Euclidean group of motions. In fact one thus obtains the group of congruent transformations of a slightly more general space, namely of a projective space endowed with a Cayley metric. That group contains a numerically indeterminate parameter λ, the constant space curvature, of which nothing but the sign is essential. According as λ is positive, zero, or negative, the resulting space is of the elliptic, parabolic (i.e. Euclidean), or hyperbolic type. These then are the only homogeneous spaces, in which all points are equivalent, likewise all directions at a point, and so on.

[It is hard to talk intelligently about these problems without an exact description of the Euclidean groups Γ and Δ^+ before our eyes. A Cartesian frame of reference in three-dimensional Euclidean space consists of a point O, the origin, and three mutually perpendicular vectors $\vec{e_1}$, $\vec{e_2}$, $\vec{e_3}$ of equal length. The coordinates x_1, x_2, x_3 of a point P are defined by

$$\vec{OP} = x_1 \vec{e_1} + x_2 \vec{e_2} + x_3 \vec{e_3}.$$

81

Relative to such a frame a similarity mapping the point (x_1, x_2, x_3) into the point (x'_1, x'_2, x'_3) is represented by a linear transformation

$$(1) \qquad S: \quad x'_i = a_i + a_{i1}x_1 + \cdots + a_{in}x_n \qquad (i = 1, 2, \cdots, n)$$

with constant coefficients a_i, a_{ik} that satisfy the following condition: $(x'_1 - a_1)^2 + \cdots + (x'_n - a_n)^2$ is a constant positive multiple a of $x_1^2 + \cdots + x_n^2$. (Here the number n of dimensions has been left indeterminate.) The similarity is 'non-enlarging' and called an orthogonal transformation if $a = 1$. The orthogonal transformations form an invariant subgroup Δ of Γ. The condition mentioned above, which is satisfied by every similarity, implies the equation $d^2 = a^n$ for the determinant d of the a_{ik}. Hence an orthogonal transformation is either of signature $+$ $(d = +1)$ or of signature $-$ $(d = -1)$. The orthogonal transformations of signature $+$ form the group Δ^+ of Euclidean motions. Δ^+ is a subgroup of Δ of index 2, i.e., if S_1, S_2 are any two transformations of Δ of signature $-$ then $S_1^{-1}S_2$ has the signature $+$. (The fundamental fact of the distinction of left and right: two screws oppositely winding to a given screw turn in the same sense.) It makes little difference whether we claim Δ^+ or Δ as the group of congruent mappings. Assume we decide in favor of the larger group Δ. Then the continuous motion of a rigid body would be represented by an orthogonal transformation $S(t)$ depending continuously on the time parameter t and reducing to the identity I at the initial moment $t = 0$. Since the determinant of $S(t)$ is capable of the two values $+1$ and -1 only, since it equals $+1$ at the beginning $t = 0$ and varies continuously with t, it must always remain equal to $+1$. Hence even if we had admitted arbitrary orthogonal transformations, the requirement of continuity for $S(t)$ automatically eliminates those of signature $-$; a rigid body could go over into its mirror image only by a discontinuous jump.}

A far deeper aspect of the group Δ than that of describing the mobility of rigid bodies is revealed by its role as the group of *automorphisms of the physical world*. In physics we have to consider not only points but also various types of physical quantities, velocity, force, electromagnetic field strength, etc. But it is a fact that relative to a Cartesian frame, not only points but all physical quantities can be represented by numbers; e.g. a force by its components f_i $(i = 1, 2, \cdots n)$, an electromagnetic field strength by a set of skew-symmetric components $F_{ik} = -F_{ki}$, etc. And under the influence of any orthogonal mapping S, (1), of the points of space they undergo a related transformation that is uniquely determined by S; e.g. the force components transform according to the equations

$$f'_i = \sum_\lambda a_{i\lambda} f_\lambda \quad (i, \lambda = 1, 2, \cdots, n),$$

the components of the electromagnetic field strength according to the rule

$$F'_{ik} = \sum_{\lambda,\mu} a_{i\lambda} a_{k\mu} F_{\lambda\mu} \quad (i, k, \lambda, \mu = 1, 2, \cdots, n)s$$

etc.   All the laws of nature are invariant under the transformation thus induced by the group Δ.   It is not true however that they are invariant under all similarities, although it seems so on a certain level of natural phenomena.   But the facts of atomism teach us that *length is not relative but absolute*.   The atomic constants of charge and mass of the electron and Planck's quantum of action h fix an absolute standard of length, that through the wave lengths of spectral lines is also made available for practical measurements.   Thus we no longer depend on the preservation of the platinum-iridium meter bar that is kept in the vaults of the Comité International des Poids et Mesures in Paris.   We now prescribe the absolute length 1 for the basic vectors of a Cartesian frame of reference.   The orthogonal transformations of signature $-$ must be included in Δ.   For there is no indication in the laws of nature of an intrinsic difference between left and right. Now it is clear why a body all of whose places undergo a transformation $S(t)$ of the group Δ depending continuously on the time parameter t and whose physical characteristics change accordingly, has a perfectly good claim to say of himself: I have remained physically the same during my motion.

[The extensive medium of the external world is one of time as well as space.   How time is included as a fourth coordinate in the above scheme will be discussed in Section 16.   It was in preparation for this step that we left the number n of dimensions indeterminate.   For physics the case $n = 4$ is even more important than $n = 3$.   At present however we shall limit ourselves to space.]

We summarize: The group of physical automorphisms in space is the group Δ of orthogonal transformations.   The group of geometric automorphisms, by virtue of the very meaning of this term, is the normalizer Γ of Δ.   It is larger than Δ, inasmuch as it includes the dilatations $x'_i = ax_i$ with any constant $a > 0$.   This divergence between Δ and Γ proves conclusively that *physics can never be reduced to geometry* as Descartes had hoped.

{*Left and right.* Were I to name the most fundamental mathematical facts I should probably begin with the fact (F_1) that the counting of a set of elements leads to the same number in whatever order one picks up its elements, and mention as a second the fact (F_2) that among the permutations of n (≥ 2) things one can distinguish the even and the odd ones. The even permutations form a subgroup of index 2 within the group of all permutations. The first fact lies at the bottom of the geometric notion of dimensionality, the second of that of 'sense.' Consider affine vector geometry. A basis for its vectors consists of n vectors $\vec{e}_1, \ldots, \vec{e}_n$ such that every vector can be uniquely expressed as a linear combination $x_1\vec{e}_1 + \cdots + x_n\vec{e}_n$, and the theorem of the invariance of dimensionality states that every basis necessarily consists of the same number n of vectors. This assertion clearly implies the fact (F_1); for by any regrouping of the basic vectors one passes to a new basis. Vice versa, the theorem of invariance is an algebraic proposition easily deduced from the fact (F_1) in conjunction with the rule for addition and multiplication of numbers. Any arrangement of n given linearly independent vectors fixes a 'sense,' and two arrangements fix the same sense provided they arise from each other by an even permutation (definition by abstraction). An odd permutation changes the sense into its opposite. That is clearly the combinatorial root of the distinction between left and right. Again, in combination with the basic operations of affine vector geometry (addition of vectors, multiplication of a vector by a number) it leads to a comparison of sense for any two bases $\vec{e}_1, \ldots, \vec{e}_n$ and $\vec{e}_1^*, \ldots, \vec{e}_n^*$. When one expresses the vectors $\vec{e}^*$ in terms of the vectors e,

$$\vec{e}_i^* = a_{1i}\vec{e}_1 + \cdots + a_{ni}\vec{e}_n,$$

the coefficients a_{ki} have a non-vanishing determinant. The senses of the two bases are the same or opposite according to whether the determinant is positive or negative. But the definition of a determinant is based on the distinction between even and odd permutations!

Kant finds the clue to the riddle of left and right in transcendental idealism. The mathematician sees behind it the combinatorial fact of the distinction of even and odd permutations. The clash between the philosopher's and the mathematician's quest for the roots of the phenomena which the world presents to us can hardly be illustrated more strikingly.}

15. RIEMANN'S POINT OF VIEW. TOPOLOGY

The notions of dimensionality and sense are not restricted to metric Euclidean or affine space. They apply to continuous manifolds in general.   Riemann was the first to analyze mathematically the general concept of an n-dimensional manifold.   A sufficiently small neighborhood of an arbitrary point in an n-dimensional manifold may be mapped one-to-one and continuously upon a region of the n-dimensional number space, the points of the latter being the n-tuples of real numbers $(x_1, x_2, \ldots, x_n)$.   Any one-to-one transformation of the coordinates

$$y_i = \varphi_i(x_1, \ldots, x_n) \qquad (i = 1, \ldots, m);$$
$$x_k = \psi_k(y_1, \ldots, y_m) \qquad (k = 1, \ldots, n)$$

yields a new coordinate assignment suitable for the representation of the same neighborhood.   Is m necessarily equal to n?   This is the question of the topological invariance of dimensionality.

[Let $P = (x_1, \cdots, x_n)$ be a given point and $P^* = (x_1 + dx_1, \cdots, x_n + dx_n)$ any point infinitely near to P.   If the transformation functions are differentiable then the components $(dx_1, \ldots, dx_n)$ of all infinitesimal vectors $\overrightarrow{PP^*}$ issuing from P transform according to linear formulas

(1) $$dy_i = \sum_k a_{ik} \cdot dx_k, \quad dx_k = \sum_i b_{ki} \cdot dy_i.$$

the coefficients a_{ik}, b_{ki} of which depend on the point P but not on P^*. (Infinitesimal quantities may be avoided by introducing an imaginary time τ and letting a point move in the manifold according to an arbitrary law $x_k = x_k(\tau)$.   Suppose the point passes P at the moment $\tau = 0$; its velocity at that moment will be a vector at P with the x-components $u_k = (dx_k/d\tau)_{\tau=0}$.   The y-components v_i of the same velocity are related to the x-components by the equations (1),

$$v_i = \sum_k a_{ik} u_k, \quad u_k = \sum_i b_{ki} v_i,$$

which hold for all possible velocities in P.)   But these linear transformations can be inverse to each other only if $m = n$ and the determinant of the a_{ik}, the so-called Jacobian, is different from zero.   Only such 'differentiable' transformations of the coordinates are now admitted at all to the totality Ω.   Under these circumstances one speaks of a differentiable manifold.   As the Jacobian varies continuously with P, it is either positive throughout the region covered by the two coordinate assignments, or negative throughout.   We give the transformation the signature $+$ in the first case, $-$ in the second. Hence a 'sense' can be fixed over the whole region.   One sees that both dimensionality and sense derive from the fact that affine geometry holds in the infinitely small. While topology has succeeded fairly well in mastering continuity, we do not yet understand the inner meaning of the restriction to differentiable manifolds. Perhaps one day physics will be able to discard it. At present it seems indispensable since the laws of transformation of most physical quantities are intimately connected with that of the differentials dx_i, (1).}

Inspired by Gauss's theory of curved surfaces, Riemann assumed that Euclidean geometry holds in the infinitely small. Then the square of the length ds of the infinitesimal vector $\vec{PP^*}$ with the components dx_i will be expressed by a positive quadratic form

$$(2) \qquad ds^2 = \sum_{ik} g_{ik}\, dx_i\, dx_k$$

of the dx_i. Its coefficients g_{ik} are independent of the vector $\vec{PP^*}$ with the components dx_i but will in general depend on the point P with the coordinates x_i and be continuous functions of these coordinates. It is clear from the invariant significance of ds^2 how the components g_{ik} of the 'metric field' will transform under transition to a new coordinate system y_i. The metric of a 3-dimensional Riemann space of this kind imposes itself upon any surface lying in it, which is thereby branded as a 2-dimensional Riemann space. For a 3-dimensional Euclidean space, however, it is not true that every surface in it is a 2-dimensional Euclidean space; rather, all possible 2-dimensional Riemann spaces occur as subspaces of a Euclidean 3-space. Thus in Euclid's geometry the space appears as something much more special (namely, non-curved) than the possible surfaces in it, while Riemann's space concept has just the right degree of generality to do away with this discrepancy.

As the true lawfulness of nature, according to Leibniz's continuity principle, finds its expression in laws of nearby action, connecting only the values of physical quantities at space-time points in the immediate vicinity of one another, so the basic relations of geometry should concern only infinitely closely adjacent points ('near-geometry' as opposed to 'far-geometry'). *Only in the infinitely small may we expect to encounter the elementary and uniform laws,* hence the world must be comprehended through its behavior in the infinitely small.

If one requires the space to be metrically homogeneous — and a space that can serve as 'form of phenomena' is necessarily homogeneous — then one is thrown back at once from the Riemannian to the classical space concept, to which Helmholtz's postulates for the group of motions lead. But Riemann had an entirely different conception of the nature and origin of the metrical properties of space. For him the metric field is not given rigidly once and for all, but is causally

86

connected with matter and thus changes with the latter. He considers the metric not as part of the static homogeneous form of phenomena, but of their ever-changing material content. Riemann asks for the inner reason of the metrical relationships in space, and having distinguished (in the words quoted on p. 43) between the cases of discrete and continuous manifolds, he continues, "Therefore, either the reality on which our space is based must form a discrete manifold, or else the reason for the metrical relationships is to be looked for externally in binding forces acting upon it." The metric field makes itself felt through the physical effects which it has upon rigid bodies, upon light rays, and all events in nature, and these effects alone permit us to ascertain the quantitative state of the metric field. But whatever acts must suffer too; it must itself be something real and cannot be enthroned in unattackable 'geometric' rigidity above the forces of matter. Thereby, in spite of the non-homogeneity of the metric field, the free mobility of bodies without changes in measure is regained, since a body in motion will 'take along' the metric field that is generated or deformed by it. Einstein, after having extended space by the inclusion of time to the full four-dimensional medium of the external world, has developed Riemann's idea into a detailed physical *theory of gravitation* and, in particular, has ascertained the laws according to which matter acts upon the metric field.

Riemann and Einstein maintain that the group of — geometric or physical — automorphisms coincides with the totality Ω of all differentiable transformations. In this respect their theories differ radically from the standpoint expounded in the previous section. Their principle of general relativity is acceptable only after inserting the metric field among the physical quantities that act upon, and are reacted upon by, matter. Nevertheless Euclidean geometry is preserved for the infinitesimal neighborhood of any given point P_0. For it is a mathematical fact that for all line elements at a given point P_0 the metric equation (2) takes on the special form

$$ds^2 = dx_1^2 + dx_2^2 + \cdots + dx_n^2$$

if appropriate coordinates x_i are chosen for the neighborhood of P_0. In this form there is no room for any indeterminacy, and we may therefore say that the *nature* of the metric is the *same* at every point. But the coordinate system in which the metrical law assumes this fixed standard form and which, as we shall say, is characteristic for the *orientation* of the metric is in general different from place to place. We use an analogous phrase in Euclidean geometry when we say that all cubes (of given size) are of the same *nature* and differ only by their *orientation*. The nature of the metric is *one*, and is absolutely given;

only the mutual orientation in the various points is capable of continuous changes and dependent upon matter. Euclidean space may be compared to a crystal, built up of uniform unchangeable atoms in the regular and rigid unchangeable arrangement of a lattice; Riemannian space to a liquid, consisting of the same indiscernible unchangeable atoms, whose arrangement and orientation, however, are mobile and yielding to forces acting upon them.

Perhaps this is brought out better by a different formulation of Riemann's conception, which has become indispensable in quantum physics when the quantities characterizing a spinning electron are to be fitted into general relativity theory. From the above illustration by velocities it is clear what is meant by the body of tangent vectors (velocities) at P. They form an n-dimensional vector space. The coordinate assignment $P \rightarrow x_i$ determines a vector basis $\vec{e}_1, \ldots, \vec{e}_n$ in this tangent vector space $V(P)$ at P such that $u_1\vec{e}_1 + \cdots + u_n\vec{e}_n$ is the vector with the x-components u_i. Assuming that the vector space at P bears a Euclidean metric (with an absolute standard of length) we can introduce in it a local Cartesian frame of reference $\mathfrak{f} = \mathfrak{f}(P)$ consisting of n mutually perpendicular vectors of length 1. The arbitrariness in the choice of this frame is expressed by the group Δ_0 of Euclidean rotations. That group consists of all linear transformations

$$S: \quad z'_\beta = \sum_\gamma a_{\beta\gamma} z_\gamma \qquad (\beta, \gamma = 1, \ldots, n)$$

for which

$$z'^2_1 + \cdots + z'^2_n = z^2_1 + \cdots + z^2_n.$$

Here the variables z_β designate the components of an arbitrary vector of $V(P)$ with respect to the Cartesian frame $\mathfrak{f}$. The numerical values $e_{i\beta}$ ($\beta = 1, \cdots, n$) of the components of each of the vectors $\vec{e}_i$ ($i = 1, \cdots, n$) with respect to $\mathfrak{f}$ describe the embedment of the frame $\mathfrak{f}$ into space. Thus the n^2 quantities $e_{i\beta}$, which depend on the choice of the coordinates x_i as well as on the Cartesian frame $\mathfrak{f}(P)$ at P and are functions of P, now serve to characterize the metric field. Riemann's g_{ik} are easily computed to have the values

$$g_{ik} = e_{i1}e_{k1} + \cdots + e_{in}e_{kn}.$$

Only after coordinates x_i and a Cartesian frame $\mathfrak{f}(P)$ at each point P have been chosen can all physical quantities be represented by numbers. The laws of nature are invariant (1) with respect to arbitrary transformations of the coordinates x_i, and (2) with respect to a rota-

88

tion S of the frame $\mathfrak{f}(P)$ that may depend in an arbitrary (continuous) manner on the point P. Hence there is this double invariance, the one described by the group Ω of all transformations of the coordinates x_i, the other by an element of the group Δ_0 that can vary arbitrarily with the position P.

What has happened in the transition from special to general relativity theory is obviously this. The physical automorphisms forming the group Δ as described in the previous section have been split into their translatory and rotatory parts. The group of translations has been replaced by that of all possible transformations of the coordinates, whereas the rotations have remained Euclidean rotations but are now tied to a center P and must be allowed to vary freely while the center P moves over the manifold. Space, the extensive medium of the material world, is clearly the seat of the group Ω of coordinate transformations; but the group Δ_0 seems to have its origin in the ultimate elementary particles of matter. The quantities $e_{i\beta}$ thus mediate between matter and space.

[The question arises for what inner reasons nature has picked Δ_0 among all possible groups of homogeneous linear transformations. One answer is provided by Helmholtz's theory, according to which Δ_0 is completely characterized by the axiom of free mobility: Any incident set σ of 1-, 2-, . . . $(n - 1)$-dimensional directions can be carried into any other such set by a transformation of Δ_0 while those transformations of Δ_0 that leave a given set σ of incident directions fixed form a subgroup containing two elements only (namely the identity and the reflection in σ). However, this characterization carries less conviction now where the group can no longer be interpreted as describing the mobility of a rigid body. (Moreover it breaks down for the Lorentz group, which in the four-dimensional world takes the place of the orthogonal group in 3-space.)

The group Δ_0 could be considered as an abstract group of which various representations by linear transformations are characteristic for various physical quantities; e.g. the representation Δ_0 by orthogonal transformations itself for the vectors, a certain 'tensor' representation for the electromagnetic field strength, and a very remarkable one, the so-called spinor representation, for the electronic wave field.]

Topology. In general a coordinate assignment covers only part of a given continuous manifold. The 'coordinate' $(x_1, \ldots, x_n)$ is a symbol consisting of real numbers. The continuum of real numbers can be thought of as created by iterated bipartition. In order to account for the nature of a manifold as a whole, topology had to develop combinatorial schemes of a more general nature. By this combinatorial approach it also got rid of the restriction to differentiable manifolds.

In order to subject a continuum to mathematical treatment it is necessary to assume that it is divided up into 'elementary pieces' and that this division is constantly refined by repeated subdivision according to a fixed scheme (which in the one-dimensional case consists in the bipartition of each elementary segment). The effect is that the continuum is spun over with a subdivision net of increasing density. Thus, properly speaking, every continuum has its own arithmetical scheme which is already completely determined by the combinatorial description of the manner in which the individual elementary pieces of the initial division border on each other; we call this the 'topological skeleton' of the manifold. The introduction of numbers as coordinates by reference to the particular division scheme of the open one-dimensional continuum is an act of violence whose only practical vindication is the special calculatory manageability of the ordinary number continuum with its four basic operations. The topological skeleton determines the connectivity of the manifold in the large. It is an important but difficult mathematical question to decide when two such skeletons are equivalent, i.e. when they represent two different ways of decomposition of the *same* continuum into elementary pieces. In the case of an n-dimensional closed manifold, the skeleton consists of a finite number of elements of rank $0, 1, 2, \ldots, n$ (vertices, edges, . . .); these elements are to be represented by arbitrary symbols. An element of the ith rank is bounded by certain elements of rank $i - 1$, and the skeleton is completely described by telling which element is bounded by which. The requirements such a skeleton has to meet, the properties which it possesses, and the question of equivalence constitute the subject of combinatorial topology.

[Topology has the pecularity that questions belonging in its domain may under certain circumstances be decidable, even though the continua to which they are addressed may not be given exactly but only vaguely, as is always the case in reality. For instance, the topological skeleton of an undamaged brick is recognizable with certainty. Or an endless thread, which determines only approximately a curve in the exact sense of geometry, is definitely either knotted or not. Whenever the possible cases form a discrete manifold, an individual case can be fixed with absolute accuracy. Thus the rational analysis of continua proceeds in three steps: (1) morphology, which operates with vaguely circumscribed types of forms; (2) topology, which, guided by conspicuous singularities or even in free construction, places into the manifold a vaguely localized but combinatorially exactly determined skeleton; and (3) geometry proper, whose ideal structures could only be carried with exactness into a real continuum after this has been spun over with a subdivision net of a fineness increasing *ad infinitum*. (Such geometrical properties of configurations in the continuum as are independent of the arbitrariness involved in the construction of the subdivision net may be conceived as based on a structural field spread over the continuum after the fashion of the metric field.) The significance which the idealizing geometry has for reality, in spite of the evident impossibility of fulfilling the above requirement for its application, will be discussed in Part II. The three steps described reveal the sensual-categorical ambivalence of geometry, which caused Plato to assign to geometrical configurations an intermediate position between ideas and sensory objects. For a more careful phenomenological analysis of the contrast between vagueness and exactness and of the limit concept, the reader may be referred to the work by O. Becker quoted at the end of Section 9. Carrying out the subdivision of the topological skeleton according to a fixed scheme implies the assumption that in dealing with a concretely given continuum we were not in error as to the topological character of the pieces generated by the first division. That is to say, we disregard the possibility that a more detailed scrutiny of a surface might disclose that, what we had considered an elementary piece, in reality has tiny handles attached to it which change the connectivity character of the piece, and that a microscope of ever greater magnification would reveal ever new topological complications of this type, *ad infinitum*.

The Riemann point of view allows, also for real space, topological conditions entirely different from those realized by Euclidean space. I believe that only on the basis of the freer and more general conception of geometry which had been brought out by the development of mathematics during the last century, and with an open mind for the imaginative possibilities which it has revealed, can a philosophically fruitful attack upon the space problem be undertaken.⊦

R E F E R E N C E S

H. von Helmholtz, Über die Tatsachen, die der Geometrie zugrunde liegen (1868), *Wissenschaftliche Abhandlungen*, II, p. 618.

B. Riemann, Über die Hypothesen, welche der Geometrie zugrunde liegen (1854), *Gesammelte mathematische Werke*, 1876 (separate edition with notes by H. Weyl, Berlin, 1923).

H. Weyl, *Mathematische Analyse des Raumproblems*, Berlin, 1923.

O. Veblen, *Analysis situs*, Am. Math. Soc. Colloquium Publications, second edition, New York, 1931.

P. Alexandroff, *Einfachste Grundbegriffe der Topologie*, Berlin, 1932.

Part Two. Natural Science

'Ο ἄναξ οὗ τὸ μαντεῖόν ἐστι τὸ ἐν Δελφοῖς, οὔτε λέγει οὔτε κρύπτει ἀλλὰ σημαίνει.

(The Lord whose is the oracle at Delphi neither reveals nor hides but gives tokens.)

HERACLITUS.

CHAPTER I

Space and Time, the Transcendental External World

16. THE STRUCTURE OF SPACE AND TIME
IN THEIR PHYSICAL EFFECTIVENESS

THE possible space-time locations or *world-points* form a four-dimensional continuum. Only to spatio-temporal coincidence and immediate spatio-temporal proximity can we assign an intuitively evident meaning. A definite *structure* is already ascribed to the four-dimensional extensive medium of the external world if one believes in a severance of the universe in the sense that it is objectively significant

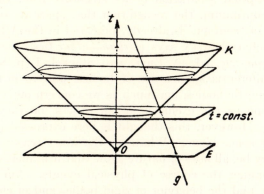

Figure 1. Graphic representation of stratification and fibration of the world. World line g of uniform translation. Light cone K.

to say of any two different events, narrowly confined in space-time, that they are happening at the same place (at different times) or at the same time (at different places). All simultaneous world-points form a three-dimensional *stratum*, all world-points of equal location a one-dimensional *fiber*. According to this view we may describe the structure of the world as possessing a stratification whose layers are traversed by fibers. (Through each world-point runs one stratum and one fiber; any one fiber intersects a stratum in but a single world-point.) Let us, for the sake of graphical representation, drop one of the spatial dimensions, thus concerning ourselves merely with the happenings on a surface, more particularly a plane. Let us represent the latter by a horizontal plane E and lay off the time t in the direction perpendicular to it. Then we are able to draw a picture of the world in intuitive space; a picture in which the layers of simultaneous world-points all appear as horizontal planes while the fibers of equally located world-points are represented by vertical straight lines.

One attributes furthermore to time and space a *metrical structure* by assuming that equality of time intervals and congruence of spatial configurations have an objective meaning. The statements of Euclidean geometry describe the spatial structure in greater detail. If in our graphical image segments of equal length on the perpendicular time axis correspond to equal time intervals, then the graphical time table of the motion of a body travelling with uniform speed along a straight line will be an inclined straight line. On this *world-line* lie those, and only those, space-time places which are occupied by the body in the course of its history. The world-lines of bodies at rest are vertical straight lines. Two bodies will meet if their world-lines intersect in some space-time point.

The conceptual separation of its structure from the underlying amorphous continuum, the recognition that space as such is merely the medium of 'contact,' is already indicated in the Aristotelian idea of space. Lobatschewsky says (*Urkunden zur Geschichte der nichteu-klidischen Geometrie*, ed. by Engel and Stäckel, I, p. 83), "Contact forms the differentiating feature of bodies, and to it they owe the name of geometric bodies, inasmuch as we concern ourselves with this property alone to the exclusion of all others, be they essential or accidental." However, this thought is here expressed not for space-time but for space alone. Whatever the inner reason of the world structure may be, all laws of nature show that it influences in the most trenchant manner the course of physical events. Among its manifestations we find the behavior of rigid bodies and of clocks; the uniform straight-line motion of a body which is free from all outside influences; the straightness of a light ray in empty space (used when sighting); the propagation in concentric spheres or circles of a light or sound wave, or of a wave in water, etc. It is our task to recognize this structure through these its physical effects. How can we, so we must ask, ascertain objectively the equi-locality or the simultaneity of events, the equality of time intervals and the congruence of spatial configurations?

Concerning the first part of the question we note that the theory of the *relativity of motion* has always been opposed to the dogma of absolute space. Aristotle designates location (τόπος) as the relation of one body to the bodies of its vicinity. Descartes (*Principia*, Chap. II) defines motion as "transportation of a portion of matter or a body from the neighborhood of those bodies which are in immediate contact with the former and which are considered at rest into the neighborhood of other bodies." A penetrating discussion of the relativity of location is given by Locke (*Enquiry concerning Human Understanding*, Book II, Chap. 13, Sections 7–10). Galileo illustrates it rather neatly with the example of the scribe who makes his notes aboard a moving vessel and who will therefore draw with his quill 'in reality,' i.e. relatively to the earth, a smooth slightly undulating line extending from Venice to Alexandrette (Dialogo, *Opere* VII, p. 198). In his controversy with Clarke (and Newton), Leibniz defends with all thoroughness, also in logico-epistemological respects, the relativity of location in space. On that occasion (Leibniz's fifth letter to Clarke, §47) he uses the happy illustration of positions in a family tree.

{Also of importance is the argumentation of Leibniz in his third letter, §5. "Under the assumption that space be something in itself, that it be more than merely the order of bodies among themselves, it is impossible to give a reason why God should have put the bodies (without tampering with their mutual distances and relative positions) just at this particular place and not somewhere else; for instance, why He should not have arranged everything in the opposite order by turning East and West about. If, on the other hand, space is nothing more than just the order and relation of things, if without the bodies it is nothing at all except the possibility of assigning locations to them, then the two states supposed above, the actual one and its transposition, are in no way different from each other. Their apparent difference is solely a consequence of our chimerical assumption of the reality of space in itself. In fact, however, each of them would be the same as the other since the two are completely indistinguishable, and therefore it is a quite inadmissible question to ask why one state was preferred to the other."[1] In contrast, Newton, the absolutist, considers motion a proof for the creation of the world out of God's arbitrary will; for otherwise it would be inexplicable why matter moves in this rather than in any other direction (Preface to the second edition of

[1] Compare with this the statement of Kant concerning left and right which was quoted at the end of Section 14. Kant has been interpreted as follows: If the first creative act of God had been the forming of a left hand, then this hand, at the time even when it could be compared to nothing else, would already have possessed that definite character of the left one (in contrast to the right one) which can only intuitively but never conceptually be apprehended. This is incorrect, as Leibniz points out, if we intend this to mean that something else would have happened had God created a 'right' hand first, rather than a 'left' hand. One must follow the process of the world's genesis further in order to uncover a difference: Had God, rather than making first a left and then a right hand, begun by making a right one and proceeded to form another right one, then He would have changed the plan of the universe not in the first but in the second act, by bringing forth a hand which was equally rather than oppositely oriented to the first-created one.

Principia by Cotes, ed. Cajori, p. XXXII, and *Principia*, ed. Cajori, p. 546). Leibniz is prevented by his theology from burdening God with such decisions as lack 'sufficient reason.' }

The body of reference upon which we rely with good reason most of the time in our daily lives when we speak of rest and motion is the "well-founded permanent earth."[2] For practical purposes this choice, suggested to us as a matter of course, is by far the most expedient. Only a sovereign imagination, breaking the bonds of sensuous appearance and freely constructing in space, could disengage itself from it. Thus Anaxagoras projected the conical shadow of the earth into space and deduced from the eclipses and the phases of the moon the correct spatial arrangement of the earth, the sun, the moon, and the stars; in the 'Moon's face' he recognized the effect of her mountains' shadows. Following the same method the Pythagoreans arrived at the hypothesis of the motion of the earth. In conscious opposition to the Pythagorean and Platonic spirit of *a priori* mathematical construction, Aristotle returned to the geocentric system. At the same time it is a definite religious attitude toward the universe that finds expression in reserving for the earth, the dwelling place of mankind, an absolute prerogative among all other bodies of reference. It is the attempt to uphold within the realm of objective reality the idealistic position, according to which I am the center of the world disclosed to me. But here where the recognition of the thou is required of the ego and the ego has to be extended so as to include the whole of mankind, the idealistic position of necessity takes on a historical and cosmo-theo-logical character. This is the reason why the book of Copernicus became a turning point of world conception; and in this direction Bruno drew the conclusions with stormy enthusiasm. The supreme act of redemption by the Son of God, crucifixion and resurrection, no longer the unique pivot of world history but the hurried small-town performance of a road show repeated from star to star — this blasphemy displays perhaps in the most pregnant manner the religiously precarious aspect of a theory which dislodges the earth from the center of the world. (Bruno had to pay for it at the stake.) "The statement, found equally with Kepler, Galileo, and Descartes, that it be foolish to think of the purpose of the universe as lying in man," says Dilthey (Der entwicklungsgeschichtliche Pantheismus, *Gesammelte Schriften*, II, third ed., 1923, p. 353), "consummates a complete change in the interpretation of the world. As these thinkers were led to an immanent teleology finding its expression in the harmony and beauty

[2] "Die wohlgegründete dauernde Erde," quotation from Goethe, Grenzen der Menschheit, verse 3.

of the universe, the character of the hitherto prevailing Christian religiosity was changed." And Goethe, in his *Geschichte der Farben-lehre* (3te Abteilung, 2te Zwischenbetrachtung), "Perhaps never before has a greater demand on mankind been made; for what did not go up in smoke with this acknowledgment: a second paradise, a world of innocence, of poetry and piety, the testimony of the senses, the conviction of a poetical-religious faith. Small wonder then that one did not want to let go of all this, that one opposed in every conceivable manner a theory which involved for him who accepted it the right and the challenge of a hitherto unknown, nay undreamed-of, freedom of thought and elevation of mind."

From the viewpoint of the relativity of motion there can be no quarrel as to the truth or falsity of the Copernican system. It is merely that the laws of planetary motion become much simpler if this motion is described as relative to the sun instead of relative to the earth.

[Newton bases the development of his mechanics in the *Principia* upon the ideas of absolute time, absolute space, and absolute motion. "Absolute, true, and mathematical time, of itself, and from its own nature, flows equably without relation to anything external. . . . Absolute space, in its own nature, without relation to anything external, remains always similar and immovable. . . . Absolute motion is the translation of a body from one absolute place into another." (*Principia*, ed. Cajori, Scholium following the Definitions, I, II and IV, pp. 6–7.) As a kinematic differentiation of the various possible states of motion of a body is undeniably impossible, Newton strives to distinguish the state of rest among all possible states of motion *dynamically*, on the basis of phenomena such as the centrifugal forces. "The causes by which true and relative motion are distinguished, one from the other, are the forces impressed upon bodies to generate motion. . . . It is indeed a matter of great difficulty to discover, and effectually to distinguish, the true motions of particular bodies from the apparent; because the parts of that immovable space, in which those motions are performed, do by no means come under the observation of our senses. Yet the thing is not altogether desperate; for we have some arguments to guide us, partly from the apparent motions, which are the differences of the true motions; partly from the forces, which are the causes and effects of the true motions. For instance, if two globes, kept at a given distance one from the other by means of a cord that connects them, were revolved about their common center of gravity, we might, from the tension of the cord, discover the endeavor of the globes to recede from the axis of their motion, and from thence we might compute the quantity of their circular motions. . . . But how we are to obtain the true motions from their causes, effects, and apparent differences, and the converse, shall be explained more at large in the following treatise. *For to this end it was that I composed it.*" (*Principia*, ed. Cajori, pp. 10 and 12.)

Newton's belief in absolute space is theologically influenced. Thus he says of God in his *Opticks* that "in infinite space, as it were in his Sensory, [He] sees the things themselves intimately, and thoroughly perceives them, and comprehends them wholly by their immediate presence to himself" (ed. Whittaker, p. 370). Newton adopts here the theology of Henry More. For More, space is the first and authentic witness for the verity and necessity of "immaterial natures"; in its properties he rediscovers the characteristics of the divine substance; space is the link between the latter and the individual objects. The nature of the world structure, that it consists of a fibration, is thus laid down by Newton in terms of an *a priori* metaphysical idea. But the actual course of the fibration in the real world has to be ascertained through its effects upon observable real phenomena. That is his scientific program. Incidentally, Newton does not succeed in mastering the problem completely. He accomplishes only the dynamic separation of uniform translation as the pure inertial motion of a body uninfluenced by external forces, from the other states of motion; but he does not succeed in isolating the state of rest among these translations. In this he *must* fail on account of the so-called *special relativity principle*, which is satisfied by the laws of Newtonian mechanics and whose validity for all natural phenomena has been confirmed today by a series of the most exact experiments: In the cabin of a ship sailing a straight course with uniform speed all events will take place in the same manner as if the vessel were at rest; given any event in nature, the one which arises from it by imparting to all participating bodies a uniform translation is equally possible. The principle has been developed by Galileo in his "Dialogo" (*Opere*, VII, pp. 212–214) in clear and lucid manner. Newton at this point resorts to a hypothesis unfounded in experience and a dialectical dodge which strike a discordant note in the midst of the magnificent and cogent inductive development of his system of the world in the third book of the *Principia*. The hypothesis states that the universe has a center and that this center is at rest. The common center of gravity of the solar system, like that of any system of bodies not subjected to external forces, moves uniformly along a straight line; thus he concludes correctly from the mechanical laws. And now we read (*Principia*, ed. Cajori, p. 419), " . . . but if that center moved, the center of the world would move also, against the hypothesis," no

100

attempt whatever being made to give a reason for this identification of the center of gravity of the planetary system with the hypothetical stationary center of the world (unless it be in the consideration that the stationary center ought to be a point, constructible on the basis of material events, whose motion according to the laws of mechanics is a uniform translation).]

The experiences which prove the dynamic inequivalence of different states of motion teach us that the world bears a structure. But in the concept of absolute space this *inertial structure* is evidently not sized up correctly; the dividing line does not lie between rest and motion but between uniform translation and accelerated motion. Referring to the graphical representation described above we can say that it is in the world as it is in space: straight lines can be objectively distinguished from curves, but in the family of all straight lines one can single out the 'vertical' ones only by a convention based on individual exhibition.

And what about the stratification, the concept of simultaneity? The trust placed in its objective significance rests on the fact that everybody considers as a matter of course the events he observes as happening at the moment of their observation. In this manner I extend my time to the whole world which enters my field of vision. Although this naive opinion lost its basis through the discovery that light has a finite velocity of propagation, there yet remained (beyond the reluctance to abandon a prejudice once held) some reason for adhering to that belief. In our graphical representation the horizontal plane passing through a world point O separates past and future as seen from O. 'Past' and 'future,' what is the reality behind these words? By shooting bullets from O in all possible directions with all possible velocities I can only hit those world points which are later than O; I cannot shoot into the past. Likewise any event happening at O has influence only upon the events at later world points; the past cannot be changed. That is to say, the stratification has a causal meaning; it determines the *causal connection of the world*. This was recognized by Leibniz, who explains in his "Initia rerum mathematicarum metaphysica" (*Math. Schriften* VII, p. 18), "If of two elements which are not simultaneous one comprehends the cause of the other, then the former is considered as *preceding*, the latter as *succeeding*." A simple method of instantaneously transferring time from one place A to another place B consists in giving a jerk to the end A of a rigid rod extending from A to B; the jerk observed at B is simultaneous with the one given at A.

But in regard to the causal structure of the world the modern development of physics has led to an essential correction. Let the segment representing one second on the *t*-axis of our graphical diagram be of the same length as the segment on the horizontal plane *E* representing the distance covered by a light ray in one second. A light signal issuing from *O* and spreading in all directions with the same velocity *c* will be received at all those world points which lie on the surface of a vertical circular cone with vertex at *O* and a vertex angle of 90°. According to Einstein's *special theory of relativity*, the 'light cone,' consisting of the above surface and its prolongation backward beyond *O*, rather than the horizontal plane through *O*, accomplishes the separation of the world into past and future. No effect is propagated at a greater speed than that of light (including the jerk given to a rigid rod for the purpose of transferring time); the velocity of any body remains of necessity below *c*. This is an inevitable consequence of the principle of special relativity and the fact that the light cone

Figure 2. Causal structure. Light cone *K*, life line *L*.

issuing from *O* depends on *O* alone and not upon the state, in particular the state of motion, of the light source which emits the signal at *O*. (Unfortunately the somewhat inadequate phrase 'constancy of the velocity of light' has been chosen to denote the latter fact.)

If I am at *O*, then *O* will divide my life line, that is the world line of my body, into two parts, past and future; in this respect nothing has been changed. But the situation is different as far as my relation to the world is concerned. In the interior of the forward part of the cone are found all those world points upon which my doings at *O* are of influence, in its exterior all those events which lie closed behind me, about which nothing can be done any more; the front cone comprehends my *active future*. In the interior of the backward part of the cone, on the other hand, are located all those events of which I either was a witness or of which I might have received some message; only these events might possibly have influenced me at all; it is the domain of my *passive past*. The two regions, active future and passive past, do not border on each other without a gap as had been the case according to the older conception.

〔It is our task, moreover, to describe in physical terms how to ascertain the equality of time intervals and the congruence of material

102

bodies. A clock is a closed material system which will return to exactly the same state S in which it found itself at some earlier instant. Let us assume the principle of causality, which asserts that the state of a system at any moment uniquely determines its entire history. Then the same process, the same cyclic sequence of states, leading from S to S will be repeated again and again, and each of these periods has by definition the same duration. What is measured in this way is the 'proper time' of the clock; it can be directly employed for all events occurring along the world line of the clock. Helmholtz says (Zählen und Messen, *Wissenschaftliche Abhandlungen*, III, p. 379), "Measurement of time presupposes that we have found physical processes, repeating themselves under equal conditions and in invariably the same manner such that if they are begun at the same instant (it would be more correct to say 'in contiguous space-time points') they also end simultaneously; such as days, the strokes of a pendulum, the running-down of sand- or water-clocks. The justification for the assumption of invariable duration rests on the circumstance that all different methods of measuring time, if carefully executed, always lead to concordant results." Concerning the empirical determination of spatial congruence he says on another occasion (*Wissenschaftliche Abhandlungen*, II, p. 648), "I call two spatial magnitudes physically equivalent if under equal conditions and in equal intervals of time the same physical events can occur within them. The process which, with appropriate caution, is employed most frequently to determine the physical equivalence of spatial magnitudes is the transfer of rigid bodies such as compasses and rulers from one place to another." The physical geometry founded on this concept of physically observable congruence is considered by Helmholtz to be an empirical science, in fact "the first and most perfect of the natural sciences." Speaking of this physical geometry in his inaugural lecture, Riemann points out what may conceivably become of major significance in the physics of the future, that "the empirical concepts upon which the spatial metric is based, the concepts of the rigid body and of the light ray, cease to be valid in the domain of the infinitely small." As a matter of fact it can be shown that the metrical structure of the world is already fully determined by its inertial and causal structure, that therefore mensuration need not depend on clocks and rigid bodies but that light signals and mass points moving under the influence of inertia alone will suffice.

A three-dimensional continuum when referred in some way to coordinates x_0, x_1, x_2 is thereby mapped upon the three-dimensional number space, i.e. upon the continuum of all number triples. Using a more familiar mode of expression, we shall replace the number space by the three-dimensional intuitive space equipped with a Cartesian coordinate system. It does little harm that, in applying the procedure to the four-dimensional world, we shall have to deprive it in imagination of one of its dimensions. A two-dimensional example is provided by the planar geographical maps. On a Mercator map, for instance, I find that San Francisco, the southernmost point of Greenland, and the North Cape, lie on a straight line, but I am not surprised to discover that on an orthographic map of the northern hemisphere this fails to be the case. Likewise a certain mapping of the world serves as the basis for the application of the customary geometrical-kinematical terms, with the x_0-axis being interpreted as the axis of time. (For instance, we shall say of a body that it is at rest if its world line is a vertical straight line, i.e. a line along which x_1, x_2, x_3 are constant and merely x_0 varies.) Only such relations will have objective meaning as are independent of the mapping chosen and therefore remain invariant under arbitrary deformations of the map. Such a relation is, for instance, the intersection of two world lines. If we wish to characterize a special mapping or a special class of mappings, we must do so in terms of the real physical events and of the structure revealed in them. That is the content of the *postulate of general relativity*. According to the *special theory of relativity*, it is possible in particular to construct a map of the world such that (1) the world line of each mass point which is subject to no external forces appears as a straight line, and (2) the light cone issuing from an arbitrary world point is represented by a circular cone with vertical axis and a vertex angle of 90°. In this theory the inertial and causal structure and hence also the metrical structure of the world have the character of rigidity, they are absolutely fixed once and for all. It is impossible objectively, without resorting to individual exhibition, to make a narrower selection from among the 'normal mappings' satisfying the above conditions (1) and (2).}

The discrepancy between the kinematical and the dynamical analyses of motion calls for a solution. Huyghens, as we know from his letters, endeavored to carry through the viewpoint of the equivalence of all states of motion even in their dynamical aspect; an attempt in this direction has been preserved in his posthumous papers (reprinted in *Jahresberichte der Deutschen Mathematiker-Vereinigung*, Vol. 29, 1920, p. 136). In our days Mach undertook the same thing in his *Mechanik* (seventh ed., 1912). He would see in the polar flattening of the earth an effect of its rotation relative to the fixed stars; the fixed stars are to hold and to carry with them the plane of Foucault's pendulum. Leibniz, on the other hand, however determinedly he rejects Newton's metaphysics of space, holding firmly to the opinion that space is nothing more than "the mere order of things among themselves," evidently agrees with Newton's mechanical program to separate true from apparent motion by dynamical criteria. (Compare the letter to Huyghens dated June 12/22, 1694, *Math. Schriften*, II, p. 184, and the explanation in In. rerum math. metaph., *Math. Schriften*, VII, p. 20: "We say that an object moves if it changes its position and if in addition the cause of this change lies within the object itself.") Euler (*Theoria motus*, 1765, especially §81) also is of the opinion that the principle of the relativity of motion, evident as it may be to our reason, has to be abandoned in the face of dynamical experiences. With some good will one may read into Kant's exposition in the *Metaphysische Anfangsgründe* a correct formulation of the problem, but they certainly throw no light on its solution.

Incidentally, without a world structure the concept of relative motion of several bodies has, as the postulate of general relativity shows, no more foundation than the concept of absolute motion of a single body. Let us imagine the four-dimensional world as a mass of plasticine traversed by individual fibers, the world lines of the material particles. Except for the condition that no two world lines intersect, their pattern may be arbitrarily given. The plasticine can then be continuously deformed so that not only one but all fibers become vertical straight lines. Thus no solution of the problem is possible as long as in adherence to the tendencies of Huyghens and Mach one disregards the structure of the world. But once the inertial structure of the world is accepted as the cause for the dynamical inequivalence of motions, we recognize clearly why the situation appeared so unsatisfactory. We were asked to believe that something producing such enormous effects as inertia — for instance, when in combat with the molecular forces it rends the cars of two colliding trains — is a rigid geometrical property of the world, fixed once and for all. Leibniz (opposing Descartes) has emphatically stressed the dynamic character of inertia as a tendency to resist deflecting forces; for instance, in a letter to de Volder (*Philosophische Schriften*, II, p. 170) he writes, "It is one thing if something merely retains its state until some event happens to change it — a circumstance which may occur if the subject is completely indifferent with respect to either state; it is another thing and signifies much more if the subject is not indifferent but possesses a power, an inclination as it were, to retain its state and to resist the causes of change." Hence the solution is attained as soon as we dare to *acknowledge the inertial structure as a real thing that not only exerts effects upon matter but in turn suffers such effects.* This step was taken by Riemann as early as the middle of the nineteenth century regarding the metrical structure of space; for indeed the inertial and the metrical structures of the world are so intimately connected (the metric after all determines the straight lines) that the metrical field will of necessity become flexible as soon as the inertial field is deprived of its geometric rigidity.

Einstein rediscovered this idea independently of Riemann, completing it by an important insight that rendered it physically fruitful. From the *equality of inertial mass and weight* — before him an enigmatic fact well-established but not understood — he concluded that, in the dualism of force and inertia, *gravitation* has to be put on the side of inertia rather than on the side of force. The phenomena of gravitation thus divulge the flexibility of the field of inertia, or, as I prefer to call it, the 'guiding field,' and its dependence on matter. The splitting of the unified guiding field into a homogeneous part obeying Galileo's law of inertia and a much weaker deviation called gravitation, which surrounds the individual stars, cannot be accomplished in an absolute manner but is relative to a system of coordinates. The laws replacing Newton's law of attraction and governing the action of matter upon the field of inertia follow conclusively from this conception. Their consequences have been fully confirmed by experience.

⟦The guiding field is (very slightly) disturbed by matter, just as the surface of a lake is disturbed by the steamships cruising on it; it will go over into the undisturbed state described by the special theory of relativity when all matter disappears, as the surface of the lake becomes a smooth homogeneous plane when the ships ride at anchor. Although Einstein, too, flirts with that idea of Mach's, it is impossible, according to an earlier remark, to eliminate the field of inertia, or the 'ether,' as an independent power from the natural phenomena. It is not the stars that guide the plane of Foucault's pendulum, but the joint motion of both — of Foucault's pendulum and of the star compass formed by the light rays reaching the terrestrial observer from the stars — is due to the overwhelming power of the ether in its interaction with matter. The old conception, separating inertia and gravitation in an absolute fashion, erred only in this point, that it saw in the actual position of all water particles of the lake, to return to our illustration, the resultant of a unique state of rest and a displacement caused by the cruising steamers. This is incorrect; indeed, as the water comes to rest at night when all ships ride at anchor, we shall undoubtedly have the same 'qualitative state' as in the morning before the ships got under way, the unruffled plane surface; but the 'material state' which is concealed behind this, i.e. the location of the various water particles, may have shifted completely. This does not contradict the principle of sufficient reason which calls for a uniquely determined state of equilibrium in the lake. For if all water particles are alike, the two states S and S' of the lake which arise from one another by having the particles interchange their locations in some arbitrary fashion are not different from each other if either is considered by itself. Only after 'a coordinate system has been introduced,' meaning in this case that numbers have been assigned to the particles, which introduce artificial differences among them and adhere to them during their motion, only then will it be meaningful to speak of the two material states S and S' as such (cf. Appendix B). In truth, however, it is not the individual material state S, the arrangement, which one can lay hands on, but only the permutation, that is the transition from the material state S to S'. This should be compared with the previously quoted remarks of Leibniz regarding the relativity of location (p. 97).}

{The group Δ_0 of the Euclidean rotations (see Section 15) in the three-dimensional space has now been replaced by the so-called Lorentz group. It consists of all homogeneous linear transformations

$$z_i' = \sum_k a_{ik} z_k \qquad (i, k = 0, 1, 2, 3)$$

which leave the indefinite quadratic form $-z_0^2 + z_1^2 + z_2^2 + z_3^2$ invariant. For each such transformation the absolute values of the coefficient $a = a_{00}$ and of the determinant d of the 3×3 coefficients a_{ik} ($i, k = 1, 2, 3$) are ≥ 1. We ascribe to the transformation the temporal signature $+$ or $-$ according to whether $a \geq 1$ or $a \leq -1$; in the same manner the sign of d determines the spatial signature. The Lorentz transformations of temporal signature $+$ form a subgroup of index 2 of the total group, and so do the transformations of spatial signature $+$. Their common part is contained in each of them again as a subgroup of index 2. The transformations of temporal signature $-$ interchange past and future, those of spatial signature $-$ interchange left and right. The most fundamental experiences of our life seem to indicate that Δ_0 should be limited to the Lorentz transformations with temporal signature $+$ (but include those of spatial signature $-$). But physics has found it rather hard to decide this question (cf. Section 23, C). A third signature, the topological signature, attaches to the coordinate transformation and is determined by the sign of its Jacobian.}

With the 'general theory of relativity' we may sum up the final result of the historical development of the structural problem of space

107

and time as follows: *The world is a four-dimensional Riemannian space.* There is associated with every line element, issuing from the point P with the coordinates x_0, x_1, x_2, x_3 and connecting it with the infinitely closely adjacent point $P' = (x_i + dx_i)$, a numerical measure

$$ds^2 = \sum_{i,k=0}^{3} g_{ik}\, dx_i\, dx_k \qquad (g_{ik} = g_{ki})$$

which is independent of the arbitrary coordinate system employed. The coefficients g_{ik} depend on the coordinates x_0, x_1, x_2, x_3 of P but not on the dx_i. The metric ground form on the right is not positive-definite but possesses one negative dimension; i.e. in an appropriate coordinate system at the point P it assumes the universal normal form

$$ds^2 = -dx_0^2 + dx_1^2 + dx_2^2 + dx_3^2.$$

In consequence of this circumstance, the 'light cone at P,' containing all line elements emanating from P that make ds^2 equal to zero, separates a domain of active future for P from a domain of passive past. The metric ground form determines, in a manner readily describable in detail, the behavior of clocks and rulers, it defines the light cones in their entire extension, and it separates the world lines of purely inertial motion (traced, for instance, by the planets) from the totality of all possible world lines. Its coefficients, the continuous functions $g_{ik}(x_0, x_1, x_2, x_3)$, describe, in terms of the chosen coordinate system, the metrical field or the 'state of the ether,' which interacts with matter.

When raising the question about the *total extent of the universe* one must distinguish between the purely topological and the metrical aspects. The transition from the Aristotelian world system, enclosed in a crystal sphere and rotating about a center, to the indifferent expanse of the infinite Euclidean space, uncentered and populated by stars throughout, was welcomed by Bruno as a mighty emancipation. Nevertheless the Aristotelian space (the interior of the crystal sphere) differs only in its metrical relations, not topologically, from the infinite one. The infinite Euclidean space leads to absurdities if we assume that the masses are on the whole uniformly distributed throughout the universe and that Newton's law of attraction is valid. Even though the gravitational force of a constant mass decreases with the inverse square of the distance, the far-off masses would then be so predominant in the entire gravitational effect that the total force exerted upon any one star would remain completely indeterminate. It is possible, however, that space is finite and yet unbounded; indeed it may be a closed manifold, like the two-dimensional surface of a sphere. It is an appealing interpretation of A. Speiser's (*Klassische Stücke der Mathematik*, 1925, p. 53) that Dante, without denying the validity of Aristotle's conception of perceptive space, assumes the real space of creation (of which the former is but an image) to be closed rather than bounded. The radii emanating from the center of the earth, the seat of Satan, converge toward an opposite pole, the source of divine force. The force of the personal God must radiate from a center, it cannot embrace the world sphere reposing in spatial quiescence like the "unmoved primal mover" of Aristotle (compare *Divina Comedia*, Paradiso, beginning with the 28th Canto). When Einstein tried, in the framework of his theory of gravitation, to carry through Mach's principle, he constructed a static universe U_a with a closed three-dimensional space in which matter is evenly distributed; the total mass in the world determines the volume of the space. Einstein's space, of course, in contrast to that of Dante, lacks a pair of distinguished opposite poles. It is as homogeneous as Euclid's space. U_a results as a possible solution of the laws of gravitation, provided they are made to include the so-called cosmological term which introduces a universal constant a of the dimension of a distance (and of the order of magnitude of the 'world radius').

Dropping two of its spatial dimensions we may picture U_a as the surface of a straight vertical cylinder of radius a and of infinite extent in both directions. This shows that U_a has two separate 'fringes,' that of infinitely remote past and that of infinitely remote future, and in this topological sense U_a extends from eternity to eternity. With the same reduction of dimensions the map of the universe U_∞ of the common Euclid-Bruno conception, i.e. of an empty world whose metrical structure is described by special relativity, is a vertical plane, and it therefore has but one connected fringe. It is this topological difference between U_a and U_∞ (two fringes as opposed to one fringe) to which in the last analysis the terms closed and open space allude.

[In Einstein's cosmology the metrical relations are such that the light cone issuing from a world point is folded back upon itself an infinite number of times. An observer should therefore see infinitely many images of a star (unless they are washed out by rarefied clouding media in interstellar space or by diffraction), showing him the star in states between which an 'eon' has elapsed, the time needed by the light to travel around the sphere of the world. The present would be shot through with the ghosts of the long ago. Moreover the solution is unstable. Yet de Sitter found that the laws of gravitation also admit the possibility of a mass-free world extending from eternity to eternity in which the domain of the future emanating from a world point does not overlap with itself. The systematic shift of the spectral lines of the most remote celestial objects, the spiral nebulas, to the red side of the spectrum has been interpreted in terms of an expanding universe, of which de Sitter's construction provides the simplest model (Weyl, Friedmann, Lemaître, H. P. Robertson, and others). For a one thus obtains a value $\sim 10^{27}$ cm. Incidentally the behavior of every world satisfying certain natural homogeneity conditions in the large (whether it is void or carries mass) follows this model asymptotically when, in the process of expansion, the world radius becomes essentially larger than a. (Compare also Section 23 C.) The postulate that for each world point O the two world domains of active future and passive past are disjoint (not only in the immediate vicinity of O but in their entire extent) rules out the possibility of a world which is closed in its spatial as well as its temporal dimensions. In such a world, that which happened once would, to the tradition handed down from generation to generation, appear as an eternal recurrence of the same events (Nietzsche's 'ewige Wiederkunft').}

REFERENCES

I. NEWTON, *Philosophiae naturalis Principia Mathematica.*
L. LANGE, *Die geschichtliche Entwicklung des Bewegungsbegriffs*, 1886.
Das Relativitätsprinzip (collection of the most important papers of H. A. LORENTZ, A. EINSTEIN, H. MINKOWSKI, H. WEYL), edited by O. Blumenthal, fifth edition Leipzig 1923; English edition, London 1923.
A. S. EDDINGTON, *Space, Time, and Gravitation*, Cambridge, 1920; *The Mathematical Theory of Relativity*, Cambridge, 1923.
H. WEYL, *Raum Zeit Materie.*
A. S. EDDINGTON, *The Expanding Universe*, Cambridge, 1937.
H. P. ROBERTSON, Relativistic Cosmology, *Reviews of Modern Physics* 5, 62–90, 1933.

17. SUBJECT AND OBJECT (THE SCIENTIFIC IMPLICATIONS OF EPISTEMOLOGY)

The doctrine of the subjectivity of sense qualities has been intimately connected with the progress of science ever since Democritus laid down the principle, "Sweet and bitter, cold and warm, as well as the colors, all these things exist but in opinion and not in reality (νόμῳ, οὐ φύσει)"; what really exists are unchangeable particles, atoms, which move in empty space. Also Plato (*Theaetetus*, 156e) holds that "properties such as hard, warm, and whatever their names may be, are nothing in themselves," but arise in the encounter of "motions" originating in the subject and in the object. Reality is pure activity; only in the "image," in the consciousness suspended between the motions is suffering. Galileo may be mentioned as another witness, "White or red, bitter or sweet, noisy or silent, fragrant or malodorous, are names for certain effects upon the sense organs." He holds that they can no more be ascribed to the external objects than the titillation or the pain which might be felt when things are touched. A detailed discussion of this is given in the final sections of Descartes' *Principia* and in his *Traité de la Lumière* (the theory of optical perception is indebted to him for important advances), likewise in Locke's *Enquiry Concerning Human Understanding* (Book II, Chap. 8, §§15–22). The subjectivity of sense qualities must be maintained in two regards, one philosophical, the other scientific. In the first place, such a quality by its very nature can only be given in our consciousness through sensation. One sees in it either an inherent attribute of sensation itself or, upon deeper analysis, an entity belonging to the intentional object which the act of consciousness puts before me. But it remains manifestly incomprehensible how quality disjoint from consciousness can be attributed as a property as such to a thing as such. This is the fundamental tenet of epistemological idealism. In the second place, the qualities in which the objects of the outer world garb themselves for me do not depend on the objects alone. They also depend quite essentially upon the concomitant physical circumstances, for instance, in the case of color, on illumination and on the nature of the medium between the object and my eye, and furthermore upon myself, on my own psycho-physical organization. My sense of vision does not grasp the objects where they are; rather, what I see is determined by the condition of the optical field in its zone of contact with my sensuous body (the retina). These are scientific facts which even the realist cannot deny. How differently the world would appear to our vision if the human eye were sensitive to other wave lengths or if the physiological processes on the retina were to transform the infinite-dimensional realm of composite physically different colors not merely into a two-dimensional but into a three- or four-dimensional manifold!

{To Locke we are indebted for the classical distinction of 'secondary' and 'primary' qualities; the primary ones are the spatio-temporal properties of bodies — extension, shape, and motion. Democritus, Descartes, and Locke held them to be objective. Locke expresses himself as follows: "The ideas of primary qualities of bodies are resemblances of them, and their patterns do really exist in the bodies themselves; but the ideas produced in us by the secondary qualities have no resemblance of them at all" (*op. cit.*, Book II, Chap. 8, beginning of §15). Although Descartes teaches that between an actual occurrence and its perception (sound wave and tone, for instance) there is no more resemblance than between a thing and its name, he yet maintains that the ideas concerning space have objective validity because in contrast to the qualities we recognize them clearly and distinctly. And a fundamental principle of his epistemology claims that whatever we comprehend in such a way is true. In support of this principle, however, he has to appeal to the veracity of God, who does not want to deceive us. Obviously one cannot do without the idea of such a God who guarantees truth, once one has grasped the principle of idealism and yet insists on building up the real world out of certain elements of consciousness that for one reason or another seem particularly trustworthy. "He is the bridge . . . between the lonely, wayward and isolated thinking, which is certain only to its own selfawareness, and the external world. The attempt turned out somewhat naive, but still one sees how keenly Cartesius measured out the grave of philosophy. It is strange, though, how he uses the dear God as the ladder to climb out of it. Yet even his contemporaries did not let him get over the edge" (quotation from Georg Büchner's philosophical notes, G. Büchner, *Werke*, Inselverlag Leipzig, 1922, pp. 268–269). Hobbes in his treatise *De Corpore* starts with a fictitious annihilation of the universe (similar to Husserl's "epoché") in order to let it rise again by a step-by-step construction from reason. But even he uses as building material the general notions which form the residue of experience, in particular those of space and time. This viewpoint has its counterpart in the physics of Galileo, Newton, and Huyghens; for here all occurrences in the world are constructed as intuitively conceived motions of particles in intuitive space. Hence an absolute Euclidean space is needed as a standing medium into which the orbits of motion are traced. Well-known is Galileo's pronouncement in the "Saggiatore" (*Opere*, VI, p. 232), "The true philosophy is written in that great book of nature (*questo grandissimo libro, io dico l'universo*) which lies ever open before our eyes but which no one can read unless he has first learned to understand the language and to know the characters in which it is written. It is written in mathematical language, and the characters are triangles, circles, and other geometric figures."}

Leibniz seems to have been the first to push forward to a more radical conception: "Concerning the bodies I am able to prove that not only light, color, heat, and the like, but motion, shape, and extension too are only apparent qualities" (*Philos. Schriften*, VII, p. 322). Also Berkeley and Hume are to be named here. For d'Alembert, the justification for using the "residue of experience" in the construction of the objective world no longer lies in the clarity and distinctness of the ideas involved as it did for Descartes, but exclusively in the practical success of this method. According to Kant, space and time are merely forms of our intuition. Stumpf (*Über den psychologischen Ursprung der Raumvorstellung*, 1873, p. 22) finds it impossible to imagine the atoms as spatial bodies without color, whose play of motion only engenders those oscillations of the ether which are the carriers of color by virtue of their wave lengths; for no more than color without spatial extension could space (according to Berkeley's and Hume's doctrine) be imagined without the raiment of some quality of color. Intuitive space and intuitive time are thus hardly the adequate medium in which physics is to construct the external world. No less than the sense qualities must the intuitions of space and time be relinquished as its building material; they must be replaced by a four-dimensional continuum in the abstract arithmetical sense. Whereas for Huyghens colors were 'in reality' oscillations of the ether, they now appear merely as mathematical functions of periodic character depending on four variables that as coordinates represent the medium of space-time. What remains is ultimately a *symbolic construction* of exactly the same kind as that which Hilbert carries through in mathematics.

The distillation of this objective world, capable only of representation by symbols, from what is immediately given to my intuition, takes place in different steps, the progression from level to level being enforced by the fact that what exists on one level will reveal itself as the mere apparition of a higher reality, the reality of the next level. A typical example of this is furnished by a body whose solid shape constitutes itself as the common source of its various perspective views. This would not happen unless the point from which it is viewed could be varied and unless the different viewpoints actually taken present themselves as instances of an infinite continuum of possibilities laid out within us. We shall return to this in the next section. A systematic scientific explanation, however, will reverse the order; it will erect the world of symbols as a realm by itself and then, skipping all intermediate levels, attempt to describe the relation that holds between the symbols representing objective conditions on the one hand and the corresponding data of consciousness on the other.

⟦Thus *perspective* teaches us to derive the optical image from the solid shape of a body and from the observer's location relative to the body. A physical example, taken from among the upper levels, is the constitution of the concepts 'electric field' and 'electric field strength.' We find that in the space between charged conductors a weakly charged 'test particle' experiences a certain force $\vec{F} = \vec{F}(P)$ when put at a given place P. Well determined as to size and direction, the force turns out to be the same whenever the test particle is brought back to the same place P. Employing various test particles we find that the force depends on the latter, yet in such a manner that $\vec{F}(P)$ may be split up into two factors:

$$\vec{F}(P) = e \cdot \vec{E}(P),$$

where the vectorial factor $\vec{E}(P)$, the 'electric field strength,' is a point function independent of the state of the test particle, while the scalar factor e, the 'charge' of the test particle, is determined exclusively by the inner state of the particle, depending neither on its position nor on the conductors, and is thus found to be the same no matter into what electric field we may place the particle. Here we start from the force as the given thing; but the facts outlined lead us to conceive of an electric field, mathematically described by the vectorial point function $\vec{E}(P)$, which surrounds the conductors and which *exists, no matter whether the force it exerts on a test particle be ascertained or not.* The test particle serves merely to render the field accessible to observation and measurement. The complete analogy with the case of perspective is obvious. The field $\vec{E}$ here corresponds to the object there, the test particle to the observer, its charge to his position; the force exerted by the field upon the test particle and changing according to the charge of the particle corresponds to the two-dimensional aspect offered by the solid object to the observer and depending on the observer's standpoint. Now the equation $\vec{F} = e \cdot \vec{E}$ is no longer to be looked upon as a definition of $\vec{E}$ but as a *law of nature* (to be corrected if circumstances warrant it) determining the ponderomotoric force which an electric field $\vec{E}$ exerts on a point charge e. Light, according to Maxwell's theory, is nothing but a rapidly alternating electromagnetic field; in our eyes, therefore, we have a sense organ capable of apprehending electric fields in another manner than by their ponderomotoric effects. A systematic presentation will introduce $\vec{E}$, the electric field strength, in a purely 'symbolic' way without explanations and then lay down the laws it satisfies (for instance, that the line integral of $\vec{E}$ extended over a closed curve in space is zero) as well as the laws according to which ponderomotoric forces are connected with it. If forces are considered observable, the link between our symbols and experience will thus have been established.}

One may say that only in the general theory of relativity did physics succeed in emancipating itself completely from intuitive space and time as means for the construction of the objective world. In the framework of this theory (which by the way includes all previously adopted standpoints either as particular or as limiting cases), the relation of subject and object may be illustrated by means of a typical example, the observation of two or more stars. By way of simplification we assume the apprehending consciousness to be a *point eye* whose world line may be called B. Let the observation take place at the moment O of its life. The construction is to be carried out in the four-dimensional number space, only for the sake of readier intelligibility we shall use a geometrical diagram instead. Let Σ be the world lines of two stars. The rearward light cone K issuing from O will meet each of the two star lines Σ in a single point, and the world lines

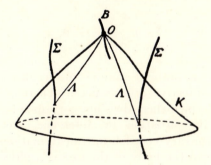

Figure 3. Data on which observation of angular distance of two stars depends.

Λ of the light signals which arrive at O from the stars join these points to O on the cone K. With the help of a construction, describable in purely arithmetical terms, it is possible to determine from these data the numerical measure of the angle ϑ under which the stars will appear to the observer at O. This construction is invariant, that is, of such a kind as to lead to the same numerical measure ϑ if, after an arbitrary deformation of the entire picture, it is carried out anew on the deformed image according to the same prescription. And everything is contained in it — the dependence of the angle on the stars themselves, on the metrical field extending between the stars, on the observer's position in the world (spatial perspective), and on his state of motion (i.e. on the direction with which the line B passes through O; this is the velocity perspective, known under the name of aberration). The angles ϑ between any two stars of a constellation determine the objectively indescribable, only intuitively experienced, visual shape of the constellation, which appears under the equally indescribable assumption that I myself am the point eye at O. If they coincide with those

115

of a second constellation, then both constellations at the moment O appear to be of equal, otherwise of different, shapes.

The objective world simply *is*, it does not *happen*. Only to the gaze of my consciousness, crawling upward along the life line of my body, does a section of this world come to life as a fleeting image in space which continuously changes in time.

{An important role in the construction of the angles ϑ is played by the 'splitting' of the world into space and time carried out at every moment O of my consciousness. Objectively this is to be described as follows: If e_0, e_1, e_2, e_3 are the components of a vector indicating the direction of B at O, then my immediate spatial vicinity will be spanned by the totality of all line elements $(dx_0,\ dx_1,\ dx_2,\ dx_3)$ issuing from O which are orthogonal to e, i.e. which satisfy the equation

$$\sum_{i,k=0}^{3} g_{ik}\, dx_i\, e_k = 0, \qquad g_{ik} = g_{ik}(O).\}$$

Thus the objective state of affairs contains all that is necessary to account for the subjective appearances. There is no difference in our experiences to which there does not correspond a difference in the underlying objective situation (a difference, moreover, which is invariant under arbitrary coordinate transformations). It comprises as a matter of course the body of the ego as a physical object. The immediate experience is *subjective and absolute*. However hazy it may be, it is given in its very haziness thus and not otherwise. The objective world, on the other hand, with which we reckon continually in our daily lives and which the natural sciences attempt to crystallize by methods representing the consistent development of those criteria by which we experience reality in our natural everyday attitude — this *objective* world is of necessity *relative;* it can be represented by definite things (numbers or other symbols) only after a system of coordinates has been arbitrarily carried into the world. It seems to me that this pair of opposites, subjective-absolute and objective-relative, contains one of the most fundamental epistemological insights which can be gleaned from science. Whoever desires the absolute must take the subjectivity and egocentricity into the bargain; whoever feels drawn toward the objective faces the problem of relativity. This thought is vividly and beautifully developed in the introduction of Born's book on relativity theory, quoted earlier.

Within the natural sciences the conflicting philosophies of idealism and realism signify principles of method which do not contradict each other. We construct through science an objective world which, in order to explain the sense data, must satisfy the following fundamental principle that was already mentioned on p. 26: *A difference in the perceptions offering themselves to us is always founded on a difference in the real conditions* (Helmholtz). Lambert, in his *Photometria* (1760), enunciates as an axiom the following special case: "An appearance is the same whenever the same eye is affected in the same way." Here the natural sciences proceed realistically.

{For as long as I do not go beyond what is given, or more exactly, what is given at the moment, there is no need for the substructure of an objective world. Even if I include memory and in principle acknowledge it as valid testimony, if I furthermore accept as data the contents of the consciousness of others on equal terms with my own, thus opening myself to the mystery of intersubjective communication, I would still not have to proceed as we actually do, but might ask instead for the 'transformations' which mediate between the images of the several consciousnesses. Such a presentation would fit in with Leibniz's monadology. Instead of constructing the perspective view which a given body offers from a given point of observation, or conversely constructing the body from several perspective images, as it is done in photogrammetry, we might eliminate the body and formulate the problem directly as follows: let A, B, C each represent a consciousness bound to a point body, and let K be a solid contained in their field of vision. The task is to describe the lawful geometrical connections between the three images which each one of the three persons A, B, C receives of K and of the locations of the other two persons. This procedure would be more unwieldy; in fact, it would be bound to fail on account of the limitations and gaps in any single consciousness as compared to the complete real world. At any rate, there can be no doubt that in this respect science proceeds in tune with a realistic attitude.}

On the other hand science concedes to idealism that its objective reality is not given but to be constructed (nicht gegeben, sondern aufgegeben), and that it cannot be constructed absolutely but only in relation to an arbitrarily assumed coordinate system and in mere symbols. Above all the central thought of idealism comes into its own in the converse of the above fundamental principle: *the objective image of the world may not admit of any diversities which cannot manifest themselves in some diversity of perceptions;* an existence which as a matter of principle is entirely inaccessible to perception is not admitted. Leibniz says concerning the fiction of absolute motion (Leibniz's fifth letter to Clarke, §52): "I reply that motion is indeed independent of actual observation, but not of the possibility of observation altogether. Motion exists only where a change accessible to observation takes place. If this change is not ascertainable by any observation then it does not exist." To be sure, many physically different colors will produce the same sensation of red; but if one sends all these various reds through the same prism, then the physical differences will manifest themselves in the perceptible differences between the streaks of colored light emerging from the prism. The prism, so to speak, unfolds the hidden differences to our senses. A difference which can in no way be broken down for our perception is non-existent. This is of great importance as a methodical principle of theoretical construction.

[The formula customarily given (Schwarzschild formula) for the metrical field surrounding a mass, such as the sun, can be interpreted as follows, if the coordinates occurring in it are taken to stand

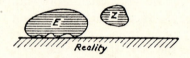

Figure 4. Schematic representation of a theory with a redundant part Z.

for a mapping of the real space into a Euclidean one: '(I) In reality Euclidean geometry holds. But the spherically symmetric field of gravity surrounding the mass center O acts upon rigid bodies in such a fashion that a radially directed rod at P is foreshortened in the ratio $\sqrt{1 - 2\alpha/r}:1$ (where $r = \overline{OP}$ and α is a constant determined by the mass), while a rod perpendicular to OP remains unchanged.' Rods after all are known to change their length with changing temperatures, why should they not react in a similar way to a gravitational field? But in making use of a certain other coordinate system one arrives at the following description: '(II) In reality Euclidean geometry holds. But the rod at P, no matter what its direction, will be changed in the ratio $(1 + \alpha/2r)^2:1$ by the field of gravitation.' Both descriptions express the same factual situation. To every possible coordinate system there corresponds a corrective prescription salvaging Euclidean geometry. Yet one is as good as the other. Each introduces into the factual state of affairs an arbitrary element which has no perceptually confirmable consequences and which therefore must be eliminated. And it can be eliminated by employing, with Einstein, none but the physical geometry as it is defined by the direct comparison of measuring rods. (That geometry is, of course, not the Euclidean one.) Each of the two theories can, if properly formulated, be split into two parts:

118

the theory (E) of Einstein, and an addition (Z) which is neither connected with (E) nor touching on reality and which must therefore be shed (compare the schematic diagram in Figure 4).

In Bohr's model of the hydrogen atom the period of the emitted light has nothing to do with the time in which the electron completes a revolution around the nucleus. Though it explains the spectrum as satisfactorily as could be wished, this lack of observable data corresponding to the period of revolution of the electron is felt as a disturbing feature which ought to be removed. In order to clarify the idea of relativity, Poincaré once set up the fiction that overnight, while all consciousness was asleep, the world with all its bodies in it, including my own, had been magnified in a definite ratio; awake again, neither I nor anyone else would notice the change in any way. In the face of such an event science makes common cause with the idealist; for what on earth could be meant under such circumstances by the statement that the world was magnified? A difference may be posited only where the assumption of equality would conflict with the principle that equals under equal conditions (especially equal objective characteristics, location, and motion, of the observer) will be perceived as equal[3] (and with the principle of causality).

Between the real world and the given there is a correspondence, a mapping in the mathematical sense. Yet, while on one side there is the one, quantitatively determined, objective world, we have on the other not only what is actually given at the moment but also the *possible* perceptions (perhaps remembered, or expected in response to definite intentions of will) of an ego; and further there enter into this correspondence, besides the unique objective state of the world, the *possible* objective states of this perceiving ego (world line of its body, *etc.*). Helmholtz sets up the principle of the "empiristic view" (*Physiologische Optik*, III, p. 433): "The sensations are signs to our consciousness, and it is the task of our intelligence to learn to understand their meaning." In this one may agree with Helmholtz as he means it and yet be of the opinion with Husserl that the spatial object which I see, notwithstanding all its transcendency, is perceived as such bodily in its concreteness (Husserl, Ideen zu einer reinen Phänomenologie, in his *Jahrbuch für Philosophie*, Vol. 1, 1913, pp. 75, 79); for within the concrete unity of perception the data of sensation are animated by 'interpretations,' and only in union with them do they perform the 'function of representation' and help to constitute what we call 'the appearance of' color, shape, etc. A dog approaching

[3] This principle cannot be taken as a definition of objective equality but only as an implicit requirement, because the concept of equality occurs in it twice: *equals* under *equal* conditions. . . .

another dog will see and smell 'a fellow dog,' an integrated whole that is more than a 'bundle of sensations.' We merely describe here one of the levels through which the process of constitution of the external world will pass. And there is no denying that the definite manner in which a thing is bodily put before me by means of those animating functions will be directed by a multitude of previous experiences. For how else should we describe this than by saying that "we always imagine such objects to be present in our field of vision as must be present in order to bring about the given impression under ordinary normal conditions for the use of our eyes" (Helmholtz, *Physiologische Optik*, III, p. 4). Helmholtz speaks here of "unconscious inferences." This sounds somewhat questionable; yet he stresses explicitly that only in their result do they resemble inferences, more accurately inferences by analogy, although the underlying psychic acts probably are quite different from acts of conscious inference and although their effects cannot be annulled by better knowledge. The sense impression of a mirror image, or of a broken rod immersed in water, or of the rainbow, does not deceive; only the bodily object which, as Husserl says, is put before me by this impression is an error. What truly exists can be ascertained only by taking into consideration all sensuous signs, which in the examples adduced above will soon reveal the prevalence of 'abnormal' conditions. Only imagine our eyes to be sensitive to light whose wave length is of the order of magnitude of the atomic distances in solids; how difficult it would then become for us to interpret the optical 'signs' (the Laue interference patterns)! In the ultimate description of the connection between appearance and reality one therefore does better to ignore all intermediary levels of constitution. }

And what significance does this objective world, representable only in symbols, have for the everyday life of man, taking place as it does in the sphere of integrated data of perceptions? Helmholtz answers (*op. cit.*, p. 18), "Once we have learned to read those symbols correctly we shall be able with their help to design our actions so that they yield the desired result, namely, that the expected new sensations will arise. A different comparison between conceptions and things not only does not exist in reality — all schools agree on this point — but a different manner of comparison is inconceivable and would be devoid of meaning. . . . Thus such a presentation (*Vorstellung*) of a single individual body is indeed already a concept (*Begriff*) which comprises an infinite number of intuitions in temporal sequence all of which can be derived from it.[4] The presentation of a single

[4] In agreement with a number of philosophers writing in English the term *presentation* has been chosen here as the equivalent of Kant's and Helmholtz's *Vorstellung* and Locke's *idea*. [Translator's note.]

individual table that I carry within me is correct and accurate if I am able to derive from it correctly and accurately the sensations I shall experience when I bring my eyes or my hands into this or that definite position with respect to the table. What other kind of similarity there can subsist between such a presentation and the object represented by it I cannot comprehend" (*op. cit.*, p. 26). In the same sense, Leibniz remarks concerning the Cartesian principles (*Philosophische Schriften*, IV, p. 356), "Of the sense data we cannot know more, nor do we have to require more, than that they are in agreement with each other as well as with the indisputable dictates of reason and that thus to a certain extent the future may be predicted from the past. To search for a truth or reality other than thus vouched for would be futile — the sceptic may not demand, the dogmatist not promise more." Or Husserl (*Ideen*, p. 311), "To the essence of a thing-noema there belong ideal possibilities of unlimited development of concordant intuitions that follow, moreover, prescribed directions of determinate type." But in the erection of empirical reality discrepancies will occur which will force us to make "corrections." Owing to its empirical character cognition of reality must of necessity pass through errors. "What is given never implies material existence as certain and necessary but merely as presumptive reality. This means that it can always happen that the further course of experience will force one to abandon what with good empirical justification had earlier been posited." (Husserl, *Ideen*, p. 86.) It might well be within the range of possibility that in the moving picture of perceptions every beginning of concordance would irreparably "explode." In that case the attempt to harmonize them according to principles of reason would fail, and no real world would be constituted.

The requirements which emerge from our discussion for a correct theory of the course of the world may be formulated as follows:

1. *Concordance*. The definite value which a quantity occurring in the theory assumes in a certain individual case will be determined from the empirical data on the basis of the theoretically posited connections. *Every such determination has to yield the same result.* Thus all determinations of the electronic charge e, that follow from observations in combination with the laws established by physical theory, lead to the same value of e (within the accuracy of the observations). Not infrequently a (relatively) direct observation of the quantity in question (for instance, of the location of a comet among the stars at a certain moment) is compared with a computation on the basis of other observations (for instance, the location at the desired moment computed according to Newton's theory from the locations on previous

days).   The demand of concordance implies consistency,[5] yet transcends the latter in that it brings the theory in contact with experience.

2. It must in principle always be possible to determine on the basis of observational data the definite value which a quantity occurring in the theory will have in a given individual case.   This expresses the postulate that the theory, in its explanation of the phenomena, must not contain redundant parts.

Hume attempted to uphold with inexorable consistency the viewpoint that the given is the whole of reality.   Since it became apparent through him that this viewpoint fails completely in the explanation of those cognitive positions which play a basic role in everyday life and in science, he was indeed the first to reveal the problem of reality in its full difficulty.   Reason in its function of constituting reality is described by him as the faculty of imagination.   With complete sincerity he confesses the irreconcilable conflict between thought and life, into which he finds himself thrown.   To carry his approach through is as impossible as to found arithmetic on nothing but the concretely existing numerals.   The positivism of a Mach or Avenarius appears to me merely as a less consistent renewal of Hume's attempt; for in their systems theoretical hypostases, strictly avoided by Hume, play once more a considerable role.   But then we are back in the midst of theoretical construction, which supplements the given in the interest of totality, and we are no longer forced to use sense data as our building material.   Kant's transcendental idealism reestablished the insights already gained by Leibniz.   The content of this Section may be considered as an elucidation of Kant's concept of reality as "that which is connected with perception according to laws."   He advances beyond Leibniz in transmuting the old metaphysical ontological concepts of substance and causality into methodical principles for the construction of empirical reality.

[In the part on logic we had insisted that existence could not be stated about something exhibited, that the logical symbol Σ_x carries an index x which refers to a blank.   This seems to be contradicted by a proposition such as 'this chair is real.'   But the assertion of real existence contains either, idealistically interpreted, the prediction of a multitude of concordant impressions expected in response to certain intentions of will, or, realistically interpreted, the statement that a thing x exists which stands in a certain metaphysical relation to the given chair phenomenon.

[5] Indeed in an inconsistent theory the formula $e = 2e$ would be deducible, and hence the actual value e as well as $2e$ for the electronic charge could be derived from such a theory in combination with the observational data.

122

Concerning the problem of realism versus idealism we find a striking analogue in geometry, which has a factual connection with it in so far as in the objective world the coordinate system is, as it were, the residue of the annihilation of the ego. As in Section 12, we consider vectors $\overrightarrow{x}$ in a plane and represent them in terms of a basis consisting of two linearly independent vectors $\overrightarrow{e_1}$, $\overrightarrow{e_2}$, thus: $\overrightarrow{x} = \xi_1 \overrightarrow{e_1} + \xi_2 \overrightarrow{e_2}$ (compare p. 69). The numbers ξ_1, ξ_2, which are uniquely determined by $\overrightarrow{x}$ and $\overrightarrow{e_1}$, $\overrightarrow{e_2}$, are called the coordinates of $\overrightarrow{x}$ with respect to the basis $(\overrightarrow{e_1}, \overrightarrow{e_2})$. We construe the vectors in our plane as analogues of the objects in the real world, bases as analogues of real observers, the numbers as analogues of subjective phenomena, and thus speak in our analogy of the pair of coordinates (ξ_1, ξ_2) as the 'appearance of the object $\overrightarrow{x}$ for the observer $(\overrightarrow{e_1}, \overrightarrow{e_2})$.' For the geometric vector plane we can construct an algebraic model, defining a vector $\overrightarrow{x}$ as a pair of numbers (x_1, x_2) and the operations of adding two vectors $\overrightarrow{x}$, $\overrightarrow{y}$ and of multiplying a vector $\overrightarrow{x}$ by a number α as follows:

$$(x_1, x_2) + (y_1, y_2) = (x_1 + y_1, x_2 + y_2), \quad \alpha(x_1, x_2) = (\alpha x_1, \alpha x_2).$$

Calling the numbers x_1, x_2 the absolute coordinates of $\overrightarrow{x} = (x_1, x_2)$ and $\overrightarrow{i_1} = (1, 0)$, $\overrightarrow{i_2} = (0, 1)$ the absolute basis in our model, we realize at once that the absolute coordinates of a vector are its relative coordinates with respect to the absolute basis. Transition from the geometric vector space, in which all bases are equally admissible, to its algebraic model is effected by assigning, as it were, to an arbitrarily chosen basis the role of absolute basis. On the other hand, the individual character of the various bases in the model can be extinguished and all bases put on an equal footing by ascribing objective significance only to such properties and relations as are definable in terms of the two fundamental operations, or, what is the same, as are invariant under arbitrary linear coordinate transformations. The relation of a vector to its absolute coordinates is not objective, but is a special case of the objective relation prevailing between a vector $\overrightarrow{x}$, a basis $(\overrightarrow{e_1}, \overrightarrow{e_2})$, and the coordinates ξ_1, ξ_2 of $\overrightarrow{x}$ relative to $(\overrightarrow{e_1}, \overrightarrow{e_2})$. Our analogy assumes that only the realm of numbers (the appearances) but not the geometrical space (the things themselves) is open to our intuition. Hence the model is the world of my phenomena and the absolute basis is that distinguished observer 'I' who claims that all phenomena are as they appear to him: on this level, object, observer and appearance all belong to the same world of phenomena, linked however by relations among which we can distinguish the 'objective' or invariant ones. Real observer and real object, I, thou, and the external world arise, so to speak, in unison and correlation with one another by subjecting the sphere of 'algebraic appearances' to the viewpoint of invariance. On this issue our theory bears out Leibniz (compare, for instance, *Nouveaux Essais*, Libre IV, Chap. 11) as opposed to Descartes, who through his "cogito ergo sum" assigns to the reality of the ego a precedence in principle over the reality of the external world. The analogy renders the fact readily intelligible that the unique 'I' of pure consciousness, the source of meaning, appears under the viewpoint of objectivity as but a single subject among many of its kind.

Yet in truth, the absolute subject, I, remains forever unique, notwithstanding the *objective* equivalence of the various subjects. This is in agreement with the facts as I find them. On purely cognitive grounds conscientalism is irrefutable, it can be carried through completely. But for all this the recognition of the 'thou' is demanded of me not only in the sense that in my thinking I yield to the abstract norm of 'objectivity,' but in an absolute sense: Thou art for thyself once more what I am for myself, conscious-existing carrier of the world of phenomena. This step can be taken in our analogy only if we pass from the algebraic model of affine vector geometry to its axiomatic description, where the concepts of a vector and of the two fundamental operations enter as undefined terms. In the axiomatic system it is no longer necessary to enforce the equivalence of all coordinates by abstraction, for in it a definite pair of vectors $(\vec{e_1}, \vec{e_2})$ can be distinguished only by an individual act of exhibition. Pattern and source of any such demonstrative act is the word 'I.' Thus axiomatics reveals itself once again (compare p. 66) as the method of a purified realism which posits a transcendental world but is content to recreate it in symbols.

The postulation of the ego, of the 'thou,' and of the external world is without influence upon the cognitive treatment of reality. It is a matter of metaphysics, not a judgment but an act of acknowledgment or belief (as Fichte emphasizes in his treatise *Über die Bestimmung des Menschen*). Yet this belief is the soul of all knowledge. From the metaphysico-realistic viewpoint, however, egohood remains an enigma. Leibniz (Metaphysische Abhandlung, *Philosophische Schriften*, IV, pp. 454–455) believed that he had resolved the conflict of human freedom and divine predestination by letting God (for sufficient reasons) assign existence to certain of the infinitely many possibilities, for instance to the beings Judas and Peter, whose substantial nature determines their entire fate. This solution may objectively be sufficient, but it is shattered by the desperate outcry of Judas,

"Why did *I* have to be Judas?" The impossibility of an objective formulation of this question is apparent. Therefore no answer in the form of an objective insight can ensue. Knowledge is incapable of harmonizing the luminous ego with the dark erring human being that is cast out into an individual fate.

Postulation of the external world does not guarantee that it will constitute itself out of the phenomena through the cognitive work of reason as it attempts to create concordance. For this to take place it is necessary that the world be governed throughout by simple elementary laws. Thus the mere positing of the external world does not really explain what it was meant to explain, the question of the reality of the world mingles inseparably with the question of the reason for its lawful mathematical harmony. The latter clearly points in another direction of transcendency than that of a transcendental world; towards the origin rather than the product. Thus the ultimate answer lies beyond all knowledge, in God alone; emanating from him, the light of consciousness, its own origin hidden from it, grasps itself in self-penetration, divided and suspended between subject and object, between meaning and being. ⎬

18. THE PROBLEM OF SPACE

⎱A. Origin of the Presentation of Space. A detailed investigation into the psychological origin of the presentation of space was not undertaken until the 19th century. The sense regions which contribute above all to the constitution of space are the visual and tactile impressions. Bain added to these the sensations of motion and the muscular feelings.

A single eye sees the qualities spread out in a two-dimensional field of vision. The latter is two-dimensional because it is dissected by any one-dimensional curve which runs through it. It is a basic physiological fact that the place in the field of vision at which we localize a visual impression is determined by the portion of the retina that is stimulated. We have here a one-to-one continuous 'mapping' in the mathematical sense. The places in the field of vision are continuously connected in the same way as the places on the retina to which they correspond. J. Müller, the originator of the law of the specific sense energies, even says (*Zur vergleichenden Physiologie des Gesichtssinnes*, p. 54), "In any field of vision, the retina sees only itself in its spatial extension during a state of affection. It perceives itself as spatially dark when the eye is at rest and completely closed." A great step forward is marked by Helmholtz's *Physiologische Optik*, as he no longer speaks of identity but of correspondence. The same remark clarifies the famous problem as to why things are seen upright (in the objective space, the image on the retina is turned through 180° as compared to the original). If we confront the 'objective' space on one side and my intuitive space on the other, and if we assume both to bear a Euclidean metrical structure, then the utmost in faithfulness that could be demanded of the correspondence between objective thing and its image given in my intuition is an isomorphic (or similar) mapping in the sense defined in Section 4. Such an isomorphism would mean that all geometrical characteristics of the thing, describable in terms of the metrical concepts of objective space, are reflected in geometrical characteristics of the image expressed in terms of the synonymous metrical concepts of intuitive space. But it is nonsensical to ask questions which would be meaningful only if the thing as well as its image were located in the same space.

The field of vision has indeed a metrical structure; the resting eye undoubtedly is capable of apprehending something like shape, which here appears as a quality of what is seen, and of distinguishing different shapes. Such shape, however, is similar neither to the thing seen nor to the objective image produced on the retina. (The deformation with respect to the configuration formed by the rays of vision is described by Helmholtz in *Physiologische Optik*, III, pp. 151–153.) A more detailed shaping and partial correction of this metrical structure is achieved through the movements of the eyes; if a shift in the direction of my glance has the effect of changing image I into image II (i.e., objectively, if the same portion of the retina stimulated prior to the movement by the visual impression I is stimulated after the movement by the visual impression II), then I and II are mutually congruent. Consequently, in the domain of ocular movements, shape is no longer a given quality but a concept obtained by abstraction from the relation of congruence (compare Section 2).

Lotze demands the existence of 'local signs' on the basis of his physiological principle "that only the qualities of sensation may be considered as directly perceptible and intellectually differentiable" (*Wagners Handwörterbuch der Physiologie*, III, Section 1, 1846, p. 183) and that from these qualities the mind has to build up the presentation of spatial extension. The local signs are sensations whose qualitative gradations form the basis of the different locations in the field of vision. He has attempted to characterize them more precisely as impulses to move the eyes so as to bring the place in question into the center of the field of vision, and the feelings in the ocular muscles that would accompany such a movement. But here that is taken as basic which itself calls for a basis. For the eye at rest has its continuously spread field of vision, independently of the ocular movements, which belong to the next higher level in the constitution of space. An attempt by Wundt to stamp color gradations as local signs is better passed over in silence. Helmholtz, though accepting Lotze's thesis, admits that the local signs are *qualitates occultae*. In view of the indissoluble connection between color and extension in the field of vision, Lotze's thesis leads to the problem (cf. Poincaré, *La valeur de la science*, ed. Flammarion, p. 91) of explaining how the one sensation can split up into the two components of color and extension — how, in other words, it is possible that two sensations of the same red produced at two different points P, Q of the retina should have a close affinity, which is absent in the case of a red at P and a green at Q. But if one understands well the punctiform character of a simple sensation, one will hardly be inclined to consider that which gives the red its extensiveness again as a continuously graded sensation, but will acknowledge with Kant and Fichte (*Bestimmung des Menschen*, ed. Medicus, III, p. 326): "I am originally not only sentient but also intuiting." There *is* something for me only inasmuch as a continuum of quality covers a (temporal or spatio-temporal) continuum of extension. This conceded, sensations as local signs become redundant.

How about the visual impressions of rest and motion? When I glance up and down I have the impression that the things in my field of vision stay at rest, although their images are produced at varying places of the retina. But this is true only when the ocular movement is produced voluntarily by the motor apparatus of the eye. Thus it is here not a question of the muscular feelings connected with the ocular movement but of the voluntary intentions. There exists an original impression of rest and motion (change). *A thing gives the impression of being at rest, if its retina image does not shift about and if at the same time no ocular movements are intended.* Between the displacements of the retina image and the voluntary intentions directing the ocular movements there exists a system of compensations which experience apparently has developed to considerable refinement. The task is simplified by a circumstance known as Listing's law. It states that the eye cannot be voluntarily rotated while fixing a definite point in the field of vision, but that any one such point determines (except for minute fluctuations) only one corresponding position of the eye. Thereby the three degrees of freedom of the eye ball are reduced to two. The possibility of turning freely the direction of sight in response to the will is made use of when certain changes in the field of vision are 'interpreted' as motions.

The immediate impression of rest may not be invoked as 'testimony of the senses' for a refutation of the physical theory of relativity. For we saw that the objective 'explanation' of this phenomenon (which after all can only consist in exhibiting an objective difference where a difference exists among intuitive data) takes recourse only to the idea of relative motion of physical entities covering each other (displacement of the image on the retina) and to the dynamical concept of motion (voluntary intentions, which, through muscular forces, cause the eye ball to deviate from its natural movement as it is conditioned by the field of inertia and by the eye's imbeddedness into the human body). The same holds for the motory sensations of our body; they do not tell of 'absolute motion' but are invariably sensations of acceleration, indicating that the body or part of it is torn out of its natural inertial motion and registering the ensuing dynamical disturbances.

The optical perception of depth, as Wheatstone has shown strikingly by the stereoscope, is closely tied up with binocular vision. (In addition, sensations produced by the accommodation effort come into play.) The positions in the fields of vision of the two eyes are in one-to-one correspondence, with the effect that the images formed at corresponding places are seen as one. The stereoscopic apperception of depth depends on the disagreement between the two images, which results when the same color quality and in particular the same contours do not appear at corresponding places of the two fields of vision. The details form a matter of dispute between two rival theories, a "nativistic" theory, represented especially by Hering, and the "empiristic" theory of Helmholtz. The former places all responsibility on the sensations, maintains that the stimulation of corresponding points on the two retinas, e.g. of the two retinal foveas, produces a simple sensation, and ascribes to the places on the retina, in addition to local signs indicating direction, a depth value modifying the sensation. Helmholtz's theory, on the other hand, considers optical depth as the result of a constitutive process. Only the latter theory is easily reconcilable with the facts. Yet it must be added, in the sense of the nativistic theory, that with the dimension of depth something new and original emerges. With its help the material of the two preceding levels — the two-dimensional purely visual field and the field of ocular movement — serves to constitute the centered three-dimensional space in which the body of the ego finds its position, though still the distinguished position of the center. (On the two previous levels we evidently do not yet have such a body-ego.) In the case of the (involuntary or voluntary) 'reversal' of the perspective interpretation of a plane figure (compare, for instance, Helmholtz, *Physiologische Optik*, III, p. 239), the 'animating' or 'integrating' function which converts the figure in the field of vision into the appearance of an object hit by the visual ray in centered space is felt particularly clearly. It is on this level, too, that the tie-up with the localization field of the sense of touch and of the movements of limbs occurs. The grasping for the seen object is constantly used as a control in the pertinent psychological experiments on vision. Husserl emphasizes that "all these facts, allegedly mere contingencies of spatial intuition that are alien to the 'true,' 'objective' space, reveal themselves, except for minor empirical particularities, as essential necessities" (*Ideen*, p. 315); and in this sense O. Becker has given a more detailed phenomenological description of the constitutive levels of spatiality.

By walking toward the indefinitely far horizon of the centered space and by the displacements connected therewith, by the feeling of the free possibility of bodily movement in response to voluntary intentions, the homogeneous space arises from the centered one. Only at this stage the body becomes an equal among other spatial objects, and we become capable of adopting in imagination someone else's standpoint. Only this space can be conceived as being one and the same for several subjects; it is the presupposition for the construction of the intersubjective world. And thus the ascertainment of the orientation of objects in it is capable of intersubjective control and correction.

As opposed to Aristotle, who held that space is an $\alpha i\sigma\vartheta\eta\tau\grave{o}\nu$ $\kappa o\iota\nu\acute{o}\nu$, Berkeley has taken the view that there are only distinct sense spaces. Stumpf (*op. cit.*, p. 287) objects to this by asking, "Are we to believe that also the duration of a tactile sensation and that of a visual sensation are heterogeneous contents?" Berkeley may be right in that the pre-spatial localization fields (of the first and second levels) are separate ones for the senses of touch and vision. But beginning with the third level it can only be a question of one space, which comprehends the sense data of touch as well as vision. Thus space becomes the connecting link between the various sense domains. Bain's association theory of space aims at bringing out this function of it. In more precise form, such a theory has been developed by Poincaré. He first distinguishes qualitative changes and motions by pointing out that the latter can be reversed by a movement of the ego-body, which betrays itself by voluntary intentions and accompanying kinesthetic sensations (*La valeur de la science*, Chap. IV, §§1-4). He then attempts to set up criteria for the coincidence of two points in space arrived at by different series of kinesthetic sensations and voluntary intentions; and finally he investigates the 'mapping' upon one another — usually interpreted as identification — of the spaces appertaining to different sense organs (for instance, to the two finger-tips, or to the visual sense of the left eye and the tactile sense of the right thumb). According to this view, the statement that the sense of vision, but not that of touch, reaches into the distance merely brings out the fact that two places in the space of any sense organ must be coincident if they correspond to two coincident places in the space of a tactile organ; while to two non-coincident places in a tactile space there may correspond two coincident places in the visual space. J. S. Mill accepted Bain's view, except that his presentation of space is not made up of Bain's sensations and their associations but emerges from them by creative synthesis ('psychic chemistry'). All these theories ignore the undeniable data on the lowest levels of constitution that do not possess the character of sensations, such as juxtaposition in the pure field of vision. }

B. THE ESSENCE OF SPACE. The penetration of the This (here-now) and the Thus is the general form of consciousness. A thing exists only in the indissoluble unity of intuition and sensation, through the superimposition of continuous extension and continuous quality. Phenomenologically it is impossible to go beyond this. If, meta-physically, with Plato, one lets the passive consciousness spring from the encounter of two 'motions,' one originating with the ego, the other with the object, then one will tend to relegate quality to the sphere of the object, extension to that of the ego (and not vice versa, since extension is the qualitatively undifferentiated field of free possibilities, while the concrete variety resides in the qualities). "Translucent penetrable space pervious to sight and thrust, that purest image of my knowledge (*Wissen*)," so Fichte says (*Werke*, ed. Medicus, III, p. 325), "is not seen but intuited, and in it my seeing itself is intuited. The light is not without but within me, and I myself am the light." But the manner in which this intuition as an integrating force penetrates the sense data and utilizes their material is largely conditioned by experience.

The fact that both constituents, extension and quality, are bound to each other is the root of Aristotle's thesis of the impossibility of empty space. Thus Hume interprets it (*Treatise*, Book I, Part II) (in which connection it must be remembered that the spatial — or, more exactly, the spatio-temporal — separation is a fact as immediately ascertainable as spatio-temporal contact). But only through a *metabasis eis allo genos* can this essential epistemological fact be turned about into an assertion concerning substantial-physical events, leading to such conclusions as Descartes drew (and which were ridiculed by Hume), namely, that the walls of a box would have to touch if the latter were pumped empty. Leibniz denies emptiness on grounds connected with the perfection of the world and the principle of sufficient reason. He explains the fact that space is bound to the sensuous qualities by denoting space, together with time, as the order of the phenomena. Stumpf (*op. cit.*, pp. 15, 26) objects, "When we distinguish different orders, we have to acknowledge in each case a specific absolute content with respect to which the order takes place," and consequently he asserts that "space denotes, rather, that positive absolute content upon which order is based." He demands that positional relations between points in space must be founded in a 'position' of the individual points severally, and by adopting this logical principle of the self-insufficiency of relations (which he may have taken over from F. Brentano, Zur Lehre von Raum und Zeit, *Kant-Studien*, XXVI) he bars himself from an understanding of the relativity of position.

Since the mere Here is nothing by itself that might differ from any other Here, space is the *principium individuationis*. It makes the existence of numerically different things possible which are equal in every respect. That is why Kant contradistinguishes it as the *form of intuition* from "the matter of phenomena, i.e. that which corresponds to sensation." Here lies the root of the concepts of similarity and congruence. Leibniz infers from this the ideality of space and time; for they violate the principle of the identity of indiscernibles, which — along with Spinoza — he postulates as necessary in the domain of substances (namely as a consequence of the principle of sufficient reason).

The dual nature of reality accounts for the fact that we cannot design a theoretical image of being except upon the background of the *possible*. Thus the four-dimensional continuum of space and time is the field of the *a priori* existing possibilities of coincidences. That is why Leibniz calls the "abstract space the order of all positions assumed to be possible" and adds that "consequently it is something ideal" (Leibniz's fifth letter to Clarke, §104).

[If we state the distance of the earth from the sun in yards, this statement acquires a meaning verifiable through what is given only if a rigid ladder, on which a scale has been marked off by means of a movable yardstick, is placed with one end upon the earth and with the other against the sun. The physically clearest realization of a rigid body is the crystal. If coordinates are to have an immediately ascertainable meaning, we must imagine the whole world to be filled out by a crystal. Among the motions of the crystal lattice that carry it into itself (covering motions) we can distinguish the translations by their peculiar properties; the covering translations can be used (by actually carrying out the translative motions) to introduce number triples as coordinates for the atoms of the lattice, and these can then be employed as position marks in the entire space. But that ladder joining the earth and the sun is non-existent, its mensuration by a rigid yard stick is not actually carried out. Similarly the 'coordinate crystal' fails to exist and the covering translations are not carried out. Indeed their ideality is essential, for their existence would produce real forces which would influence the course of world events. As to the structure discussed in Section 16, we may assert only the *possibility* of ascertaining it from events producible by the experimenter's free interference. The geometrical statements, therefore, are merely ideal determinations, which taken in individual isolation lack any meaning verifiable by what is given. Only here and there does the entire network of ideal determinations touch upon experienced reality, and at these points of contact it must 'check.' That, expressed in the most general terms, may well be called *the geometrical method.* "It must be admitted that he who undertakes to deal with questions of natural sciences without the help of geometry is attempting the unfeasible," Galileo says (Dialogo, *Opere*, VII, p. 229). Enemies of this method are, on the one hand, the empiricists, because any aprioristic construction is a thorn in their flesh; they fondly imagine it to be possible to grasp reality as a thing of one stratum, as it were, without aprioristic ingredients, by a purely descriptive approach (Bacon versus Galileo, Hume versus Kant, Mach versus Einstein). On the other hand, out of hatred for the freedom, the open field of geometrical construction, those metaphysicians oppose the method who build up a rigid dialectical world of concepts as the true reality (Hegel versus Newton). From both angles Aristotle (versus Archytas-Plato) is the great anti-mathematician. ⎬

C. A Priori or A Posteriori? The belief in the aprioristic character of geometrical cognition, in particular of Euclidean geometry, had taken deep roots in former times. Thus Kepler says (in his famous letter to Galileo, April 1610; Galileo, *Opere*, X, p. 338), "The science of space is unique and eternal and is reflected out of the spirit of God. That men may partake of it is one of the reasons why man is called the image of God." Leibniz has tried to show that the geometrical truths are analytic. With respect to geometry Kant raises the problem of the Critique of Pure Reason: How are synthetic judgments *a priori* possible? And he believes that he has answered this question for geometry by his thesis that space is pure non-empirical intuition. "That in which sensations are merely arranged, and by which they are susceptible of assuming a certain order, cannot itself be sensation; hence indeed the matter of all phenomena is given to us *a posteriori* only, while its form must lie ready *a priori* in the mind and therefore must be capable of investigation independently of all sensation. . . . Hence our explanation alone renders comprehensible the possibility of geometry as synthetic knowledge *a priori*." This certainty is shaken by the development of non-Euclidean geometry.

{Proclus already, in his commentary on Euclid, sounded a warning in connection with the axiom of parallels not to make undue use of intuitive evidence. Gauss writes to Olbers (1817, *Werke*, VIII, p. 177), "I am coming more and more to the conviction that the necessity of our geometry cannot be demonstrated, at least neither by, nor for, the human intellect. Perhaps in some other life we may arrive at other insights into the nature of space that are at present inaccessible to us. Until such time geometry should be ranked, not with arithmetic, which is purely aprioristic, but with mechanics." Or, in 1830, to Bessel (*op. cit.*, p. 201), "We must admit humbly that, while the number is a product of our intellect alone, space has a reality beyond our mind whose rules we cannot completely prescribe."}

Helmholtz shows that the two parts of the Kantian doctrine of space, namely, (*i*) that space is pure form of intuition, and (*ii*) that the science of space, Euclidean geometry, holds *a priori*, are not so closely connected that (*ii*) follows from (*i*). He is willing to accept (*i*) as a correct expression of the state of affairs; but nothing can be inferred from that, according to him, beyond the fact that all things of the external world have spatial extension. In accord with Riemann he points out the empirical physical content of geometry and refers to Newton, who in the introduction to *Principia* had declared, "Therefore geometry is founded in mechanical practice, and is nothing but that part of universal mechanics which accurately proposes and demonstrates the art of measuring." If there were, aside from the "physical equivalence" of spatial quantities (cf. p. 103), an equality given by immediate transcendental intuition, then the agreement of the two concepts could after all be only a matter of experience, while in the case of conflict the transcendental equality "would be degraded to the level of a sense illusion, i.e. an objectively false semblance" (Helmholtz, *Wissenschaftliche Abhandlungen*, II, p. 654). Against the argument that non-Euclidean geometry is devoid of intuitivity (Anschaulichkeit), he sets up a definition of intuitivity. The latter consists, he says, in "the complete imaginability of those sense impressions which the object would produce in us according to the known laws of our sense organs under any conceivable observational conditions and by which it would differ from other similar objects." We may refer to the description given in Section 17 of the relation between the objective world and its subjective image as conceived by the point eye moving along a world line. Against the argument that an attempted experimental test of geometry always involves physical statements about the behavior of rigid bodies and light rays it may be pointed out that the individual laws of physics no more than those of geometry admit of an experiential check if each is considered by itself, but that a constructive theory can only be put to the test *as a whole*.

Under the influence of modern mathematical axiomatic investigations one has come to distinguish the 'mathematical space,' whose laws are logical consequences of arbitrarily assumed axioms, from the 'physical space,' the ordering scheme of the real things, which enters as an integral component into the theoretical construction of the world. With regard to this distinction Einstein says (*Geometrie und Erfahrung*, p. 3), "As far as the propositions of mathematics refer to reality they are not certain, and in so far as they are certain they do not refer to reality." The general philosophical development, on the other hand, has since taken a course that led to a split of Kant's judgments *a priori* into two directions. On the one hand, there are the non-empirical laws (Wesensgesetze), which express the manner in which data and strata of consciousness are founded upon each other, but do not claim to involve statements of fact; this line of pursuit culminated in Husserl's phenomenology, in which the *a priori* is much richer than in the Kantian system. On the other hand, principles of theoretical construction are formulated, which according to the most extreme point of view (Poincaré) rest on pure convention.

After what has been said in Part I we need not enter here into a detailed discussion of the general mathematics of continua and of the more important structures with which they can be endowed. In the case of physical space it is possible to counterdistinguish aprioristic and aposterioristic features in a certain objective sense without, like Kant, referring to their cognitive source or their cognitive character. In fact, according to the Riemann-Einstein view, we may contrast the one absolutely given Euclidean-Pythagorean *nature* of the metric, which does not participate in the irradicable vagueness of that which occupies a variable place on a continuous scale, with the mutual *orientation* of the metrics in the various points, i.e. the quantitative course of the metrical field; the latter is accidental, dependent on the distribution of matter, ever-changing, and ascertainable only approximately and with the help of immediate intuitive reference to reality. Thus the general theory of relativity does not altogether deny that there is in this sense something aprioristic to the structure of the extensive medium of the external world, but the line between *a priori* and *a posteriori* is drawn at a different place. (To be exact, this juxtaposition, or separation, must be understood as meaning — as always in cases of this kind — that the aprioristic factor can be isolated from the whole without thereby exhausting the latter; there is no residue of purely *a posteriori* character, however, that would be left after the first part has been 'subtracted' from the whole.) Among the aprioristic features of the world, beside and above the one nature of the metrical field, there is the topological connectivity, which is fixed once and for all, especially the dimension number 4. The quantitative course of the metrical field obeys exact natural laws, namely, the Einstein laws of gravitation, which resemble the Maxwell laws of the electromagnetic field. Within the *a posteriori* one has thus to make yet another distinction, between what is necessitated by natural law and what even under their rule remains free and thus appears as contingent. The binary gradation is replaced by a ternary one.

{In addition to the physical space one may acknowledge the existence of a *space of intuition* and maintain that its metrical structure of necessity satisfies Euclidean geometry. This view does not contradict physics, in so far as physics adheres to the Euclidean quality of the infinitely small neighborhood of a point O (at which the ego happens to be at the moment). For the angles which are formed by the spatial directions of the light beams issuing from the various stars and striking the point eye do indeed fulfil the laws of spherical trigonometry in Euclidean space. But then it must be admitted that the relation of the intuitive to the physical space becomes the vaguer the farther one departs from the ego center. The intuitive space may be likened to a tangent plane touching a curved surface (the physical space) at a point O; in the immediate vicinity of O the two coincide, but the larger the distance from O the more arbitrary will the one-to-one correspondence between plane and surface become that one tries to establish by continuing the relation of coincidence near O. This does not mean that the intuitive space as such must necessarily be of a vague character. The intuitive space after all does not overcome the discrepancy created by binocular sight by vacillation or compromise (provided extreme circumstances, or attention directed toward the visual perceptions as such, do not cause a contest between the fields of vision to break out) but is intuitively of unobscured clarity, though in the objective construction the state of affairs can only be represented as a compromise.}

Regarding the aprioristic features of space the task arises to understand on rational grounds the peculiarities that give them their distinctive position within the range of the more general possibilities revealed by formalized mathematics. Thus there are three different possibilities as to the nature of a four-dimensional Riemann manifold, according as its fundamental metrical form possesses 0, 1, or 2 negative dimensions. If the world corresponded to the case 0, no propagation of effects from a world point O would be possible, while in the case of 2, past and future would be melted into one world domain. Thus it can be argued that the middle case of 1 negative dimension is realized by the metrical field of the real world because of the necessity of a causal structure by virtue of which an ego may be actively and passively connected with the world in such a manner as to separate past from future, what is known from what is planned. Likewise it must be asked in connection with n-dimensional Euclidean or Riemannian geometry, which resulted by cogent formalization from the three-dimensional one (Section 12), what inner reasons there are for the distinction of the case $n = 3$ realized by the actual space. Aristotle gave several answers to this, which still move in the sphere of mythical thought. Galileo discusses and rejects them at the beginning of his "*Dialogo*." The solution which he himself proposes is merely a clearer formulation of the problem but is no answer. The best chances for success seem to me to lie in theoretical physical construction.[6] Thus it can be shown by means of the wave equation of light (which can be immediately extended to n dimensions) that only in a space of an odd number of dimensions is the extinction of a candle followed by complete darkness about the candle (within a radius that increases as rapidly as light travels). This, at least, shows up an important inner difference regarding the propagation of effects between *even* and *odd* numbers of dimensions. Those particularly simple and harmonious laws which Maxwell had developed for the electromagnetic field in empty space are invariant with respect to an arbitrary change of the standard unit length at every world point, provided the world is four-dimensional. This principle of 'gauge invariance' holds for no other number of dimensions.

[The group structure of the Euclidean group of rotations (which still dominates the metrical nature of the world even if the Riemann-Einstein infinitesimal geometry is adopted) is decidedly different for the various numbers of dimensions. This circumstance suggests that the mathematical and physical laws may cease to be indifferent to the number of dimensions on some deeper level that has hardly been touched by the physics of today. There is thus good reason

---

[6] One blushes at the thought of the naive geometrical blunders committed again and again in an attempt to solve this deep problem. A recent example of this can be found in Natorp's *Logische Grundlagen der exakten Wissenschaften*, pp. 303 ff.

to hope that our problem will one day find a cogent solution along such lines. An attempt to make the three-dimensionality of space comprehensible through its role in the constitution of the external world for the consciousness was made by Bolzano (*Abhandlungen der Böhmischen Gesellschaft der Wissenschaften*, 1843). A more recent attack by O. Becker in the same direction is less absurd, but still far from satisfactory.

A way to understand the Pythagorean nature of the metric (which finds its expression in the Euclidean group of rotations) exactly through the separation of *a priori* and *a posteriori* has been pointed out by the author. Only in the case of this particular group does the contingent quantitative distribution of the metrical field, however that distribution is chosen within the framework of its *a priori* fixed nature, uniquely determine the infinitesimal translation, the non-rotational progression from a point into the world. This assertion involves a rather deep group-theoretical theorem which was proved by the author. The space problem, thus solved, plays a similar part within the Riemann-Einstein theory as the Helmholtz-Lie problem (Section 14) plays for the rigid Euclidean space. It may be that the postulate of the unique determination of 'straight progression' can be justified on the basis of the requirements posed by the phenomenological constitution of space. Becker persists in attempting to base the significance for intuitive space of the Euclidean group of rotations upon Helmholtz's postulate of free mobility. If in agreement with a remark made in Section 15 the transformation group Δ_0 in 3 or 4 dimensions is considered as representation of an abstract group, then more emphasis should be placed on the distinctive features of the structure of this abstract group than on the special concrete representation Δ_0. }

REFERENCES

R. DESCARTES, *Principia philosophiae.*

G. GALILEO, *Il saggiatore.*

T. HOBBES, *De corpore.*

G. W. LEIBNIZ, *Philosophische Schriften*, ed. Gerhardt, VII, pp. 352–440 (Discussion between Leibniz and Clarke).

J. LOCKE, *An Enquiry concerning Human Understanding.*

D. HUME, *Treatise of Human Nature.*

I. KANT, *Kritik der reinen Vernunft.*

H. VON HELMHOLTZ, *Handbuch der physiologischen Optik*, III.

—— Über den Ursprung und die Bedeutung der geometrischen Axiome, in *Vorträge und Reden*, fourth edition, II, 1896.

—— Über den Ursprung und Sinn der geometrischen Sätze, *Wissenschaftliche Abhandlungen*, II, p. 643.

E. MACH, *Analyse der Empfindungen*, third edition, Jena, 1903.

R. LUNEBURG, Metric Methods in Binocular Visual Perception, *Courant Anniversary Volume*, New York, 1948, pp. 215–239.

E. Study, *Die realistische Weltansicht und die Lehre vom Raum*, Braunschweig, 1914.

M. Schlick, *Allgemeine Erkenntnislehre*, Berlin, 1918.

E. Husserl, Ideen zu einer reinen Phänomenologie und phänomenologischen Philosophie, *Jahrbuch für Philosophie und phänomenologische Forschung*, I, pp. 1–323, 1913. (English translation under the title *Ideas* in the Library of Philosophy, London and New York, 1931.)

O. Becker, Beiträge zur phänomenologischen Begründung der Geometrie und ihrer physikalischen Anwendungen, *ibid.*, 6.

H. Weyl, *Mathematische Analyse des Raumproblems*.

R. Carnap, Der Raum, *Kant-Studien*, Ergänzungsheft, 56, 1922.

CHAPTER II

Methodology

19. MEASURING

THE opinion that cognitive connections can be found in the real world only in so far as qualitative determinations are reduced to *quantitative* ones, which asserted itself in modern times in opposition to Aristotle's philosophy, has assumed fundamental importance for natural science. This is Kepler's succinct formulation, "Ut oculus ad colores, auris ad sonos, ita mens hominis non ad quaevis sed ad quanta intelligenda condita est." The standard of our knowledge is found in its approximation to the "nudae quantitates." Galileo enunciates the principle, "to measure what is measurable and to try to render measurable what is not so as yet." A splendid illustration of the second part of this postulate is his invention of the thermometer.

But what does the process of measuring consist in? Let us take *inert mass* as an example.

According to Galileo the *same* inert mass is attributed to two bodies if neither overruns the other when they are driven against each other with equal velocities (they may be imagined to stick to each other upon colliding). This is a definition by abstraction. The physically defined equality of mass is a relation of the character of equality (see Section 2), as can be confirmed by experience. Experiment must show, in addition, that equality is independent of the attendant circumstances of the defining process, such as the speed of collision. Equality, this first requisite of all mensuration, usually carries with it the relation of 'smaller' and 'larger.' In our case: that body has the larger mass which, at equal speeds, overruns the other. Finally a process of addition must be given; in the case of masses this consists simply in joining the two bodies. By assuming certain axioms concerning these fundamental concepts (which Helmholtz for instance discusses in his repeatedly quoted essay on *Zählen und Messen*) one can establish a measuring scale which characterizes every value of the quantity in question by a number. It may be necessary to fix arbitrarily a certain unit of measure (herein lies a new component of relativity, and this is actually the case with line-segments and masses, for example); while under other circumstances a natural unit of measure exists, like the complete rotation (360°) in the realm of angles. From a practical viewpoint the unit must fulfill the requirement that it be reproducible everywhere and at all times as accurately as possible.

A type of quantities different from the 'additive' quantities just characterized are the absolute and material constants that occur in such functional relations between additive quantities as are accepted as laws of nature. In this category belongs the coefficient of refraction n, whose significance is evinced by Snell's law of refraction: the sine of the angle of incidence equals n times the angle of refraction (the two angles are the additive quantities put in relationship to each other by this law). Helmholtz calls constants of this sort "intensive" quantities, in contrast with the additive or extensive quantities. In particular all numerical valuations of properties are intensive quantities.

{A good example is the measuring of temperature. Bodies have equal temperature if they produce no change in each other when in contact. It is by no means a self-evident fact, but one to be confirmed through experience, that when A and B, and B and C, possess equal temperature, A and C also have equal temperature. An addition that would lead to a definite measuring scale does not exist in the field of temperatures. Yet on the basis of the experience that bodies of unequal temperature cause changes of length between one another, one proceeded to define temperature by means of the length of a standard body which is brought into contact with the body to be measured. This determination of the temperature is always reproducible and independent of past history, while to our sense of temperature a body of physically constant temperature feels warm or cold, according to the degree of warmth to which our skin was exposed immediately before. Wood and iron of equal temperature feel different to the touch — when warm, iron feels warmer; when cold, iron feels colder. The external conductivity of heat is a codeterminant for the resultant sensation. The objective concept of temperature is thus pretty far removed from the sense data of heat perception. The temperature scale is dependent on the choice of the standard body. At least all gases, however, react approximately alike, and their behavior can be described within relatively small errors by a simple law, which in turn is laid down as characteristic for an 'ideal gas.' But only by deriving from the ideal-gas law the so-called second theorem of thermodynamics, which holds for all bodies, did it become possible to realize in an unambiguous way the temperature scale of the ideal-gas thermometer. The *absolute temperature* is characterized, aside from the statement that bodies of equal temperature have the same temperature value T, by the following law: the integral of dQ/T over any cycle of virtual states is zero. Here T is the temperature of the individual state σ and dQ the infinitesimal increment of heat that takes place in passing on from σ to the next state along the cycle. The heat is measured as energy and thus an additive quantity. Consequently the temperature T is an intensive quantity in the Helmholtz sense. Its definition is an implicit one and as such presupposes the validity of certain natural laws. It leaves arbitrary only the unit of measure, but not the zero point. T is by necessity always positive, and there exists an *absolute zero point* of temperature. (On defining as 100° the difference between the boiling and freezing temperatures of water under atmospheric pressure these temperatures turn out to be 373° and 273° respectively in the absolute thermodynamic scale.)

The laws of 'mechanical similarity,' of which Galileo speaks on the second day of his *"Discorsi,"* are based on the relativity of quantitative determinations in regard to arbitrarily chosen standards. These laws make it possible to use small models in order to study real events, just as the proportion of the sides of a triangle whose angles are known can be found from a small model. If, in a problem of floating or flying, the viscosity of the medium (water or air) must be taken into account, then, when changing over to the model, the medium must generally be replaced by one whose viscosity is changed according to the size of the model. Yet the physical laws of similarity have their limits. Thus, according to the set-up of the special theory of relativity, only one arbitrary unit of length for time and space remains, the velocity of light c becoming the absolute standard of velocity. However, the existence of an absolute unit for velocities is no more extraordinary in relativity theory than the existence of an absolute angular unit in geometry. It is merely a consequence of the metrical structure of the four-dimensional world. If the gravitational constant is added, there remains just one unit for all physical mensuration that has to be chosen arbitrarily, say the time unit of the second. Thus far one can get without taking the atomistic structure of matter into consideration. As for the atomic theory and the absolute constants of nature to be obtained from the atomic laws, see Section 22 E (p. 184) and Appendix F.}

The theory of mensuration involves the question of *how it is possible to determine quantities much more accurately than the differentiating capacity of our senses permits.* What good is it to differentiate between two shades of yellow (such as the yellows of the two adjacent D-lines in the sodium spectrum), if they are sensually indiscernible? A simple example is the exact determination of the duration of a pendulum oscillation: one waits for, say, 1,000 oscillations and divides the entire time by 1,000. The accuracy has thus been increased a thousandfold compared to that obtainable by observing a single oscillation. To be sure, a theoretical assumption has been made here; namely, that all single oscillations are of equal duration. For the intuitionist, who respects the limits of sensory accuracy and does not want to transgress them a thousand times, this assumption is just as meaningless as the indirectly obtained assertion concerning the duration of a single oscillation. Yet the assumption can be confirmed to some extent by observing that the ratio of the duration of m successive oscillations to that of n oscillations (m and n being large integers) is $m:n$, of course within the limits of accuracy of direct observation. (The test is carried out with several series of oscillations picked at random.) In general the situation is as follows: By virtue of the exact laws of the basic theory, the quantity x to be determined is functionally dependent on several others. By observing the latter, one can arrive at conclusions regarding the value of x by which x can be determined more exactly than by direct observation. The basic theories are confirmed if within the expected margin of error all indirect methods for the determination of x lead to the same result. In particular, a fact is determined the more exactly the further its causal consequences continue to develop in time. A deviation in direction of two missiles which at first may not be noticeable eventually leads to the most obvious difference; one hits the mark and the other misses. It must be remembered, however, that any such indirect quantitative determination and any establishment of a difference not manifest to the senses is possible only on the basis of *theories*. Their verification takes place by testing them in all their numerical consequences and finding that they yield *concordant* results. (Otherwise the observations enforce modification of the theory.)

[In this field belong all the indirect methods of experimental physics, beginning with the simplest tools — the vernier, the mirror reflections for the measurement of small deviations, the rotating mirrors which help to resolve the sound-generating vibrations of luminous bodies, the microscope — up to the experimental and instrumental refinements of modern atomic research that aim at making the single atomic particle visible through its effects. Mach, in the chapter "Das physische Experiment und seine Leitmotive" in *Erkenntnis und Irrtum* (1905), makes an interesting survey of, and an attempt to organize, the various methodical principles involved. Here is a wide field for the inventiveness of the experimenter.]

Even if the opinion can thus be justified that the world is far more accurate than it appears to the senses, or even that it is absolutely accurate, nevertheless this absolutely accurate state could only be ascertained by me as the observer if I waited for the resulting develop-ments till the end of time (as well as for the perfection of theoretical physics which has to provide the exact laws). Complete accuracy is therefore a limiting idea and by no means immediately given. Leib-niz's thought of preestablished harmony — which he himself illus-trates by the example of two entirely independent clocks that are synchronous, not because they exert a regulating influence upon each other but because they are identically constructed — contradicts, therefore, the nature of the continuum. In his *Treatise*, Hume states that the refinement of mensuration is based on repeated mutual cor-rection, but that "the notion of any correction beyond what we have instruments and art to make is a mere fiction of the mind, and useless as well as incomprehensible" (Book I, Part II, Section 4). Even so one can understand the necessity and expedience of exact mathe-matics: the exact theory provides the framework for approximate verifications. If, for example, we adopt Euclidean geometry as the theory of space, then, with the theorem that the diagonal of a square is to the side as $\sqrt{2}:1$, we are prepared for all future refinements in direct or indirect methods of mensuration; it will lead us again and again to new predictions (approximative in character) or to ever finer criteria by which to check the standard measuring bodies as to whether they satisfy the ideal assumptions of Euclidean geometry up to the degree of accuracy attainable at each step.

⟦Recently the Danish geometer Hjelmslev has espoused a purely approximative geometry (*Abhandlungen des Mathematischen Seminars der Universität Hamburg*, Vol. 2, p. 1), with the same arguments as Hume, who remarked, "Our ideas seem to give a perfect assurance, that no two right lines can have a common segment; but if we consider these ideas, we shall find, that they always suppose a sensible inclina-tion of the two lines, and that where the angle they form is extremely small, we have no standard of right line so precise as to assure us of the truth of the proposition" (*Treatise*, Book I, Part III, Section 1). But when Hjelmslev continues by formulating the Pythagorean theorem thus, "In a right triangle, numbers can be assigned to the sides in such a manner that the square of the number assigned to the hypotenuse is equal to the sum of the squares of the numbers assigned to the other sides," then one can already see the tendency to reduce the range of indeterminacy of directly observed measurements by declaring the functional relationship enunciated in the ordinary theorem of Pythag-oras to be an inviolable exact law. If one keeps in mind that the same segment which here functions as the hypotenuse may be a con-stituent in infinitely more figures, with whose remaining parts it is linked by similar functional relations, one reaches that concept of an exact theory which dominates constructive physics. Hjelmslev, incidentally, is far too concerned with figures drawn on a blackboard and is apt to forget that geometry must also serve as an ideal basis for astronomy and atomic physics. Constructive science can sustain the intuitionism of Brouwer; but the sensualism of Hume and Hjelmslev — which on principle would recognize as real only the immediately given, without being able to carry this through — is deadly for science.]

Measuring, as we have considered it up to now, was based on the fact that in many cases physical quantities are subject to the notions of equality and addition with their characteristic axioms, by virtue of which their values are projected on a numerical scale. "Thus," says Maxwell (*Scientific Papers*, I, p. 156), "all the mathematical sciences are founded on relations between physical laws and laws of numbers." However important the particular way discussed here of introducing numerical symbols into natural science may be, it seems nevertheless not to be the decisive feature of quantitative analysis. If a basis for an arithmetical differentiation of the individual places in a continuum is created by spreading a division net over the continuum, with a wide margin of freedom in all its steps of refinement and sharpening (though bound by a fixed combinatorial scheme), then the procedure is different and much looser, as it were, than it is in the case of mensuration proper. Moreover, the measuring of many physical observables (which are not scalar but vector or 'tensor' quantities, such as the metrical field) is possible only relative to a coordinate system thus arbitrarily introduced into the world. This free insertion of coordinates and that mensuration based on the addition of equal elements may be typical for the different levels on which the two methods are applied: the first to the form, the latter to the content of the world. However, the only decisive feature of all measurements is, it seems, *symbolic representation;* even numbers are in no way the only usable symbols. Measurement permits things (relative to the assumed measuring basis) to be presented *conceptually*, by means of symbols. If a part of the infinite Euclidean plane is materialized by a flat metal disc, we can at first fix places within the domain of the metal disc only, using material markings on the disc that are qualitatively different and permanently recognizable. But once two rectangular axes and a unit length have been scratched onto the disc, then we are not only able ('ideally' and on the basis of a theory concerning the behavior of rigid measuring rods) to spread over it an arbitrarily fine net of well-characterized locations by means of coordinate assignments, but this indirect method enables us to put such ideal 'numerical marks' even beyond the boundaries of the disc. It is thus that we use the earth as a basis to plumb the sidereal space.

Finally, in carrying out measurements there is a tendency to reduce the immediate sensory observations, which of course can never be eliminated, to the safest and most exact among them, namely spatio-temporal coincidences (in particular, one tries to do without the subjective comparison of colors and light intensities). Any mensuration should ultimately ascertain, so one wishes, whether a mark on one scale (a movable pointer or such) coincides with a certain mark on another scale. In the case of an astronomical observation the reading of the graduating circle is done in just that way, while the training of the instrument on a star utilizes a coincidence modified by the intercalation of light, namely the 'coincidence' of star and cross threads.

20. FORMATION OF CONCEPTS

Dilthey, in his essay on the autonomy of thought in the 17th century (*Gesammelte Schriften*, II, third ed., 1923) describes the development of mechanics up to Galileo. "Galileo came, and with him there followed an actual *analysis of nature*, after more than two thousand years of mere description and consideration of form in nature, that had culminated in Copernicus's picture of the world." For this analysis it is decisive to isolate simple occurrences within the complexity of facts, and to dissect the course of the world into simple recurrent elements. Bacon already devised the formula "dissecare naturam." "Only the mathematicians contrived to reach certainty and evidence, because they started with the easiest and simplest" (Descartes, *De Methodo*). In no small measure is the strength of natural science based upon its renunciation of designing a 'system of nature' in one draft, its condescension to deal with the small individual problems and its boundless patience in submitting them to a detailed analysis. Descartes himself still sinned heavily against his own methodical remark. Galileo's superiority over him in the field of natural science is partly attributable to the fact that Galileo, in his research into the laws of falling bodies, strictly exercised that "restraint which proves the master."[7]

We can distinguish the following phases of dissection into simple elements, the first three of which still belong to the pre-scientific stage.

[1. Dissection of the three-dimensional spatial reality into single partial systems (bodies or things), each forming an intuitive spatially

---

[7] "In der Beschränkung zeigt sich erst der Meister" is a familiar line from Goethe's sonnet *Natur und Kunst*.

isolated and relatively constant unit. In its behavior each is considered as independent of the others, unless progressive analysis calls for corrections. Closely connected with this is the dissection of the four-dimensional spatio-temporal reality into single isolated events that form natural intuitive units.

2. The conception of an intuitively experienced event as having come about by spatio-temporal coincidence and amalgamation of several simple phenomena (each of which would produce other perceptions than the phenomenon as a whole if the others were 'erased' or replaced by 'normal conditions'; e.g. the sun setting behind a gold-edged cloud).

3. Apperception of the 'being-so,' bringing out the characteristic features (self-insufficient parts) of the phenomena. Upon this procedure is based the grouping together of similar things, the subordination under concepts, in one word: classification. Such classification will correct itself as the wealth of our experience increases. It will thus learn to distinguish better and better.the truly essential from the inessential and progress to the formation of more and more 'natural' classes. A concept is the more essential the more connotations it entails according to the evidence of experience, namely the more characteristics not contained in the concept itself are empirically found to be common to the objects falling under the concept.

4. We are not satisfied with intuitively isolable elements but interpret a series of properties which always appear together as an indication of a concealed something. This leads to hypothetical elements, such as atoms, forces, electro-magnetic field, etc. Moreover, we learn to interpret not only the observable properties but also the reactions that occur if one system is brought together with another as manifestations of such hypothetical elements and of their intensive and quantitative values. (Reactions instigated at will are the essence of experiments.) Finally, we do not hesitate to dissect hypothetically even the intuitively simple, e.g. the white sunlight into the colors of the spectrum, or the acceleration of a planet into the partial accelerations brought about by the sun and the other planets. It is evident that along with the dissection the synthetic principles also have to be established according to which the elements unite into a whole (e.g. formation of the resultant of forces). ⎬

Starting everywhere with the simplest, we find, among the recurring elements thus obtained and the variations of their values, constant lawful relationships which can be quantitatively explored and expressed by mathematical functions. What is decisive is this: the farther the analysis progresses, the more detailed the observations become and the finer the elements into which we dissect the phenomena, the simpler — and not the more complicated, as might be expected — become the basic laws, and the more completely and accurately do they explain the factual course of events. And only by way of this analysis do the right constructive concepts evolve which serve to describe objective nature; they are bound up throughout with definite facts and valid natural laws.

{What is it that compels us in physics to think of the uniform white color as something composite? It is the causal law, asserting that equals, under equal conditions, produce equal reactions. It requires that two colors, which to the senses appear as the same white, contain 'hidden' differences, since, upon passage through the same prism, they yield different spectra. (In principle, what happens here is no different from the case of two spherical balls of identical appearance but different inertia and weight, one of which when cut open is seen to contain a core of gray lead.) It will be found that the variety hidden in the white light can be described most expediently in terms of the spectrum and its intensity distribution. At first the apparatus used in the reaction, the prism with its special properties, will get mixed in, and it will be necessary by varying the form and substance of the prism to learn to separate the two influences. In this way one will arrive at the wave-length scale of spectral colors, which is independent of the prism. In the earlier example of the ponderomotoric force suffered by a test particle in a field generated by charged conductors, we have explained in detail how such a separation can be effected. Polarized light of a certain spectral color and intensity, on the other hand, proves to be something simple, because its behavior in all reactions is completely determined by the characteristics mentioned.}

A typical example of the formation of physical concepts is Galileo's mass concept. We mentioned above the criterion of mass equality. Here the concept of *momentum* appears as prior to that of mass. Two bodies moving toward each other (each undergoing uniform translation in accordance with the law of inertia) have oppositely equal momenta if neither overruns the other upon collision. We repeat Galileo's criterion by saying that two bodies have the same mass if, at equal velocities, they possess equal momenta. We are thus dealing with a constructive concept in the sense of the description on p. 37. Instead of, or besides, purely intellectual manipulations in the realm of numbers, we have here, in the material sphere, real (or at least really possible) experiments, the results of which are used for the numerical determination of characteristics. This is a step of great importance. After matter was stripped of all sensory qualities, it seemed at first as though only geometrical properties could be attributed to it. In this respect Descartes was wholly consistent. But it now appears that other numerical characteristics of bodies can be gathered from the laws to which changes of motion in a reaction are submitted. Thus the sphere of properly mechanical and physical concepts is opened up beyond geometry and kinematics. Basically Galileo's definition of mass implies the law of momentum: "An isolated body (moving uniformly) has a certain momentum, $\vec{I} = m\vec{v}$, which is a vector having the same direction as the velocity $\vec{v}$. The sum of the momenta of the individual bodies of an isolated system prior to a reaction is the same as after the reaction." By subjecting the observed motions to this law, it is possible to obtain data for the numerical evaluation of the ratios of the masses m of the individual bodies before and after the reaction. *Constructive natural science has the general task of assigning to the objects such constructive quantitative characteristics* (dependent only on the object though not necessarily directly observable) *as will make their behavior, under circumstances described by characteristics of the same kind, completely determinate and predictable on the basis of the natural laws.* The implicit definition of characteristics is tied to these laws. In this way science complies with the postulate (which fails to be satisfied if nothing but sensory qualities are admitted) that "all changes which bodies undergo have their cause in the nature and the qualities of the bodies themselves" (Euler, *Anleitung zur Naturlehre*, Chap. I, §2, *Opera postuma*, II, 1862). The fact that we do not find but enforce the general principles of natural knowledge was particularly emphasized by the conventionalism of H. Poincaré.

Turning to the temporal analysis of the process of reaction, and considering that for an isolated body k the momentum $\vec{I}$ is constant in time, we take the change of this quantity per unit of time, $d\vec{I}/dt$, called *force*, as a measure of the effect which other bodies k_1, k_2, . . . have upon k. Indeed Newton recognized that the force is composed additively (according to the parallelogram law of vector addition) of individual forces exerted upon k by each of the bodies k_1, k_2 . . . , and that this occurs in such a manner that, for example, the force exerted by k_1 on k at a certain moment depends solely on the condition of these two bodies (location and velocity) at that instant. This is the real meaning of the decomposition of the one force into several component forces. Looking at these facts one cannot escape the conclusion that the definition 'force = time-derivative of momentum' does not reflect the nature of force adequately but that the real state of affairs is the other way round: force is the expression of an independent power that connects the bodies according to their inner nature and their relative position and motion, and that power *causes* the change of momentum with time. Thus the living metaphysical interpretation conforms to the theoretical construction. Through the basic mechanical law of motion, physics is given the task of exploring the forces operating among bodies in their dependence on position, motion, and inner condition. The latter will enter the laws of force by way of numbers characteristic of the inner state of the reacting bodies, like the electrical charge in the case of Coulomb's law of electrostatic attraction and repulsion. Thus the concept of force becomes a source of new measurable physical characteristics of matter.

While the metaphysical conception of nature is modified by the results of theoretical construction, which should find in that conception a suggestive and fruitful expression, there usually is already at the bottom of concrete research a preconceived idea that is in happy consonance with the facts. In the process of motion Galileo sees the dynamic intensity, the driving push, the impetus or momentum. Motion to him depends on the struggle of two tendencies, inertia and force, force that deflects the body from the path dictated by inertia. Mass is the dynamic coefficient according to which inertia resists the deflecting force. With reference to Galileo, Goethe remarked in his *Geschichte der Farbenlehre* (Section 4, Galileo Galilei): "In science all depends on what is called an *aperçu*, on a recognition of what is at the bottom of the phenomena. And such a recognition is infinitely fruitful." Given the right basic aspect, the right basic concepts will emerge in the course of detailed research conducted under its guidance.

{In his book *Substanzbegriff und Funktionsbegriff* (1910) E. Cassirer has endeavoured to show that the formation of concepts in mathematics and physics in no way corresponds to Aristotle's logical scheme. In plane analytic geometry an ellipse is defined by its equation, by setting a positive quadratic form of the coordinates equal to unity. The individual ellipse is obtained by substituting specific values for the coefficients of the quadratic form (which vary over a predetermined range, namely the continuum of real numbers). We cannot agree with Cassirer's remark that in this procedure the more general concept is the richer; for the properties of the individual ellipse depend, in addition to the general form of the equation, on the specific values of the coefficients. It is true, though, that the special cases are obtained from the general one by assigning definite values to the 'variables' — within a range which is completely given or open to free construction. Aristotle ascends from the single object to the concept by isolating individual features of the object and by "abstracting" from everything else. Thus every other object which exhibits those same features falls under the same concept, or into the same class. In this procedure (as in descriptive botany or zoology) only the *really existing* objects are concerned, and classes are formed preferably in such manner that, according to the testimony of experience, the concepts entail as many 'connotations' as possible. In the mathematical-physical or 'functional' formation of concepts, on the other hand, no abstraction takes place, but we make certain individual features variable that are capable of continuous gradation (such as the coefficients of the quadratic form in the case of the ellipse), and the concept does not extend to all actual, but to all *possible* objects thus obtainable. "The possibility of arbitrary refinement, the easy survey and the facility in handling a whole continuum of cases with the assurance of completeness," according to Mach (*Prinzipien der Wärmelehre*, third ed., 1919, p. 459), "warrant the preference placed on such quantitative constructions." In this connection it is essential, though, that the continuum is not a closed aggregate but a field of determinations open to infinity; for otherwise we would return after all to the Aristotelian scheme of characteristic features ("a set of points (x, y) is an ellipse if numbers a, b, c *exist* such that all points of the set and no others satisfy the equation $ax^2 + 2bxy + cy^2 = 1$"). Thus the individual objects falling under the functional concept have to be *generated*, and the question whether a given object falls under it must not be asked in the expectation that the 'facts as they are' will necessarily answer with a clear-cut yes or no.

By means of the Platonic diagram on p. 53, which is identical with the division net of the one-dimensional continuum, Plato assigns their places to all beings by proceeding from the general to the specific by bipartition (diaeresis). This scheme, as well as the Platonic conception of ideas as numbers based on it, is not so far removed from the modern mathematico-physical conception of the world as might appear at a first glance. The former would merely have to be modified to the extent that firstly some but not all of the levels and divisions — as Plato maintains (dissection of the sacrificial animal, *Phaedrus*, 265c; *Politicus*, 287c) — are prescribed by the facts and capable of exact execution (for this possibility ceases whenever a uniform connected continuum is present); and secondly that the process continues *ad infinitum* and the individual thing appears but at the horizon as a limiting idea. It is characteristic of Aristotle that he reverses this diagram and begins at the bottom, with the individual beings, while Plato starts with the 'one.'

In particular, the concepts obtained by mathematical abstraction in accordance with the rule given at the end of Section 2 are of 'functional' nature.]

21. FORMATION OF THEORIES

The constructive character of the natural sciences has become obvious through what has been said above. Individual scientific statements cannot be ascribed an intuitively verifiable meaning, but truth forms a system that can be tested only in its entirety. Hobbes developed the view (*English Works*, VII, pp. 183 ff.) that we cognize with certainty only in those sciences which construct their objects on the basis of the structural conditions resting within the cognizant subject. Reality, to him, does not reside in the images of consciousness but in that content of theirs that makes a construction of objects possible. In contrast with the mere *cognitio* he sees in this synthetic process of generating the phenomenon from its origin the *scientia* in the strict sense. This, he claims, takes place within the natural sciences as far as mathematical deduction reaches. "Thus the sovereign consciousness of the autonomy of the human intellect and its power over the physical things," Dilthey says (*Gesammelte Schriften*, II, p. 260), "was definitely established by the great discoveries of Copernicus, Kepler, and Galileo, and the accompanying theory of the construction of nature by logico-mathematical elements of consciousness given *a priori* became the dominant conviction of the most progressive minds." In modern physics the building material is no longer the elements of consciousness abstracted from reality, but purely 'arithmetical' symbols. Dingler (*Die Grundlagen der Physik*, p. 305, 1923) in fact defines physics as that scientific domain in which the principle of symbolic construction is carried through completely. But, coupled with aprioristic construction, we have experience and the analysis of experience by the experiment. "The scientific imagination of man was regulated by the strict methods which subjected the possibilities that lay in mathematical thinking to experience, experiment, and confirmation by facts. . . . The results thus obtained have made possible a regular and connected progress in scientific research by the common efforts of the various countries. It may be said that now only did human reason become a unified force working concordantly within the various civilized nations. The most difficult work of the human mind on this planet was accomplished by this regulation of scientific imagination, which subordinated itself to experience." (Dilthey, Der entwicklungsgeschichtliche Pantheismus, *op. cit.*, II, p. 346.)

[Let us illustrate what has just been said by the theory of electromagnetic phenomena. Since we only want to bring out the essentials, it may be permissible in order to spare the reader the difficulties of relativistic physics to assume the velocity of propagation as infinite. We suppose that there are elementary material quanta to which fixed masses and charges are attached once and for all. Position and velocity of these electrons at a moment t uniquely determine the electromagnetic field by virtue of certain generating laws. This field, in accordance with further laws, is connected with spatially distributed momentum and energy, and exerts, by virtue of the flux of momentum, certain ponderomotoric forces upon the generating electrons. The force, finally, produces the acceleration of the electrons, according to the fundamental law of mechanics; but velocity and acceleration give us the change in position and velocity during the next time interval dt, thus determining from position and velocity at the time t these same data at the time $t + dt$. By iterating this infinitesimal transition $t \rightarrow t + dt$ again and again the entire motion is obtained through the mathematical process of integration. Only this complete theoretical context, in which also geometry plays its obvious part, is capable of an experimental test; and even this only under the idealizing assumption that the motion of the electrons is what we are able to observe directly. An individual law taken out of this theoretical context, however, hangs in the air. Thus, in the last analysis, all parts of physics and geometry grow together into one indivisible unit.

For the same reasons a theory develops by way of continual correction, as it is driven on by the ever-growing richness and precision of experience. "Thus the progress of science is dependent upon science itself, it is an extension and not a creation" (Enriques, *Problems of Science*, translated by Royce, Chicago and London, 1914, Chap. III, §37, p. 165). When the Kepler-Newton theory of planetary motion was established by observation, each event was tacitly assumed to take place at the instant of its perception. Only later, Roemer discovered the finite velocity of propagation of light through the apparent deviations of the motion of the Jupiter satellites from the motion required by theory. Thus a theory is employed (instantaneous propagation of light) which is later proved to be false. But the assumption of its rough correctness (together with other premises taken from experience) leads to the recognition of its finer inaccuracy and to its correction. But without the preliminary assumption not even the first step could have been taken. Newton's fourth rule for the study of nature refers to this (*Principia*, ed. F. Cajori, p. 400): "In experimental philosophy we are to look upon propositions inferred by general induction from phenomena as accurately or very nearly true, notwithstanding any contrary hypotheses that may be imagined, till such time as other phenomena occur, by which they may either be made more accurate, or liable to exceptions."}

To facilitate the task of the theorist, the experimenter endeavours to arrange the experiment in such a way that it is most sensitive to one law and as insensitive as possible to all others that play a part, namely by dampening the influence of such circumstances as are governed by the latter. This accounts, among other things, for the tedious efforts involved in screening off all kinds of 'sources of error.' All the same, the influence of certain elements such as the metrical field can never be eliminated. If a fact is in discord with the entire theoretical stock of science, it is finally left to the theorist to find the place where the theory is to be modified. It is hardly possible to formulate general rules for this, nor for the relative weight with which the known facts should bear upon the theoretical interpretation; this must be left to the discretion of the genius. Thus the general theory of relativity came about because Einstein realized the fundamental nature and the particular import of one fact, the identity of heavy and inert mass. The possibility must not be rejected that several different constructions might be suitable to explain our perceptions; in this recognition of the 'ambiguity of truth,' Hobbes and D'Alembert preceded the modern positivists. In an address on the occasion of Max Planck's sixtieth birthday in 1918, Einstein described the real epistemological situation with great justice as follows: "The historical development has shown that among the imaginable theoretical constructions there is invariably one that proves to be unquestionably superior to all others. Nobody who really goes into the matter will deny that the world of perceptions determines the theoretical system in a virtually unambiguous manner, although no logical way leads to the principles of the theory."

At any given stage of the theoretical construction there exists a hierarchy of laws, inasmuch as different degrees of stability are ascribed to the different laws. Certain ones among them are adhered to as principles with great tenacity. For a long time the laws of Euclidean geometry were held to be sacrosanct. The principles of the conservation of energy and momentum are of comparable, if not higher, stability. It is certain that a considerable portion of the theoretical system can be maintained in the face of any experiences as long as modifications of the remainder are permitted. Thus in the practice of scientific research the clear-cut division into *a priori* and *a posteriori* in the Kantian sense is absent, and in its place we have a rich scale of gradations of stability. It is the simple form and the instinctively convincing character of a law, together with its decisive significance for an extensive domain of facts, which gives it the rank of a principle. For instance, the convincing and simple law of inertia, which at first appears to be sufficiently confirmed by our experiences regarding motions relative to the earth, is maintained even when more refined experiences (Foucault's pendulum experiment) contradict it, by resorting to the 'subterfuge' of claiming the law not for motions relative to the earth but for an 'absolute motion' that is to be determined from the phenomena. The law of momentum is based on the 'evident' fact that a system of bodies, originally at rest, cannot be set into a progressive translatory motion under its own force; more exactly, interior reactions of an isolated system of bodies at rest are incapable of imparting to a portion of the system a common uniform translatory motion while the remainder stays put. When we are laughing at Münchhausen's tale of having extricated himself from the swamps by his own pigtail, we betray our intuitive knowledge of that fact. Further examples are the rule for the composition of velocities, taken for granted almost unnoticedly by Galileo (Discorsi, 4th day), and the energy principle.

{In the special form stating that the bodies of a system in a homogeneous gravitational field cannot lift themselves under their own force to a higher level, the latter is employed by Galileo and Stevin to derive the law of the inclined plane, by Huyghens to reduce the compound to the 'mathematical' pendulum (*Horologium oscillatorium*, 1673). Huyghens already conceived the general idea of the energy principle. He says (*op. cit.*, p. 95), "If only the inventors of new machines, who vainly endeavor to build a perpetuum mobile, would follow this hypothesis [!] of mine, they would soon recognize their error and would see that their goal is wholly unattainable." Leibniz reads the energy principle into the formula "causa aequat effectum" and considers it a special consequence of the principle of sufficient reason required by the "logic of quantity"; he bases on it his measure of the "*force vive.*" Much greater weight than by a single confirming experiment is carried by a universal negative experience of the type, 'This will never happen, whatever the circumstances.' Thus the energy principle is supported by the failure of all attempts at constructing a perpetuum mobile; and of the same character are the principles of relativity theory, the special relativity principle and the principle of the 'constancy of the velocity of light.' Attacking scholastic philosophy, Newton says (*Opticks*, ed. Whittaker, pp. 401–402), "To tell us that every Species of Things is endow'd with an

154

occult specifick Quality by which it acts and produces manifest Effects, is to tell us nothing: But to derive two or three general Principles of Motion from Phenomena, and afterwards to tell us how the Properties and Actions of all corporeal Things follow from those manifest Principles, would be a very great step in Philosophy, though the Causes of those Principles were not yet discover'd."}

Simplicity is considered as *sigillum veri*. "Nature loves simplicity and unity," Kepler says (*Opera*, ed. Frisch, I, p. 113). The same principle is formulated by Aristotle as follows (*De coelo*, I, 4, 217a): "At deus et natura nihil prorsus faciunt frustra," and it is held as an axiom: "frustra fit per plura quod potest fieri per pauciora." Galileo, on the third day of the "Discorsi," reconstructs the chain of thoughts which led him to the laws of falling bodies (*Opere*, VIII, p. 197): "When, therefore, I observe a stone initially at rest falling from a considerable height and gradually acquiring new increments of speed, why should I not believe that such increases come about in the simplest, the most plausible way? On close scrutiny we shall find that no increase is simpler than that which occurs in always equal amounts." He goes on to formulate the definition of uniformly accelerated motion, develops its consequences regardless of experience, and then finds them, as far as he can observe them with the means at his disposal, confirmed for the 'naturally accelerated' motion of falling bodies. Among Newton's rules for the study of nature, the first is to the effect that no more causes of natural things should be admitted "than such as are both true and sufficient to explain their appearances. . . . For Nature is pleased with simplicity, and affects not the pomp of superfluous causes." What matters is not that the absolutely simplest principles be established (as Dingler demands in his *Grundlagen der Physik*) — for then the world, for instance, would have to be attributed one dimension rather than four — but rather that the whole breadth of up-to-date experience be taken into account and that the explanation be sought which is simplest relative to the known phenomena. It often happens that for some partial domain an explanation A is simpler than B; but while A becomes increasingly complicated as the circle of experience widens, the same does not apply to B, with the result that eventually B emerges as the superior theory. Furthermore the required simplicity is not necessarily the obvious one, but we must let nature train us to recognize the true inner simplicity.

{The problem of simplicity is of central importance for the epistemology of the natural sciences. Since the concept of simplicity appears to be so inaccessible to objective formulation, it has been attempted to reduce it to that of probability, which has already been incorporated to a large extent into mathematical thought. If, for example, 20 corresponding pairs of values (x, y) of a functional connection $y = f(x)$, with the accuracy to be expected, lie on a straight line when plotted in a rectangular coordinate system, then a strict natural law will be surmised to the effect that y depends linearly on x. And this because of the *simplicity* of the straight line, or also because it would be so extremely *improbable* for the 20 points of observation to lie (nearly) on a straight line if the law in question were a different one. If one now uses the straight line for inter- and extra-polation, one arrives at predictions which go beyond the content of the observations. However, this analysis is open to criticism. Certainly functions $y = f(x)$ could be defined mathematically in many ways that are satisfied by the 20 observational data; among them such as will deviate considerably from a straight line. For each of these one might claim that it would be extremely improbable for the 20 observational points to comply with it if it did not represent the true law. It is thus essential, after all, that the function, or rather the class of functions, be held ready by mathematics *a priori* because of its mathematical simplicity. Here the class of functions must not depend on as many parameters as there are observations to be satisfied (e.g. the class of linear functions $f(x) = ax + b$ depends only on two parameters a, b, whose values may be fitted to the observational data). An important confirmation of the theory is obtained if it remains in accord with the facts which it was intended to explain even after the observational accuracy has been improved (and the number of observational points increased). An outstanding example is Euclidean geometry, which was proved by geodetic and astronomical precision measurements to be much more exactly valid than could have been conjectured on the basis of the experiences which had led to its erection. But this is far from being the only example of such a confirmation of the principle of simplicity. There is an abundance of similar cases in physics. Conversely it is a sure sign of being on the wrong scent if one's theory suffers the fate of the epicycles of Ptolemy whose number had to be increased every time the accuracy of observation improved. The three laws of Kepler were much simpler and yet agreed noticeably better with the observations than the most complicated system of epicycles that had been dreamed up. But Kepler's astronomical discovery would have been impossible without the Greek geometer's preceding discovery of the ellipses as a mathematically simple class of curves. Newton's law of attraction, especially in its formulation as a nearby action law, again is simpler than the Keplerian theory of planetary motion. The latter can be regained from the former if nothing but the attractive force of the sun is taken into consideration, while the 'perturbations' coming from the remaining planets are disregarded. And again, a splendid confirmation of Newton's 'Tieferlegung' of the theoretical foundation must be seen in the perfection with which the perturbations computed on the basis of his law have checked with innumerable observations, the accuracy of which has again been enormously improved since the time of Tycho Brahe. It should be added that the law of gravitation proved to be valid even outside the circle of experiences for which it had originally been designed, namely, for the motion of double stars about each other.

If experience has suggested a *hypothesis*, it is necessary to develop its consequences deductively, always with a view to inferring statements which are amenable to experimental test. Huyghens describes the method in the introduction to his *Traité de la lumière* (written in 1678, published in 1690): it differs greatly from geometry, he says, "because here the principles are confirmed by inferences drawn from them. . . . It is nevertheless possible to achieve a degree of probability which often is hardly inferior to a strict proof. In fact, this is the case if the consequences arrived at under the assumption of these principles are in perfect accordance with the phenomena known from experience; especially if their number is large, and all the more if new phenomena are designed and predicted that follow from those assumptions and if it is found that the result agrees with our expectation." He thus finds his wave theory of light confirmed by the discovery of the law of the double refraction of calcite. This is too complicated to be found purely empirically; but if the simplest assumption is made with regard to the propagation of light waves in calcite beyond that of a spherical wave, laws of refraction are obtained that are in accord with experience. It must be put down as a success of a theory if it reduces the complicated dependencies among directly observable quantities to simple relations among the fundamental quantities of the theory. Galileo's discovery of the law of falling bodies is based on a similar procedure.

"The essential function of a hypothesis," according to Mach (*Erkenntnis und Irrtum*, p. 237), "consists in the guidance it affords to new observations and experiments, by which our conjecture is either confirmed, refuted, or modified, by which — in short — our experience is broadened." "The seafarer, in whose imagination the objects thrown up by the ocean upon the beach create a vivid picture of the distant land, sets out to find that land. Whether his search will succeed or not, whether in place of the expected Indian or Chinese coast he discovers a new one, at any rate his experience has been widened" (*op. cit.*, p. 231).

For Galileo, Huyghens, and Newton, the deductive part still plays a much greater role than in modern times. Galileo is no less proud of the "abundance of theorems which flow from a single principle" than of the discovery of this principle itself (end of the third day of the "Discorsi"). The empirical attitude in physics has been accentuated progressively. The first great inroad was made by the discovery of electricity.⊦

Closely connected with the concept of simplicity is the category of *perfection*. It plays a considerable part not only as a methodical but also as an explanatory principle in Aristotelian philosophy. Thus Aristotle attributes the indestructibility and unalterability of the heavenly bodies to their perfect spherical form. Criticizing him, Galileo remarks in his "Dialogo," firstly, that from this point of view a deviation from the exact spherical shape by as much as a hair's breadth would be as inadmissible as one of the size of a mountain range. His sense of continuity revolts against the idea that, in nature, which permits no absolutely exact measurements, the exact value of a continuous quantity should confer properties upon the bearer that are basically different from those corresponding to nearby values no matter how close. Secondly he points out that even, say, a tetrahedron contains a sphere and that consequently only the residual corners of the tetrahedron could be destructible (although spheres may be inscribed to them too). He thus proves strikingly that for a property such as indestructibility it is not the geometrical form that matters but only the boundary surface across which the physical quantities (in this case the material density) undergo a discontinuous jump and which thereby may become the seat of special surface forces. (In fact, such capillary forces play a role in imparting spherical form to rain drops.) We here witness more clearly than anywhere else in the "Dialogue" the radical change in the interpretation of nature brought about by Galileian as opposed to Aristotelian thinking. Characteristic for Galileo's attitude is his exuberant praise of changeability in contrast to that crystalline perfection (Dialogo, *Opere*, VII, pp. 83–84); he points to the blossoming flower as something incomparably more magnificent than Aristotle's celestial bodies in their aloofness from all changes. In Kepler's work considerations of perfection still occupy a good deal of space. He is concerned with the "rank of the earth." Being convinced of the perfection of the circle, he has to go through a hard struggle before he gives up, as Brahe's measurements force him to do, the circular orbit of Mars. At first he still clings to static conceptions; he sees the harmony of the planetary system expressed in the regular Platonic solids. Only with effort he wins through to a more dynamic interpretation of the world.   Even Galileo, at a remarkable place in the "Dialogue," succumbs to the magic of deriving an explanation from geometrical perfection, when he bases upon the latter the circular (not straight!) path of purely inertial motion.   But on the whole he has already completed the turn-about much more decisively than Kepler.   He seeks perfection no longer in the fixed configurations and in the individual objects but in the dynamic relationships, the natural laws (which leave a large amount of play to contingency).   The notion of perfection is to him no longer a factual constituent of the theory, but it has become a heuristic principle, a belief which stimulates research.   "Kepler, Galileo, and Bruno," says Dilthey, "share with the ancient Pythagoreans the belief in a universe ordered by most perfect and rational mathematical laws and in divine reason as the source of the rational in nature, to which at the same time human reason is related."   On the long path of experience during the succeeding centuries this belief has always found new and surprising partial fulfilments, the most beautiful perhaps in Maxwell's theory of the electromagnetic field in empty space.   But again and again nature still proved itself to be superior to the human mind and compelled it to shatter a picture held final prematurely in favor of a more profound harmony.

Two strict requirements, according to Section 19, have to be made of any theory: (*i*) *concordance*, which implies consistency, (*ii*) the *absence of redundant* purely dogmatic *constituents*, which are without influence upon observable phenomena.   Furthermore, the principle of sufficient reason must never be violated.   In simple cases it may lead, as a principle of symmetry, to the establishment of definite laws. Thus it is used by Archimedes when he bases his theory of the lever on the theorem that equal weights attached to equally long arms of a lever are in equilibrium.   The entire configuration, including the gravitational direction, is transformed into itself by reflection with respect to the plane perpendicular to the horizontal lever at the point of support.   The notion of spatial similarity is the basis of the conclusion.   If a configuration of masses and forces, or a state uniquely determining the subsequent course of events, is mapped into itself by a similarity transformation, then the events must also be invariant with respect to this transformation.   For this reason the lever cannot lean to one side under the condition described above.   In conjunction with the general mechanical axiom that a balanced system remains in equilibrium if a balanced partial system is split off, Archimedes then derives from that special case the general law of the lever.

The same train of thoughts leads to the theorem that equal bodies have equal inertial masses; i.e. if they are propelled against each other with oppositely equal velocities, neither overruns the other. If this should happen nevertheless with two bodies of equal appearance, we infer a hidden inner difference. Although under unfavorable circumstances it might admittedly reveal itself only in the difference in mass, it would at any rate cause us to search for other differences in the physical behavior of the bodies. Frequently the principle of sufficient reason has been relied on to prove the law of inertia by inferring from it that the state of a body left to itself must remain unchanged. But what is meant by 'state'? Scholasticism interpreted it as position and thus believed that a body must remain at a standstill if not subjected to any outside influences. Galileo, on the other hand, construes it as velocity, both in magnitude and direction. Evidently only experience can decide which opinion is right. It must also inform us, in the cases mentioned above, of the 'relevant,' the determining circumstances. The argument of Leibniz in his controversy with Clarke and Newton over the relativity of motion (p. 97) is a typical example of the application of the principle of sufficient reason. Undoubtedly, however, its import as a source of factual truths has been grossly overestimated by Leibniz.

{Mach, who fights the *a priori*, the endeavour to turn, as he says, "the instinctive in science into a new mysticism and to hold it infallible," points out in his *Mechanik* (seventh ed., 1912, p. 27) that "even instinctive insights of such great logical force as the symmetry principle employed by Archimedes may be misleading. Many a reader may remember the intellectual shock at learning for the first time that a magnetic needle lying in the magnetic meridian can be deflected from the meridian by a current running parallel to the needle." However, the principle of symmetry is satisfied if we assume that a reflection with respect to the plane in which current and needle lie maps the current into itself, but interchanges the north and south poles of the magnet. Admittedly this view is possible only because positive and negative magnetism are inseparable and of equal nature. We form a theoretical conception of the nature of magnetism — namely that it is caused by molecular cyclic electric currents perpendicular to the needle — by which those facts are deprived of their astounding character, nay become necessities. }

Another guide of the theorist is the principle of *continuity*, first formulated in general terms by Leibniz. It rests upon the impossibility of proper division of a uniform continuum. It is scientifically unsound to exclude, as Euclid does, the null angle and the straight angle from the notion of an angle. Rest is not contradictory to

motion, but a limiting or special case of motion.   Leibniz says that by
virtue of that principle "the law of bodies at rest is, so to speak, only
a special case of the general rule for bodies in motion, the law of
equality a special case of inequality, the law for the rectilinear a sub-
species of the law for the curvilinear," and he calls manifolds "homo-
genous if one can be transformed into the other by a continuous
change" (Initia rerum Mathematicarum metaphysica, *Mathematische
Schriften*, VII, pp. 25, 20).   By means of the *lex continui* he disproves
the laws of impact which had been laid down by Descartes but formulated differently for a whole series of different cases.   In deriving
the law of inertia Galileo (Dialogo, *Opere* VII, pp. 171/174) starts
with the fall of a body on an inclined plane, for which he knows the law,
and then lets the inclination against the horizontal decrease to zero;
inertial motion thus is the limit of falling motion.   This origin makes
it understandable why Galileo, as it seems, recognized the law of
inertia in its classical form as true only for motions perpendicular to
the direction of gravity (an opinion with which one can agree in a sense
from the point of view of general relativity theory).   Mach (*Mechanik*,
p. 131) gives the following directive: "After having reached an opinion
for a special case, one gradually modifies the circumstances of this
case in one's imagination as far as possible, and in so doing tries to
stick to the original opinion as closely as one can.   There is no procedure which leads more safely and with greater mental economy to
the simplest interpretation of all natural events."   On the other hand,
in order to test an overall view tentatively adopted, it is common
practice in mathematics and physics to examine limiting and special
cases for which the results are pretty obvious.

The principle of *analogy* is closely akin to that of continuity.
Newton formulates it in the second of his rules concerning the study
of nature (*Principia*, ed. Cajori, p. 398): "Therefore to the same
natural effects we must, as far as possible, assign the same causes."
We meet the principle of analogy in perhaps its most significant
application in the establishment of the atomic theory.   The mechanical laws, which had been derived from the behavior of ordinary visible
bodies and had been most precisely confirmed by the planets, are
carried over to atoms.   One anticipates that the facts may later
enforce corrections, but without this preliminary adoption of the
mechanical laws no beginning of atomic research is thinkable.   Even
the most recent quantum mechanics of atoms, which deviates so
radically from tradition as to renounce any kind of a spatial picture of
the atomic events, still is based on the old mechanical laws in their
most transparent form, namely the Hamiltonian equations.   H. A.
Lorentz arrived at the fundamental electromagnetic laws of the theory

of electrons by taking the phenomenological Maxwell equations, which had been derived from observation and with which the electrical engineer works, and crossing out all quantities in which the influence of matter manifests itself in the form of material constants, such as conductivity, electrical polarization and magnetization. Under the assumption that the true 'microscopic' electromagnetic field obeys these simplified harmonic laws, in conjunction with certain ideas on the atomic structure of matter, he was able to obtain once more the old phenomenological laws for the macroscopic field by identifying the macroscopic field quantities with certain average values of the microscopic field quantities.

[The exact laws of nature must not contain any material constants; the latter should be derived from those laws on the basis of the atomic structure of the material under investigation. Since the phenomenological laws are apt to fail wherever the finer internal structure of matter is relevant, the atomic theory must at the same time disclose the limits of their validity and yield the atomic laws which, beyond these limits, take the place of the macroscopic laws. Thus Maxwell had assumed that the electric polarization is proportional to the field strength. This is correct for static and for slowly changing fields, and even for the fields of wireless telegraphy which carry out more than a million oscillations per second. But in the domain of the much more rapid optical oscillations we encounter the new phenomenon of dispersion; the proportionality factor taken as constant by Maxwell — that is, the constant of dielectricity, which equals the square of the coefficient of refraction — turns out to be dependent on the frequency of oscillation, and this according to laws which are closely connected with the atomic structure of the refracting medium and can only thus be understood. (In particular, charge and mass of an electron enter the dispersion formula in such a manner that one can derive from optical observations a definite value for their ratio.)]

What is the ultimate purpose of forming theories? H. Hertz describes the process as follows in his *Prinzipien der Mechanik* (p. 1): "We form images or symbols of the external objects; the manner in which we form them is such that the logically necessary (denknotwendigen) consequences of the images are invariably the images of materially necessary (naturnotwendigen) consequences of the corresponding objects." In the 19th century under the influence of sceptical epistemology it had become the fashion, especially among British physicists, to search only for images, for analogies covering

162

narrowly circumscribed domains of facts, and to construct mechanical models which rendered certain features of the phenomena in question but which could not possibly be taken seriously as 'explanations.' One suffered no longer from the 'delusion' of having to explore a uniquely determined reality. But the procedure proved to be singularly sterile as long as the only deliberate aim was the design of images and models. To Maxwell the physical analogies were expedients that avoid the disadvantages of a purely mathematical theory (which obscures the empirically important consequences) and of a physical hypothesis proper (which is apt to blind one to the facts).

⌈"By a physical analogy," he says, "I mean that partial similarity between the laws of one science and those of another which makes each of them illustrate the other." He mentions the analogy between gravitation and the stationary heat distribution in a medium — an analogy based on the fact that the Laplace equation holds for both processes — and confronts it with the analogy between light and the oscillations of an elastic medium. The latter "extends much farther, but, though its importance and fruitfulness cannot be over-estimated, we must recollect that it is founded only on a resemblance *in form* between the laws of light and those of vibrations. By stripping it of its physical dress and reducing it to a theory of 'transverse alternations,' we might obtain a system of truth strictly founded on observation, but probably deficient both in the vividness of its conceptions and the fertility of its method." (Maxwell, *Scientific Papers*, I, p. 156.) The example, especially in view of the further development of the theory of light inaugurated by Maxwell himself, very suitably illustrates the advantage of this standpoint, namely of affording protection against dogmatism.⌋

Mach speaks of a progressive "adaptation of thoughts to facts." The justification for the formation of theories he sees in the ensuing economy of comprehending and communicating facts and procedures (cf. *Mechanik*, Introduction). Others have adhered to the belief that reason is here at work, reason which strives according to immanent principles to construct symbolically its correlate, *transcendent reality*. Without this belief science, to them, seems an empty shell. But all are of one opinion as to the ultimate goal, the prediction of events. In how far do the economic principles or principles of reason, by which a theory comes about, guarantee the fulfilment of its predictions? This is a last fact, which points beyond knowledge — Hume's problem: the trust in induction, if it is to be justified, can only be justified by the principle of induction itself. But trust in the world and in oneself is in no need of justification; it is the natural attitude of the mind's life, especially as it manifests itself in thetic acts of reason.

Kant, in his transcendental logic, made the attempt to ascertain by a systematic procedure the aprioristic principles for the construction of empirical reality. His work deserves credit for elevating into philosophical consciousness the conception of reality which dominated the sciences since Galileo, for liberating it from the metaphysical ballast with which it was still loaded down by the Leibnizian system, and for safeguarding it against Hume's brand of sensualism that had grown out of the natural sciences. Yet the natural scientist will find it difficult to be satisfied with his attempt. What was stated by Kant is not nearly sufficient and is tied too closely to the particular form of contemporary physics; on the other hand it contains superfluous components, which got in only through the rigid logical schematism of "the great Chinese from Königsberg"[8] and his peculiar predilection for trichotomy. The ideas of *substance* and *causality*, to which the last section of this book is devoted, emerge as the really useful nucleus. Besides the two "analogies of experience," which refer to them, Kant places a third that deals with community (*Wechselwirkung*). It is preceded by the "Axioms of Intuition" ("All intuitions are extensive quantities") and the "Anticipations of Perception" ("In all phenomena, sensation, and the Real which corresponds to it in the object, has an intensive quantity, that is, a degree"). He follows up the first three groups by the "Postulates of Empirical Thought," which refer to the concepts of possibility, existence, and necessity. Kant's problem, for the solution of which a few fragments have been assembled here, remains open for the future, presumably as an infinite task. Kant, however, considered metaphysics, particularly as it strives for the solution of this problem, as "the only one of all sciences which, through a small but united effort, may count on such completion in a short time, so that nothing will remain to posterity" (Preface to the *Critique of Pure Reason*, ed. M. Müller, p. XXV).

REFERENCES

P. Duhem, *La théorie physique, son objet et sa structure*, Paris, 1906.
E. Mach, *Erkenntnis und Irrtum*, 1905.
H. Poincaré, *La science et l'hypothèse*.
E. Cassirer, *Substanzbegriff und Funktionsbegriff*, Berlin, 1910.
B. Russell, *Our Knowledge of the External World*, Chicago, 1914.
A. S. Eddington, *The Nature of the Physical World*, Cambridge, 1929.
J. Jeans, *The New Background of Science*, Cambridge, 1933.
P. W. Bridgman, *The Nature of Physical Theory*, Princeton, 1936.
K. Popper, *Logik der Forschung*, Vienna, 1935.

[8] Nietzsche's nickname for Kant; see e.g. *Jenseits von Gut und Böse*, 6tes Hauptstück, aphorism 210. [Translator's note.]

CHAPTER III

The Physical Picture of the World

22. MATTER

A. THE SUBSTANCE THEORY OF MATTER. The 17th and 18th centuries are dominated by what I should like to call the substantial conception of matter. The bodily thing contains an immutable substantial nucleus: it is the carrier of the changing sensory qualities that are inherent in the thing for our perception, but is itself unaffected by all these changes; "the continued body," Locke says, "that considered in any instant of its existence is the same with itself" (*Enquiry concerning Human Understanding*, second book, Chap. 27, §3). Because of this constancy, the changing sensuous phenomena must be *effects* on our sense organs caused by the *motions* of the substantial elements. The basic features of this conception go back to Democritus. In grandiose abstraction from sensory appearance he assumes as the only differentiation, from which all variety springs, the absolute distinction between the "empty" and the "full" — the $\mu\grave{\eta}$ ὄν of empty space as opposed to the $\pi\alpha\mu\pi\lambda\tilde{\eta}\rho\epsilon\varsigma$ ὄν of matter. That is the ultimate explanatory principle for the phenomena. At the beginning of the 17th century this theory of Democritus was revived by Gassendi. But also Galileo declares: "The variety exhibited by a body in its appearances is based on dislocation of its parts without any gains or losses. . . . *Matter is unchangeable and always the same, since it represents an eternal and necessary form of being.*" The decisive feature in this concept of substantial matter is that in principle the same substantial place can be recognized at any moment in the course of the history of a bodily system; it preserves its identity in time. The scientific justification of the concept, therefore, will depend on the development of exact methods by which in practice to follow a substantial place within the flux of movement. The four-dimensional world continuum appears dissolved into individual world lines, the world lines of the individual substantial places. This was the salient point whenever in physics a substantial medium was hypothetically introduced as the 'carrier' of certain phenomena, e.g. the ether in the mechanical theory of light. Thereby the possibility of objective differentiation between rest and motion of a body relatively to that medium was obtained.

But in a completely homogeneous substance without any quality, the recognition of the same place is as impossible as that of the same point in homogeneous space. For this reason Democritus's idea

165

necessarily leads to *atomism* and to the recognition of *empty space*. It is also their atomic constitution that explains the different density of bodies, their capacity of rarefaction and condensation — namely by a mixture of atoms and empty space in changing proportions of volume. A body occupies a certain portion of space; the total volume of that part of it that is 'covered' by atoms is to be set down as the mass of the body. The space which is required here is the Euclidean space with its rigid metrical structure and its 'far-geometric' relations. For all possible changes in the world must be temporal changes of spatial relations among the distant atoms. The atoms are *indivisible* and *rigid*, that is, they remain perpetually congruent with themselves. Moreover, they are *impenetrable;* the portions of space occupied by two atoms never overlap. *Solidity*, which includes impenetrability and rigidity, has been emphatically described, especially by Gassendi and Locke, as the basic feature of matter; as opposed to Descartes, in whose corpuscular theory the elementary bodies deform and pulverize one

Figure 5. Atom consisting of two separate parts.

another. Solidity must not be construed as the sensory property of hardness, for this would amount to excepting the qualities of the tactual sense from the subjectivity of sense qualities. Nor must it be construed dynamically as a firmness based on mutual forces of the substantial places. It is, according to its definition, an abstract geometrical property. The elastic firmness of the visible bodies is founded on this absolute property of the atoms. This point of view is defended by Huyghens, the mechanic, who thinks geometric-kinematically and in terms of principles, in his exchange of letters with Leibniz, the metaphysician, who thinks intuitive-dynamically. To be sure Huyghens himself speaks of a resistance against breakage or compression. But these terms, chosen for their greater expressiveness, must not be misunderstood; for "one must," as he says, "assume this resistance to be infinite, as it would seem absurd to ascribe to it a certain degree, say equal to that of a diamond or of iron; for no reason could be found for this in matter of which nothing but extension is presupposed. The hypothesis of infinite firmness therefore seems to me very necessary, and I fail to understand why you find it so strange, as if it introduced a permanent miracle." Possibly Huyghens would have understood the objections of Leibniz more easily if he had realized the following consequence of his 'substantial' viewpoint: even if the shape of an atom were not a connected portion of space, as indicated in Figure 5, it would always have to remain congruent to itself because of its geometric rigidity, the 'square' part would not be freely movable with respect to the 'circular' part; for 'God has willed' that this whole be a unit.

Regarding the *shape of the atoms* the spherical form is generally preferred as being the simplest. But protuberances in the form of hooks are also in vogue, by means of which the atoms supposedly cling to one another when they combine to form a solid body breakable only by force. The ideal solution would be an atomic shape of such a kind that all points in it are geometrically indistinguishable. For then we would have, on the one hand, the possibility of observing an atom as a whole during its motion, and, on the other, the impossibility of considering parts of the atom as remaining identical with themselves. The sphere is evidently chosen as the closest approximation to this ideal.

A mechanical atomistic explanation of the phenomena, reducing all processes to the motion of substantial particles, requires that the *laws of motion* of the atoms are known. First of all it must be ascertained how an atom moves freely when other atoms do not prevent it from penetrating into the adjacent portions of space. Secondly it is necessary to find out what effects the atoms exert upon one another, how their motions are modified when, in the state of contact, they are in one another's way. Epicurus considers the downward fall as free motion. Since Galileo, the fall in the field of gravitation is, of course, replaced by uniform translation in accordance with the law of inertia. Atoms act upon one another by '*impact.*' The latter, however, is not understood dynamically, the statement means nothing but that the movement of two atoms after their collision is determined by their movement before. Huyghens succeeded in establishing the relevant principles; they are the laws of conservation of energy and momentum, which are fundamental to the whole of physics. They determine the motion uniquely in conjunction with the assumption that an exchange of momentum occurs only in the direction perpendicular to the common tangent plane of the colliding atoms. "Thus the whole of natural science consists in showing in what state the bodies were when this or that change took place, and that, on account of their impenetrability, just that change had to take place which actually occurred" (Euler, *Anleitung zur Naturlehre*, Chap. VI, §50).

This is the *mechanical picture of the world* in its pure form. Euler (*op. cit.*, Chaps. 1–6) mentions as the fundamental properties of matter: extension, mobility, inertia, and impenetrability. In the concluding considerations of his *Opticks*, Newton says, "All these things being consider'd, it seems probable to me, that God in the beginning form'd Matter in solid, massy, hard, impenetrable, moveable Particles, of such Sizes and Figures, and with such other Properties, and in such Proportion to Space, as most conduced to the End for which he formed them; and that these primitive Particles being Solids, are incomparably harder than any porous bodies compounded of them; even so very hard, as never to wear or break in pieces; no ordinary Power being able to divide what God himself made one in the first Creation. . . . And therefore, that Nature may be lasting, the Changes of corporeal Things are to be placed only in the various Separations and new Associations and Motions of these permanent Particles."

Through Huyghens the atomistic substance theory had attained that degree of precision which made strict conclusions possible. As can be shown by statistical methods, spherical atoms of equal size which move according to the laws established by him form a body that has all the properties which we empirically associate with a *gas*. The manifestations of heat are due to the lively movements of the atoms. Huyghens' theory, however, has been incapable of going beyond the explanation of the gaseous state, and even in this respect it failed in one decisive point. For it was possible to derive from observation in combination with the mechanical theory rather reliable values for the magnitude of the radii as well as for the inert masses of atoms, which enter into the expressions for energy and momentum; and it transpired that for the various chemical elements the atomic masses are far from proportional to the atomic volumes. This shattered the basic conception of one matter, the conception of a homogeneous dough of substance out of which the Creator, with the help of a set of baking moulds, at the beginning of time had carved the little atom cakes, and had then given them absolute rigidity and sent them off into space with varying initial momenta. The mass ratios however, proved to be in accordance with the relative atomic weights, as derived from the actual quantitative analysis of innumerable chemical compounds. The law of multiple proportions, on which the atomistic interpretation of the findings of chemistry is based, was for a long time by far the most convincing empirical proof for the atomic constitution of matter.

B. Matter and Field. Ether. Beginning with Newton, dynamic conceptions enter the physics of substance. The main impetus to this development was given by his discovery of gravitation. At the place quoted above, at the end of his *Opticks*, he continues as follows: "It seems to me farther, that these Particles have not only a *Vis inertiae*, accompanied with such passive Laws of Motion as naturally result from that Force, but also that they are moved by certain active Principles, such as is that of Gravity, and that which causes Fermentation, and the Cohesion of Bodies. These Principles I consider, not as occult Qualities, supposed to result from the specifick Forms of Things, but, as general Laws of Nature, by which the Things themselves are form'd." Until the most recent times, various hybrid combinations of substance and dynamics were developed, but gradually the constructive dynamic properties of matter displaced its substantial ones and rendered them superfluous.

[Fundamentally, *mass* has already been introduced by Galileo as a dynamic coefficient appearing in the law of momentum; yet, along with it, the definition of mass as 'quantum of matter' stubbornly persists. Hardness and impenetrability of the atoms get replaced by the repulsive force with which they interact and by the law according to which this force depends on distance. Newton repudiates the hook-shaped atoms as an explanation which explains nothing and continues, "I had rather infer from their Cohesion, that their Particles attract one another by some Force, which in immediate Contact is exceeding strong, at small distances performs the chemical Operations above-mention'd, and reaches not far from the Particles with any sensible Effect" (*Opticks*, ed. Whittaker, p. 389). The atoms become "centers of force." Boscovich, Cauchy, and Ampère clearly profess the view that the centers are points in the strict sense. Kant in his *Metaphysische Anfangsgründe der Naturwissenschaft* constructs matter out of the equilibrium of attractive and repulsive forces. The purely mechanical interpretation of nature is replaced by the physics of central forces.[9] Berzelius first conceives the idea that the chemical affinity is of an electrical nature. Today we have succeeded to a considerable extent in explaining the structure of bodies and their elastic, thermic, electrical, magnetic, optical, and chemical behavior on the basis of the forces acting among the atoms. This applies in particular to the two extreme states of matter, the gaseous and the crystalline.

Modern physics speaks of the radius of an electron and ascribes to it a value of the order of magnitude 10^{-13} cm. This number, how-

[9] In opposition to Kant it must be said, though, that a decomposition of the uniform central force into two partial forces would be purely arbitrary unless the laws determining the two components in terms of distance each contained a parameter ('attractive' and 'repulsive' mass) which varied independently from one body to another. Thus electrical and gravitational force are separable because charge is not determined by mass. But since Kant only speaks of a single mass density (= intensity of fulness), supposedly arising from the equilibrium of repulsive and attractive force, his theory of matter hangs in the air.

ever, must be interpreted as the distance (entering into Coulomb's law of force) up to which two electrons approach each other if one is propelled against the other with a velocity comparable to that of light.]

With Newton, gravitation still appears as an instantaneous action into distance. When only nearby action is considered admissible, ether theories of gravitation arise, which at first however are still under the pressure of the purely mechanical interpretation of nature. Of course Newton too was aware of the difficulty, but declined to "frame hypotheses" about the cause of gravitation. (Apparently he thought of a non-material transmission by virtue of a 'spiritual substance' or of the all-penetrating space filled with the omnipresence of God.) The difficulty was overcome by physical means after Faraday had developed the idea of a *field* for the electric phenomena. Maxwell found that the field propagates from the centers of excitation not instantaneously but with the velocity of light. Nearby action laws, in the form of differential equations, connect the physical quantities characteristic of matter and field, namely charge and current densities and electrical and magnetic field strengths. The force, which with Newton is not an activity determined by and emanating from a single body k but a bond between two bodies k and k' which join hands across an abyss, is split up into an activity of k (excitation of the field determined by k alone) and a suffering of k' (temporal change of its momentum caused by that field). Between them the expanse of the field is spread out according to laws of its own of the utmost simplicity and harmony. The field transmits momentum as well as energy from one body to another; a radiating body not only loses energy, but as it radiates light in one direction it recoils in the opposite direction. In the field we therefore have spatially localized energy and momentum. The scalar densities and the components of the vectorial current densities of energy and momentum can be computed by means of simple laws from the two field strengths. The ponderomotoric effect of bodies upon one another is due to an exchange of field energy and momentum against kinetic energy and momentum of matter and vice versa; the increase or decrease in time of total energy or total momentum of any part V of the field is compensated by the current of energy or momentum going through the surface of V. If we determine the center of energy of a portion of space containing both matter and radiation, in the same way as we determine the center of gravity (mass center) of a 'ponderable' body, it turns out that the total momentum $\vec{I}$ contained in this portion has the same direction as the velocity $\vec{v}$ of the center of energy. If we set $\vec{I} = m\vec{v}$, the proportionality factor m may well again be called the *inert mass*. It is connected with the energy E by the universal relation $m = E/c^2$, where c is the velocity of light. A portion of a field such as the radiation in empty space enclosed by a massless shell (Hohlraumstrahlung) possesses inert mass like an ordinary body. Thus the strength with which a body, in the face of diverting forces, persists on its natural course as prescribed by the field of inertia depends on the energy compressed in the body. The mass of the electron certainly derives in part from the accompanying electromagnetic field. *Or even completely?* Since all physically important properties of an elementary material particle, as we have seen, belong to the surrounding field rather than the substantial nucleus at the field center, the question becomes inevitable whether the existence of such a nucleus is not a presumption that may be completely dispensed with.

This question is answered in the affirmative by the *field theory of matter*. According to the latter a material particle such as an electron is merely a small domain of the electrical field within which the field strength assumes enormously high values, indicating that a comparatively huge field energy is concentrated in a very small space. Such an energy knot, which by no means is clearly delineated against the remaining field, propagates through empty space like a water wave across the surface of a lake; there is no such thing as one and the same substance of which the electron consists at all times. Just as the velocity of a water wave is not a substantial but a phase velocity, so the velocity with which an electron moves is only the velocity of an ideal 'center of energy,' constructed out of the field distribution. According to this view, there exists but one kind of natural law, namely, field laws of the same transparent nature as Maxwell had established for the electromagnetic field. The obscure problem of laws of interaction between matter and field does not arise. This conception of the world can hardly be described as dynamical any more, since the field is neither generated by nor acting upon an agent separate from the field, but following its own laws is in a quiet continuous flow. It is of the essence of the continuum. Even the atomic nuclei and the electrons are not ultimate unchangeable elements that are pushed back and forth by natural forces acting upon them, but they are themselves spread out continuously and are subject to fine fluent changes.

[On the basis of rather convincing general considerations, G. Mie in 1912 pointed out a way of modifying the Maxwell equations

171

in such a manner that they might possibly solve the problem of matter, by explaining why the field possesses a granular structure and why the knots of energy remain intact in spite of the back-and-forth flux of energy and momentum. The Maxwell equations will not do because they imply that the negative charges compressed in an electron explode; to guarantee their coherence in spite of Coulomb's repulsive forces was the only service still required of the substance by H. A. Lorentz's theory of electrons. The preservation of the energy knots must result from the fact that the modified field laws admit only of one state of field equilibrium — or of a few between which there is no continuous transition (static, spherically symmetric solutions of the field equations). The field laws should thus permit us to compute in advance charge and mass of the electron and the atomic weights of the various chemical elements in existence. And the same fact, rather than the contrast of substance and field, would be the reason why we may decompose the energy or inert mass of a compound body (approximately) into the non-resolvable energy of its last elementary constituents and the resolvable energy of their mutual bond.

Besides the electromagnetic field, we have the metric or gravitational field as discussed in Section 16. The task of merging both into one unit arises. It has recently been attacked in different ways by Weyl, Kaluza, Eddington, and Einstein. At a certain stage of the development it did not seem preposterous to hope that all physical phenomena could be reduced to a simple universal field law (in the form of a Hamiltonian principle). ⎬

Geometry unites organically with the field theory; space is not opposed to things (as it is in the substance theory) like an empty vessel into which they are placed and which endows them with far-geometrical relationships. No empty space exists here; the assumption that the field omit a portion of the space is absurd. Just as in intuitive space extension and quality are tied to each other, so, in the field theory, the state quantities of the field or the field structure on the one hand, and its spatio-temporal medium, the structureless four-dimensional continuum on the other, depend on one another. If the latter is referred to coordinates, the state quantities appear as functions of the coordinates. But the concept of independent variable is correlative to that of function; as far as the range of existence of a function extends, so far the domain of variability of its arguments. (It should be noted here that the validity of the equation $E = 0$ in some portion of space does not mean that the electrical field E is interrupted in that portion, but merely that it is in the 'state of rest' there, which fits continuously into all other possible states.)

[Concerning the substance contained in a box, one may well ask what will happen when it is pumped out. The field, however, cannot be pumped out. Leibniz must have had something like this in mind when he refused to acknowledge the experiments of Guericke and Torricelli as proof for the existence of a vacuum, though he did so with the, at least in its wording, questionable argument that "the glass has minute pores through which the radiation of light and of the magnet and other very tiny particles [!] can penetrate" (Leibniz's fifth letter to Clarke, §34).]

While according to Democritus the distinction of full and empty forms the basis of substance theory, any field theory is founded on certain state quantities spread out in the four-dimensional space-time continuum. The laws of motion of the substance are replaced by differential equations (of simple build) in which, apart from the values of the state quantities at an arbitrary place, the derivatives of the latter with respect to the four world-coordinates appear. These are the field laws, which, in view of their objective significance, must be independent of the choice of the coordinate system.

The explanation of ponderomotoric effects through the propagation of energy and momentum in a continuous field arose in closest contact with experience, and today this conception permeates the whole of physics. It seems scarcely probable that this factor will again disappear one day from the description of nature, closely as it is tied up with the space-time continuum and its metrical structure. On the other hand, the pure field theory is hypothesis and program; in spite of its highly attractive features, the great hopes it once raised, and its development by men like D. Hilbert, M. Born, and others, it has remained in the limbo of speculative physics. But its discussion led to investigations from which the fortunate result emerged (Weyl, Einstein and Infeld) that the decisive facts concerning the interaction of the discrete material particles and the continuous field can be accounted for without commitment to any premature hypothesis about the inner structure of the particles. Proceeding in this way we reestablish the duality of field and matter. Their connection is a dynamic one; matter excites the field, the field acts upon matter. If less attention is paid to the connecting medium of the field, then matter and force appear as the interdependent constituents of the world. Helmholtz formulates this viewpoint as follows: "Science considers the objects of the external world according to two kinds of abstraction: on the one hand, according to their mere being, irrespective of their effects on other objects or on our

173

sense organs; as such, it calls them matter." On the other hand, we attribute to matter the capacity to act, for only through its effects do we know it. "Pure matter would be irrelevant for the rest of nature, because it could never cause a change in it or in our sense organs. Pure force would be something which should be there, and yet is not because being-there we describe as matter." F. A. Lange, in his *Geschichte des Materialismus*, takes a more critical view of matter and describes it as "the uncomprehended or incomprehensible residue of our analysis."

We saw that the Newtonian physics is entirely dominated by this dualism. The classical philosopher of the dynamical conception of the world, however, is Leibniz. To him, what is real in motion does not lie in the change of position as such, but in the moving force. *"La substance est un être capable d'action, une force primitive"* — transspatial and immaterial. "For not all truths relating to the world of bodies can be derived from merely arithmetical and geometrical axioms, that is, from axioms of larger and smaller, of shape and position," he says in criticism of Descartes (*Mathematische Schriften*, VI, p. 241) "but others must be added concerning cause and effect, activity and passivity, in order to give an account of the order of things." The ultimate element is the *monad*, an indecomposable unit without extension, from which the force bursts forth as a transcendental power. Only with regard to the distribution of the monads in space, which itself is merely a *phaenomenon bene fundatum*, is the body described as an *extended* agent. Pure activity, however, is all; preestablished harmony takes the place of such reciprocal effects as we think are carried by the field from particle to particle. Fichte, too, recognizes, apart from the sensation of qualities and the intuition of extension, active *thought* that, connecting them both, posits the *thing* as a *force* and thereby as the cause of my being affected (Bestimmung des Menschen, *Werke*, ed. Medicus, III, pp. 332, 333). Experiences of fundamental character seem to speak very distinctly in favor of another kind of causality than would fit into the framework of field theory, namely, that the field if left to itself would remain in a homogeneous state of rest and that something alien to it, matter, is the 'spirit of unrest' that excites it. Our voluntary acting must primarily always attack matter. The field is an extensive medium which transfers effects from one body to another by virtue of its structure expressed in the field laws.

Without making any assumptions about the inner structure of a particle we may derive its dynamically relevant properties from the local field around it. For instance, the field-generating or active charge of a particle may be defined as the flux which the electric field sends through a tiny imaginary shell surrounding the particle. The latter describes a narrow channel of one-dimensionally infinite extension in the four-dimensional world. Outside these channels we apply the classical field laws *in empty space*. For historical reasons presently to be explained these differential equations that treat of no other state quantities than the electromagnetic and gravitation potentials are called the *laws of ether*. It is irrelevant whether the particle is an actual singularity of the field or covers a small region where the laws of ether are suspended (and unknown laws take their place). The local field is uniquely determined by the particle, at least as far as the *nature* of the field is concerned, i.e. such characteristics of it as are invariant under coordinate transformations. In this respect the 'monad' preserves its pure activity and independence of anything extraneous. The particle suffers reactions from the field merely as far as its *orientation*, the embedment of the local field into the external field, is concerned. Indeed the mechanical laws of motion follow from the fact that the individual field of the electron must fit into the field distribution outside the particle that obeys the field laws of ether. Thus we can understand that charge and mass, which excite the field, at the same time appear in passive function and determine the intensity of the effect which a given field exerts upon the particle. It is no longer the energy content but the flux of the gravitational field which a particle sends through an enveloping shell that accounts for heavy mass and thus also, according to general relativity theory, for inert mass.

[The strength of this *ether theory* lies in its sober noncommittal attitude; it studies matter by its effects without attempting to penetrate into its interior. Speculation is tempted to fill the 'lacuna' left by the particle. Pure field theory of matter does it in one way; another is suggested by general relativity theory, for the latter makes it possible to entertain the hypothesis that the grooves of the elementary particles are bottomless, without forcing one to conceive of the particles as actual singularities in the space-time manifold. (I speak here of the channels in the four-dimensional world as if they were grooves in a two-dimensional surface.) Indeed general relativity does not prescribe the topology of the world, and it may therefore happen that the world has unattainable 'fringes' not only toward the infinite but also inwardly. In line with Leibniz's ideas, the material particle, although imbedded in a spatial environment from which its field effects take their start, would itself then be a *monad* existing beyond space and time. Hence one may not say, 'Here *is* a charge,' but only, 'This closed surface within the field *surrounds* a charge.' The inner fringes would be the geometrico-physical basis for the splitting of the world into space and time which takes place within our consciousness, tied as it is to a material body.

Schelling, partially under the influence of Leibniz, has expressed ideas which vaguely anticipate this development. "Thus there ought to be discernible in experience something," he says on p. 21 of his "Erster Entwurf der Naturphilosophie" (1799; *Sämtliche Werke*, III, p. 21, Cotta, 1858) "which, without being in space, would be principle of all spatiality." This "natural monad" is not itself matter but action, "for which there is no measure but its own product." Based on the thesis that "the striving of all original tendencies is toward the filling of space," he then arrives at the construction of a shapeless fluid — which we today would replace by the field.]

The naive substance theory has been followed in history by the much more refined ether theory of matter, of which we count the pure field theory and the monadic theory as two hypothetic variants. Here is the place to sketch in a few strokes the *history of the ether concept*. It can be traced back to stoic philosophy. During the epoch following Galileo the ether appears as the substantial carrier of light and gravitation, so with Huyghens and Euler. The state quantities characteristic of the ether as well as of any substance are 'density' and 'velocity.' Since it rests as a whole and is only excited into minute oscillations, it can serve at the same time to support Newton's metaphysical concept of absolute space by a physical (though hypothetical) reality. Then, as the optical phenomena subordinate themselves as a partial domain to the electromagnetic ones and the conception of electromagnetic field is developed by Faraday and Maxwell, the ether is divested of its substantio-physical character, and nothing remains but the absolute space as the medium of electromagnetic field states. It is no longer subject to excitation by matter but has become a rigid geometric entity. As a third step it is shown by the special relativity theory that the spatio-temporal structure is described incorrectly by the notion of absolute space. Not the state of rest, but the states of uniform translation form an objectively distinguished class of motions, and this puts an end to the substantial ether. Finally, and fourthly, the general relativity theory re-endows this metric world structure with the capacity of reacting to the forces of matter. Thus, in a sense, the circle is closed. Since the electromagnetic field is evidently of the same nature as the metrical field which among other things causes the gravitational phenomena, 'ether' has now become synonymous with 'field,' in the sense of a unified electromagnetic and metrical field in empty space. The state of the ether is known if, with reference to a coordinate system, the quantitative distribution of the electromagnetic field components and the Einsteinian gravitational potentials g_{ik} are given by mathematical formulas. If the ether has thus regained its physical nature, nevertheless the state quantities characteristic of it have changed completely from what they were at the beginning of the development when it entered the scene as a substantial medium.

C. Historical Connections, in particular with the Metaphysical Concept of Substance. The relation of matter to the concept of substance is clearly evidenced by the way in which Kant in the first edition of the *Critique of Pure Reason* formulates the first analogy of experience, the 'Principle of Permanence': "All phenomena contain the permanent [substance] as the object itself, and the changeable as its determination only, that is, as a mode in which the object exists." In the comments he says, "A philosopher was asked: 'How much does smoke weigh?' He replied: 'Deduct from the weight of the wood burnt the weight of the remaining ashes, and you get the weight of the smoke.' He was therefore convinced that even in the fire matter [substance] does not perish but that its form only suffers a change." The reference to quantitative measurement is even more stressed in the second edition: "In all changes of the phenomena the substance is permanent, and its quantum is neither increased nor diminished in nature." In the example quoted, the weight is assumed proportional to the quantum, but no indication is given of the principle according to which matter is to be measured. In this form the thesis of the indestructibility of matter was introduced by Lavoisier into chemistry. Wherever the individual substantial places can no longer be traced, transmittability is the criterion of substantiality. A measure of quantity must be found according to which the transmitted quantum does not change. In this sense energy may also be looked upon as substance (hypothesis of a heat substance), although there can be no question of tracing a single 'energy place' through the course of events. Hobbes considers incorporeal substance a word without meaning.

Yet the idea of substance is not so closely tied to that of physical matter as might appear from these quotations. It has its origin in the logic and metaphysics of Aristotle, and is used in a metaphysical sense by Descartes, Spinoza, and Leibniz. Today we find it difficult to grasp its meaning. Descartes defines (*Principia*, Part I, §51): "By substance we cannot understand anything but a thing which exists in such a manner that it requires no other thing for its existence." He then modifies the definition so that "created substances" also fall under it, by saying that "they require nothing for their existence but the assistance of God." Aristotelian philosophy considers matter ($\H{v}\lambda\eta$, $\tau\grave{o}$ $\H{v}\pi o\kappa\epsilon\acute{\iota}\mu\epsilon\nu o\nu$) as the determinable, in contrast to the determining form ($\epsilon\H{\iota}\delta os$). Matter is possibility of becoming formed. In a production process of several stages its matter appears 'more formed' at each step, and thereby the range of possibilities for further forming becomes more restricted. At the same time that matter, the component of merely potential rather than actual being, shrinks more and more. Substantiality is ascribed to the forms rather than to matter. The forms push matter from potentiality to actuality; the transition itself takes place in 'movement.' Natural science as conceived by Galileo had first of all to subdue this metaphysics of substantial forms. Leibniz reintroduced them through his monads, yet without wishing thereby to abandon the new 'mechanical' way of explanation. "However much I stand on the side of scholasticism in the general and, so to speak, metaphysical explanation of the principles of the physical world, I am, on the other hand, the most radical adherent of the corpuscular philosophy with regard to the explanation of particular phenomena" (to Arnauld, *Philosophische Schriften*, II, p. 58). "My opinion thus is to the effect that bodies, which are commonly considered substances, are nothing but real phenomena and are as unsubstantial as a mock sun or a rainbow. . . . The monad alone is substance" (to de Volder, *Philosophische Schriften*, II, p. 262). Admittedly "there are no material particles in which monads are not present" (to Bernoulli, *Mathematische Schriften*, III, p. 538). "It does not matter whether we denote this principle as 'form,' as '$\grave{\epsilon}\nu\tau\epsilon\lambda\acute{\epsilon}\chi\epsilon\iota\alpha$,' or as 'force.'" The essence of the monad he sees in the *law*. "That a certain law persists that includes all future states of the subject which we conceive as identical, this fact constitutes the identity of substance" (to de Volder, *Philosophische Schriften*, II, p. 264). That puts him beyond Aristotle. Most characteristic for a philosophy of nature is probably the point at which it lets the Heraclitean flux "sich zum Starren waffnen":[10] Aristotle in the immanent substantial forms, Plato in the transcendental ideas, modern natural science, like Leibniz, in the law. At the end of Newton's *Principia* (ed. Cajori, p. 547), immediately after his declaration that it suffices to know the laws of gravity and that he has no intention of devising a hypothesis as to the cause of these properties, we find the following strange words: "And now we might add something concerning a certain most subtle spirit

[10] One of Goethe's inimitable phrases, taken from the poem "Eins und Alles," verse 3:

> "Und umzuschaffen das Geschaffne,
> damit sich's nicht zum Starren waffne,
> wirkt ewiges, lebendiges Tun."

[Translator's note.]

178

which pervades and lies hid in all gross bodies; by the force and action of which spirit the particles of bodies attract one another at near distances, and cohere, if contiguous; etc."

Descartes subscribed to the doctrine, to which Plato also inclined,[11] that spatial extension is the proper substance of bodies. It fits well within the framework of the field theory, provided the contrast between substance and accident is construed as that between 'this' and 'thus.' The 'this,' which can only be given by individual exhibition but not by qualitative characterization, is here not a hidden carrier to which the qualities are inherent; it is the here-now, the individual spatio-temporal position. To use Hilbert's term, the description of the world according to the field theory consists of the here-thus relations — the here being represented by the space-time coordinates, the thus by the state quantities. If the latter are given as functions of the former, then the course of the world is completely known.

[Descartes' concept of motion seems to presuppose a substance (in the sense of Part A of this Section) which can be followed through its motion. His physics is corpuscular theory, but his corpuscles may not leave any empty spaces between one another and thus must grind and deform one another. Only his helplessness in the face of the continuum causes him to think of the discontinuities along the separating surfaces as essential for comprehending the motion. In fact, he has as the carrier of motion a fluidum which fills space continuously. For this fluidum one could, if one wished and as was actually done in later theories of matter, assume the laws of incompressible non-viscous liquids, with the modification that the dynamical variable of pressure should be eliminated from the hydrodynamical equations. This causes no difficulties. Once this is done, the substantial medium can also be discarded. One merely abstains from interpreting the vectorial state quantity $\vec{v}$ with which the differential equations deal as velocity of a substance. Thus if Descartes's basic idea is carried through consistently, a field theory results. As soon as the transition has been made from a moving substance to the spatio-temporally distributed field which no longer requires a material carrier, such theories suggested by hydrodynamics cease to have any intuitive advantage over Maxwell's field theory, which bases the choice of the state quantities to be employed on experience rather than on speculation.

[11] Compare *Timaeus*, 48 E ff.: Between the "eternal pattern" and its "imitation" in reality, which he had distinguished earlier, it would be necessary to place something which "is the recipient and, in a manner the nurse, of all generation." This third, "which like a plastic mass lies in readiness to take the imprint of anything," is *space:* inaccessible to the senses, not subject to perdition, but granting a place to all that comes into being.

Two of the most important *roots of the concept of substance* seem to be the following: (*i*) the *thing* of the external world, which is fitted as a stable factor into our world of causes and effects and which in spite of varying appearances and aspects remains the same or undergoes a familiar slow change; (*ii*) *I*, who am conscious of my identity throughout the flow of my life with its everchanging kaleidoscopic experiences. (Cf. Leibniz, *Philos. Schriften*, VI, p. 502: "Since I now see that other beings too may have the right to say 'I,' or that one may say it for them, I understand what is generally meant by substance.") Descartes, for the first time, clearly formulated the philosophical problem as to what "the wax itself" is (end of second meditation), which is different from anything within the domain of senses and remains the same in spite of any changes which taste, smell, sight or feeling may convey to me. And he finds that this cannot be in my imagination, but that I can only apprehend it by thinking. More critical are Locke's remarks about "our idea of substance in general" (*Enquiry concerning Human Understanding*, second book, Chap. 23, §2). Hume considers it altogether a misconception: "Our propension to confound identity with relation is so great, that we are apt to imagine something unknown and mysterious, connecting the parts, beside their relation" (*Treatise of Human Nature*, Book I, Part IV, sect. 6), " . . . which view of things . . . obliges the imagination to feign an unknown something, or *original* substance and matter as a principle of union or cohesion among these qualities, and as what may give the compound object a title to be called one thing, notwithstanding its diversity and composition" (*ibid.*, sect. 3).}

D. CONSERVATION THEOREMS. In view of the present state of physics, those who want to retain an aprioristic principle of conservation are liable to cling to the principle of conservation of *energy*. According to the special theory of relativity, energy is one, namely the temporal, component of an invariant objective entity, a four-vector whose spatial projection is momentum. The conservation theorems of energy and momentum therefore belong together inseparably.

{The momentum of a body moving with the velocity $\vec{v}$ had been equaled to $m\vec{v}$ and, with Galileo, we had called m the inert mass. The question arises how this inert mass depends on the velocity of the body if the velocity changes while the internal state — as judged by an accompanying observer undergoing the same motion — remains the same. The answer can be obtained from the special relativity principle, but differs according to what the causal structure of the world is

180

assumed to be. If, in line with the old view, the structure consists in a stratification $t = $ const., then the mass is independent of velocity. If however, as unquestionably is the case, it is described by the light cones, then we have

$$m = \frac{m_0}{\sqrt{c^2 - v^2}},$$

where v is the absolute value of the velocity, c is the velocity of light, and the 'mass factor' m_0 is independent of the velocity. On the basis of the relativity principle, the theorem of the conservation of energy follows from the law of momentum, and the energy of a body turns out to be $E = mc^2$. If, for instance, the energy content E of a body is increased by heating, its inert mass m increases proportionally. Not only a massive body but also a gas, consisting of whirling molecules, and even an arbitrary portion of a field possesses a certain energy E and a certain momentum $\vec{I}$. And again the laws

$$\vec{I} = m\,\vec{v}, \quad E = mc^2$$

hold, provided we understand by $\vec{v}$ the velocity of the energy center. Energy appears here as the absolute energy level of a portion of space at a given moment. It is uniquely determined by the physical state prevalent in that portion of space. }

The *phenomenological law of energy*, as it has historically emerged independently of the conservation principle for momentum, deals only with the *difference* of such energy levels, i.e., with the energy value attaching to the change from one physical state to another, $Z \to Z'$, of a given fixed system of bodies. The energy value E of the transition $Z \to Z'$ depends on the initial and terminal states, Z and Z', in such a manner that

$$E(Z \to Z') + E(Z' \to Z'') = E(Z \to Z'')$$

always holds. Leibniz proved the energy law on the basis of the axiom *causa aequat effectum*, by transforming every mechanical change of state, the cause, into a 'standard effect' with but one degree of freedom, namely, the lifting of a given weight. The lifting height is then taken as measure of the energy. Leibniz's idea really strikes at the root of the energy principle. And it will carry over to all natural phenomena as soon as the lifted weight is replaced, say, by a water calorimeter. (This generalization, however, was not conceived before the middle of the 19th century.) Indeed a non-mechanical change of state may not always be transformable into a mechanical

181

effect such as the lifting of a weight, but it is always transformable into the heating or cooling of a given standard body. The empirical facts on which the energy law is based may thus be formulated as follows: Let S be any system of bodies in which the interaction of its parts as well as the effect of arbitrary other bodies has brought about a change of state V. By connecting S with a water calorimeter at rest and with suitable auxiliary bodies we can undo this change of state in such a manner that the auxiliary bodies emerge from the process in the same state and that only the calorimeter has undergone a change of temperature (fact A_1). If the heating (or cooling) of the calorimeter has consumed w calories, i.e. if the temperature of w ccm. of water under atmospheric pressure has risen from 15°C to 16°C (or if the temperature of $-w$ ccm. has fallen from 16°C to 15°C in case w is negative), then w is the energy measure of the change V. The same value w is obtained no matter through what processes that transformation is brought about or what auxiliary bodies are employed (fact A_2).

{By virtue of the relativity principle this leads to the following consequences: (i) With every body there is associated a number m_0 dependent only on its internal state, such that the energy value of an arbitrary change of state of that body equals the difference of the values of

$$E = \frac{m_0 c^2}{\sqrt{c^2 - v^2/c^2}}$$

for the initial and terminal states. (Here, the energy differences associated with changes of state again lead to absolute energy levels of states.) (ii) Beside the energy law we have the law of momentum, the momentum being given by the following expression:

$$\vec{I} = \frac{m_0 \vec{v}}{\sqrt{c^2 - v^2/c^2}}\}$$

In a systematic treatment the energy law, of course, has nothing to do any longer with the assumption that any change of state can be transformed into a change of temperature of the standard body. In the framework of general relativity theory, the conservation theorems for energy and momentum are closely connected with the *invariance of the field laws* under arbitrary coordinate transformations. Their validity is largely independent of the particular form of the field laws of interaction. Even so, no aprioristic command would prevent physics from abandoning the strict validity of the conservation laws if that should become necessary under the pressure of new empirical discoveries. (This actually happened recently in a theory of Bohr and Slater, which however was soon abandoned again.)

[The derivation of the energy principle is somewhat surprising inasmuch as from the sole assumption that there is something conservative the measure of this something, namely the energy, is obtained. The explanation is that the basic empirical fact A_1 implies the assertion that there is altogether only one quantity in nature for which the conservation principle can hold, as long as we are concerned with changes of a given system of bodies. But the law of momentum, which follows from the energy law itself, gives that assertion the lie. Yet we are able to understand in retrospect why A_1, in spite of not being strictly valid, is found to be confirmed within the limits of accuracy attainable in experience — because of the earth, whose mass is vastly in excess of that of the reacting bodies. For in reversing a given change of state V that involves an increase of momentum in a preassigned direction, one can transmit that increase to the earth, which serves as the resting body of reference for terrestrial experiments. If a bullet is shot horizontally into a mountain we do not see what becomes of the lost momentum; experience in this case appears to invalidate the law of momentum. What we observe, is that the loss of 'kinetic' energy due to the deceleration of the bullet is transformed into thermic energy. Rigorously speaking, even the latter does not represent the full equivalent, for there is in addition the hidden kinetic energy associated with the transition of the earth from rest to the small speed imparted to it by the impact. However this kinetic energy is negligible to the same degree as is the mass of the bullet in comparison to that of the earth.

Another quantity for which a conservation law holds is the electric charge. According to the theory of electrons, charge is bound to matter, and consequently the assumption that the change V takes place in a fixed system of bodies excludes electrical discharging and recharging if rigorously interpreted. And this restriction is actually necessary for A_1 to be valid.]

Field physics extends the conservation laws of energy and momentum to radiation and thus liberates them from any tie to a definite system of bodies. Instead it must take into consideration the energy entering and leaving a given bounded portion of space by assuming an energy current (just as one needs an electrical current in addition to electrical charge). If a spatial domain D is divided into partial domains, D_1 and D_2, the total electrical charge contained in D equals the sum of the charges contained in D_1 and D_2 (law of addition). The same holds for energy and momentum. There is a difference, though, from the atomistic viewpoint, inasmuch as the field spread out between the elementary particles is free from charge but not from energy-momentum. The 'mass factor' (the 'length' of the four-dimensional energy-momentum vector) is neither dependent on the choice of coordinates, like energy, nor capable of both signs $+$ and $-$, like electrical charge. Rather it is invariant and always positive. Nevertheless it is unsuitable as a measure of quantity because it fails to satisfy the law of addition. (Just as the length of a side in a triangle is not equal to but less than the sum of the lengths of the other sides, the mass factor of a domain D is greater than the sum of the mass factors of D_1 and D_2.)

E. ATOMISM. The atomic theory originally arose from pure speculation in answer to certain epistemological requirements. Chemistry, which gave it a firm foundation and a strong empirical support by the law of multiple proportions, was interpreted atomistically by Dalton. The atoms of a chemical element must be all equal, for otherwise the constancy of the physical properties of the element would be incomprehensible. Within such a swarm of equal atoms, the identity of an atom and its discernability from other atoms cannot be warranted by its particular internal properties and the particular laws holding for them, as Leibniz had maintained with regard to the monads, but only by the continuity of motion together with the spatial separation of the atoms. Does it not in this respect resemble the ego, which also is able to maintain its identity as an individual and its distinctness from other egos no matter whether the total of its experience is completely like that of the others?

From the standpoint of a consistent substantial theory of matter there is no reason to see why, among the infinite continuous manifold of substantial spheres with all possible radii, just those few discrete possibilities are realized which correspond to the chemical elements; the mass however should be determined by the radius. We have seen before that experience is completely at variance with this requirement. The ether theory, on the other hand, imposes no restriction upon charge e and mass m of a body; here there is no collision with experience. Yet again it remains unexplained why of all these possibilities but a few are realized *for the elementary particles*. Only the pure field theory holds out some hope that it might be able to explain this basic fact. For it could happen that its (non-linear) field laws were such as to possess no more than a discrete number of regular static spherically symmetric solutions.

It is now sure that no fruitful attack on the problem could have been made before the discovery of the *electron*. Here the physicists, going beyond chemistry, had laid hands on an elementary unit of matter which is the same in all chemically different substances and occurs freely in the form of cathode rays. The 'periodic system of elements' had previously pointed, though with a somewhat vague gesture, toward a uniform structure of the various chemical atoms. But now physics entered into its golden era of atomic research. During the last half century it has provided a thorough and brilliant corroboration for the basic tenets of atomism and penetrated into ever deeper layers of the strange atomic world. To begin with, all its methods led with increasing accuracy to the same values of the charge and mass of an electron. Only through this concordance has atomistics become a well-founded physical theory. Gradually indirect methods have been replaced by more and more direct ones. Thus the Brownian motion of small suspended particles demonstrates directly to our senses the presence of a molecular thermic motion. Through ingeniously arranged experiments one has succeeded in isolating the effects of individual atomic events. Of the greatest consequence was the discovery of what we now consider the most fundamental atomic constant, Planck's *quantum of action h*. It first disclosed its existence in the thermodynamics of radiation, hence by a statistic effect depending on the disorderly cooperation of a huge number of atomic events. Planck saw himself forced to assume, contrary to classical physics, that a linear oscillator emits light of frequency ν not during its continuous oscillation but by a discontinuous 'jump' in which it loses the energy $h\nu$. Niels Bohr applied this principle to the electrons in an individual atom. By letting the frequencies thus obtained 'correspond' to the frequencies derived from the classical theory of radiation he got an approximate rule for the computation of the atomic energy levels. Thus the key was manufactured that unlocked the secret of the amazing regularities governing the series of the spectral lines which are emitted by radiating atoms and molecules. The success was most striking in the simplest case, that of the hydrogen atom. "Our spectral series, dominated as they are by integral quantum numbers," says Sommerfeld (*Die Bedeutung der Röntgenstrahlen für die heutige Physik*, Munich, 1925, p. 11) "correspond, in a sense, to the ancient triad of the lyre, from which the Pythagoreans 2500 years ago inferred the harmony of the natural phenomena; and our quanta remind us of the role which the Pythagorean doctrine seems to have ascribed to the integers, not merely as attributes, but as the real essence of physical phenomena." Thus we see a new quantum physics emerge of which the old classical laws are a limiting case, in the same sense as Einstein's relativistic mechanic passes into Newton's mechanic when c, the velocity of light, tends to ∞.

The old dream of the *unity of all matter* had certainly come a great step nearer to fulfillment by the discovery of the electron. But the positively charged nucleus of the atom about which the negatively charged electrons revolve like the planets around the sun still seemed to be a particle with a constitution of its own for each individual chemical element. It is *a priori* clear that beside the negative electron at least one positive brick is needed for the construction of all atoms. The execution of this idea of building up matter from two ultimate elementary units, the electron and the proton, presupposes of course that they are able to enter into a variety of combines that are held together by strong forces and react outwardly like solid atomic spheres. Hence sheer substance without force would never do.

[The proton is identified with the nucleus of hydrogen. The various chemical elements differ from one another by the charges of their nuclei, which are integral multiples ne of e, where $-e$ denotes the charge of the electron. If the atom is in an electrically neutral state ('non-ionized'), the factor n, called the order of the atom, coincides with the number of electrons revolving around the nucleus. There is no gap in the sequence of the orders of the various elements: $n = 1$, hydrogen; $n = 2$, helium; $n = 3$, lithium; . . . . Elements may change into one another by nuclear emission or absorption of elementary particles. In the radioactive elements, which thus betray their instability, this process goes on spontaneously, but by bombardment with elementary particles of sufficient energy all sorts of artificial nuclear transmutations have been effected. Aston found that also the *masses* of the atomic nuclei are, at least to a considerable approximation, integral multiples of the mass of the proton. If this did not become apparent in the atomic weights obtained by chemistry, the reason is to be seen in the fact that different atomic structures may belong to the same order number ('isotopes'), and that what chemists used to consider as a pure element frequently turns out to be a mixture of isotopes of different atomic weights; for these cannot be segregated by ordinary chemical means. These findings seem to corroborate the assumption of only two elementary particles, and indeed in 1926 there was no clear evidence of any other. But since then new elementary particles have made their appearance, and we now have a whole gamut: electrons, positrons, protons, neutrons, neutrinos and mesons of several kinds. Their charges are zero or $\pm e$, but quantum theory has not yet succeeded in reducing their masses to the mass of the electron or in explaining the several particles as different quantum states of one universal particle. In particular, the atomic nuclei seem to consist of densely packed protons and neutrons (two particles which have equal or nearly equal masses).⎬

The composition of the material world out of one or a few units, existing in a huge number of completely alike specimens, must surely be looked upon as one of the most fundamental features in the nature of the universe; and one that is most profoundly in need of interpretation.

⎨While this remains a task for the future there is a decisive progress to record in quantum mechanics which occurred while the MS of this book was in preparation.[12] W. Heisenberg succeeded in replacing the correspondence principle, a somewhat vague and flexible prescription that could never claim the rank of a theory and in its working had become more and more ambiguous and unsatisfactory, by a complete, simple, and consistent formulation of the quantum mechanics of arbitrary atomic systems or, what is the same, by a definite rule for the computation of its energy levels. In making the necessary modifications of classical mechanics Heisenberg was guided by a universal principle that had been abstracted from the vast empirical material of spectroscopy, the so-called combination principle of spectral lines. Even so the result compels us to abandon any spatio-temporal picture of the atomic processes. Many facts, as Bohr explains in a beautiful and generally informative article on Atomic Theory and Mechanics (*Naturwissenschaften*, 1926, p. 1; English version in *Atomic Theory and the Description of Nature*, Cambridge, 1938) have driven him and other physicists to the conviction "that, in the general problem of quantum theory, one is faced not with a modification of the mechanical and electrodynamical theories describable in terms of the usual physical concepts, but with an essential failure of the pictures in space and time on which the description of natural phenomena has hitherto been based." In particular the new quantum mechanics avoids the discrepancy mentioned on p. 119 between the frequencies of revolution of the electrons in an atom and the observed frequencies of the emitted spectral lines. While Heisenberg arrives at this new mechanics by a modification of the formal rules of computation, Schrödinger, with an entirely different viewpoint, independently reached mathematically equivalent results, his theory being based on an idea which replaces the movement of the mechanical system by a wave process. This wave process, which may be considered purely fictitious, is not itself observable, but the phenomena observable in the mechanical system are derived from

_____
[12] So written in 1926!

it by means of a 'projection' based on statistical principles. While the motion of the *electron* is thus assumed to be supported by an electronic wave, the optical waves which obey Maxwell's electromagnetic field equations regulate the statistical behavior of the 'light quanta' (*photons*). The idea of a light quantum of definite energy and momentum had already been conceived in the compromise between classical and quantum physics prevalent prior to 1925 (Einstein, 1905). It served to account for the corpuscular nature of light as evidenced by the photoelectric effect, as well as for its wave nature as manifested in diffraction and interference.}

Thus it seems clear that quantum physics cannot posit matter and ether as the basic polarity underlying all phenomena, as the ether theory had done. Light is not only ether wave but also corpuscle, an electron is not only a corpuscle but also a wave. It depends on the concrete situation of their observation, on the instruments we train on them, whether photons or electrons reveal themselves to us as ether waves of definite frequencies or as corpuscles that hit here or there. Bohr has coined the word *complementarity* for this basic feature of the new quantum mechanics that in a sense replaces the old polarity of matter and force. After the previous approaches by the substance and then by the ether conception, the problem of matter now appears to have entered an entirely new stage of its historical development.

More complete information about quantum physics is contained in Appendix C.

REFERENCES

K. Lasswitz, *Geschichte der Atomistik*, Hamburg and Leipzig, 1890.
K. Kirchberger, *Die Entwicklung der Atomtheorie*, second ed., Karlsruhe 1929.
H. Weyl, *Was ist Materie?*, Berlin, 1924.
R. A. Millikan, *Electrons (+ and −), Protons, Photons, Neutrons, Mesotrons, and Cosmic Rays*, Univ. of Chicago Press, 1947 (revised edition).
F. Hund, *Linienspektren und periodisches System der Elemente*, Berlin 1927.
P. A. M. Dirac, *The Principles of Quantum Mechanics*, third edition, Oxford 1947.

23. CAUSALITY (LAW, CHANCE, FREEDOM)

A. Causality and Law. Although the relation of cause and effect dominates our theoretical knowledge as well as our practical dealings with reality, there still is considerable difficulty in bringing out quite clearly those aspects of the causal law which actually bear on scientific research. In the first edition of the *Critique of Pure Reason* Kant says, "Everything that happens [begins to be], presupposes something upon which it follows according to a rule." This second analogy of experience he supplements by a third, however: "All substances, insofar as they are coexistent, stand in complete community, that is affect each other reciprocally." Hume was the first to analyze in detail the category of causality, which until then had been used uncritically in physics and metaphysics. He found in the first place that objects or processes which are considered causes or effects of others are spatio-temporally contiguous with these. This is the principle of nearby action. Any remote effect must be due to a continuous transmission of effects. The question 'Why?' requires the insertion of a continuous causal chain without gaps. Temporally the transmission 'cause → effect' runs parallel to the relation 'past → future.' Furthermore, it is alleged that a 'necessary connection' must exist between both. But if we define, "Of two successive events A and B, A is the cause of B if it is impossible that A takes place without B taking place subsequently," we are saying something that has no empirically verifiable meaning. For how should we recognize the required necessity, since after all we have only one world, and in it B just follows A. Hume therefore replaces the necessary by *constant* connection, by one that *recurs* under all circumstances. But even so, nothing is gained at first, since a concretely given event happens only once. Thus it is necessary to add continuity requirements, to the effect that sufficiently like causes lead to nearly like effects, and that bodies and events which are too remote have no noticeable influence, and so on. The phenomena must be subordinated under concepts, collected into *classes* according to typical characteristics. The causal relation does not hold between individual events but between classes of events. Above all — and this is a point which still escaped Hume — it is necessary to isolate generally valid connections by decomposing the unique course of the world (as described in Section 20) into recurrent elements which are capable only of a gradation representable by a few numerical characteristics.

When they are subjected to measurement it should appear that simple exact functional relations obtain among them that can be ascertained once and for all. The *natural law* thus takes the place of causation. If several quantities a, b, c are connected by a functional relation, the values of a and b may determine the value of c; but the same law may also be construed in the sense that the quantity a is determined by b and c. Thus the functional relation, unlike the causal, is indifferent to the distinction between determining and determined quantities. The abandonment of the metaphysical quest for the cause in favor of the scientific quest for the law is preached by all great scientists. The discovery of the laws of fall by Galileo is the first great example. He himself says (Discorsi, third day, *Opere*, VII, p. 202), "It does not seem expedient to me now to investigate what may be the cause of acceleration," the chief concern must be to explore the law according to which acceleration takes place. Or Newton: "But hitherto I have not been able to discover the cause of those properties of gravity from phenomena, and I frame no hypotheses . . . To us it is enough that gravity does really exist, and act according to the laws which we have explained, and abundantly serves to account for all the motions of the celestial bodies, and of our sea" (end of *Principia*, 3rd ed.). According to the teachings of d'Alembert and Lagrange, dynamics does not require any laws which reach beyond its own domain to the causes of the physical phenomena and the essence of those causes; it is self-sufficient as a description of the regularities of the phenomena. In recent times Mach has fought with particular vigor against the "fetishism" of the concept of causation.

This may be the place where a few remarks on the relation between form (*Gestalt*) and law may be inserted. Kepler still saw the rationality of the world in the *form* of the planetary system, which he associated with the Platonic solids, and thus with certain ideal configurations that are geometrically distinguished *a priori*. The idea of forms and their types plays an important part in biology (systematic morphology), though here in close connection with the teleological notion of organic function. But the idea has not disappeared entirely from inorganic natural science, crystallography providing the most brilliant example of an exact morphological system. The laws of dynamics, since they are laws of nearby action, are of a continuous infinitesimal character; they, rather than the forms, are considered as original in physics today. Typical configurations come about, however, when these laws admit of certain discrete solutions of special character, such as static or periodic solutions. As for more detailed comments on form and constitution and their relation to law and evolution, especially regarding the problems of biology, the reader may be referred to Appendix F.

{The idea of functional law, to which science seems to reduce causality, is not altogether unproblematic. Twice in its history physics believed that it had overcome in principle the decomposition of the world into individual systems (individual events and their elements, which after all are only approximately isolated from one another) and had grasped the world as "a whole in which all is interwoven."[13] The physics of central forces and later the pure field physics seemed for a moment to have reached that goal. Causal law here took the following form: the derivatives with respect to time of the state quantities at a world point are mathematical functions of

[13] "Wie alles sich zum Ganzen webt," Faust's monolog at the beginning of Goethe's *Faust*.

the state quantities themselves and their spatial derivatives at that point. Consequently, the state of the world at any moment would determine the state at the immediately following moment by means of differential laws. Thus only the world's state at a single moment would remain 'arbitrary' or 'accidental,' and from it the world's whole past and future could be computed by integration of the 'Laplacean world formula.' But here again the causal law is in danger of reverting to a triviality. For let us assume that we have only one state quantity; then that quantity together with its temporal and its three spatial derivatives will yield five definite functions of the four space-time coordinates. This makes it mathematically self-evident that one functional relation must hold among them which does not explicitly contain the space-time coordinates. The assertion of regularity becomes meaningless if complications of arbitrary degree are admitted. This was emphasized already by Leibniz in his "Metaphysische Abhandlung" (*Philosophische Schriften*, IV, p. 431). What is decisive and at the same time astounding is the fact that the laws show such a simple mathematical structure, while the quantitative distribution of the state quantities in the world continuum is incredibly complicated. This has the consequence, for our knowledge, that limited experience enables us to ascertain those laws while the unique quantitative course of events remains largely unknown. This distinction, for the naive realist only the vague one between simple and complicated, becomes one of principle, when the intuitionist or constructivist view is adopted in mathematics and physics. }

Furthermore, the causal principle is meant to postulate not merely the existence of functional dependences but also a certain density of their fabric. This is brought out in the customary formulation: "Under the same conditions the same event will repeat itself." Where should the line be drawn between conditions and events happening due to these conditions? There is no difficulty here, provided cause and effect belong to different spheres of existence; as is for instance the case when we search out the real conditions of an immanent perception and demand that each difference within the perception must be based on a difference in the corresponding real conditions. But in nature cause and effect lie within the same plane. As a dividing line separating the two one may set down a three-dimensional space-like cross section $t = $ const. through the world, which will dissect it arbitrarily into a 'past' half and a 'future' half. This leads to the formulation that the content of the past determines by law the content of the future. However this is not the causal law itself but only a special form of it which fits well into field physics. It need not for this

191

reason be unconditionally accepted. A special case of practical importance is this: Assume we have a number of permanent objects or stationary events (e.g. prism and light beam), the behavior of which is known if they are isolated; arrange them now in space-time in such manner that they penetrate each other; then a novel but predictable event will result (the light is refracted by the prism into its spectral colors). The ether theory is likely to interpret the causal relation by the contraposition of matter and field ('matter excites the field'). For the experimenter, the conditions consist of that part of the events which are under his control. (This circumstance lies at the base of a methodical remark of Stuart Mill's: if we wish to learn the effects of a cause we may experiment; but if we wish to learn the cause of an effect we have to rely purely on observation.) In our will we experience a determining power emanating from us, and were we not thus actively and passively drawn into the stream of nature (be it even merely in the role of an experimenter who creates the conditions of the experiment), we would hardly regard nature under the metaphysical aspect of cause and effect. (Hume disagrees. The compulsion our reflecting mind feels to pass from one idea to another represents to him the prototype of force.) As only an inner understanding of the word 'I' reveals to me the notion of substance, so, according to Leibniz, it is "the consideration of myself which also furnishes to me other metaphysical concepts such as cause, effect, . . . " (*Philosophische Schriften*, VI, p. 502).

The causal law asserting that "the same events take place if the same conditions obtain" is not an empirical statement, as was emphasized even by Helmholtz, who is otherwise so empiristically minded, for "to prove it by induction seems a very dubious proposition. The degree of its validity could at best be compared to that of the meteorological rules" (*Physiologische Optik*, III, p. 30). Rather "the causal law bears the character of a purely logical law even in that the consequences derived from it do not really concern experience itself but the understanding thereof, and that therefore it could never be refuted by any possible experience. . . . It is nothing but the demand to understand everything" (*op. cit.*, p. 31), a norm whose validity we *enforce* in the construction of reality. (The example should here be recalled of the two reds which appear equal to sensory perception and yet are refracted into different spectra by the same prism.) Helmholtz's conception of causality as a methodical principle is in agreement with the Kantian doctrine of categories.

{As a check on the acquired understanding of the causal principle the question has often been asked: Why is it that the relation of day and night, this prototype of a regular succession, is not a causal one? The argument of the 'irreversibility' of the causal relation is a poor way out, since it can so easily be refuted by the example of the hen and the egg. Mill states the following criteria of causal relationship: unalterability ('again and again') and unconditionality ('irrespective of the remaining conditions'). He points out that the latter requirement fails to be satisfied in the succession of day and night. Against this, it may be said that no one has actually tried out what happens when the essential subsidiary condition is changed, namely, when the sun is removed. Hume replies with the "recognized principle," which is implied in the principle of continuity, that an object which has existed unaltered during an interval of time without producing another object cannot possibly be the sole cause of the latter. But it happens with certain chemical reactions that after mixing the reagents considerable time passes during which nothing is observed, until suddenly a change in color takes place. Admittedly in such cases theory will tell us of certain continuously progressing hidden changes, which produce a noticeable optical effect only after passing a certain threshold. But the same might be the case with respect to day and night: when lightness has accumulated for twelve hours, it turns into darkness, and conversely. But then we notice that lightness is always connected with the phenomenon of the sun, that there is something wrong with the constant twelve hours, that the duration of the day changes, that even during the day it becomes darker when the sun is hidden behind clouds or eclipsed, and that the dark shadows move according to the position of the sun. Thus our attention is drawn to the sun, and we find that, if it is taken into account, the accumulating nightly darkness becomes entirely superfluous as part of the cause of day. Decisive is finally the analogy: *I* am able to produce lightness by lighting a candle. By setting up a parallel between the sun and the candle under my control, I arrive at the conception that if I blew out the sun like this candle it would become dark.

In his inductive logic (*A System of Logic, Ratiocinative and Inductive*, Book 3, Chap. 8), Mill tried to reduce the empirical ascertainment of causal relations to definite rules. They cannot be considered as more than a first rough attempt to describe the methodology of inductive research. Their main shortcoming lies in the failure to explain how the various 'instances' to which the rules refer are to be isolated from a situation given as a whole.}

Causality occurs in physics not only as a methodical principle but also as a factual component of the theory, namely as the causal structure discussed in Section 16. According to pre-relativistic theory, propagation of effects takes place instantaneously. The push given to a rigid rod can be felt everywhere along the rod at the same instant; the dislocation of masses effected by spreading out my arms influences at the same moment the motion of the planets. In relativity theory, however, the velocity of light has become the upper limit of all velocities of signals and propagation, hence the effects at a distant place really occur later than the releasing cause. Kant's third analogy of experience, the principle of community between coexistent objects, has to be abandoned. Subjectively that part of the light cone which opens toward the future plays an entirely different role from that which is directed toward the past. We travel along the world line of our body with 'screened-off consciousness.' Indeed only of the content of the rear cone can we have direct knowledge based on perception. Neither the classical laws of mechanical motion nor the field laws of electricity and gravitation can account for this difference. For instance, a spherical wave converging toward a center O is just as compatible with the field laws as a spherical wave emanating from O. And yet one must demand that that basic fact of consciousness, the one-way direction of the flow of time, have a physical foundation. Phenomenological thermodynamics finds it in the law of *entropy*, according to which the natural processes are irreversible and take place in the sense of increasing entropy. Since the atomic theory reduces phenomenological thermodynamics to the statistics of atomic motion and thereby introduces an element of chance, we postpone a discussion of the problem and turn first to an analysis of *chance*.

B. CHANCE. The judgments which implicitly or explicitly determine our actions rarely exhibit the sharp division between the alternatives of true and false as demanded by classical logic. Between black and white there are all shades of grey. In particular, questions concerning the future do not point to a verification by any reality; and yet they are discussed and judged right now, under such aspects as possible, likely, inevitable, rather than true or false. For instance, a statement about what will happen within a year from now will indeed be verifiable after one year, but then in the modified temporal form 'it happened in the past year.' We make plans by figuring out in advance future possibilities and basing our decisions on weighing them. The driver of a car has to do this almost instinctively at every moment. We strive for certain ends, run risks, dangers hang over our heads. Besides hard facts we depend on expectations which often bear the emotional accents of hope and fear. One may hesitate to speak here of knowledge and judgments, but these things have the structure of judgments and mean something vital to us.

It is an ironical historical comment that the logic of probability owes its first stimulus to gamblers seeking advice from their mathematical contemporaries. Thus Pascal and Fermat were led to a mathematical analysis of chances in a game. When it is a question of the gain or loss of money, the quantitative factor cannot be overlooked; the answer had to be sought not in a descriptive analysis of the concept of probability but in a calculus of probability. The problem was to satisfy in an exact manner the requirement of justice, by justly weighing risk against gain, under the simplest circumstances, namely for a game determined by set rules. The earliest treatise on the subject has been written by Huyghens. Also the earlier parts of Jacob Bernoulli's *Ars conjectandi* (published 1713) move within a circle of concepts of subjective nature, such as 'hope,' 'expectation,' 'conjecture.' True, the classical definition of quantitative probability coined by Laplace — the quotient of the number of favorable cases over the number of all possible cases — emphasizes the objective aspect. Yet this definition presupposes explicitly that the different cases are *equally possible*. Thus it contains as an aprioristic basis a quantitative comparison of possibilities.

[This is especially evident when it is a question not of a finite number but of a continuous manifold of possibilities, e.g. the possible positions of a particle freely mobile within a box. The probability $v(D)$ that the particle will be found in a portion D of the box is a function of D, possessing the additive property: $v(D) = v(D_1) + v(D_2)$, if the domain D is in any way divided up into the two partial domains D_1 and D_2. This requirement is identical with the one set down as a matter of course for the volume measure of arbitrary domains D in a continuum (compare the axioms on areas, p. 28), and it is therefore understandable that continuous probabilities are usually treated by mathematicians today under the title of 'measure theory.']

It is because of the arbitrariness of such a measure that Laplace, from his consistently deterministic conception of nature, is in the end unable to ascribe to probability anything but a subjective meaning; it deals with events whose premises are incompletely known, and thus is "relative to this our knowledge and ignorance." Laplace therefore calls two events equally possible if we are equally undecided as to their occurrence. However, a purely mathematical part can be split off where probabilities of events are computed not absolutely but on the basis of the given probabilities of other events with which they are causally or logically connected. If A, B are two events (or the statements that they occur), it is possible to form the logical combinations 'A or B' and 'A and B.' Their probabilities are respectively equal to the sum and the product of the individual probabilities, provided A and B exclude each other in the first case, and are independent of each other in the second. Statistical independence may be defined in the sense of the subjective Laplacean viewpoint by the requirement that the probability of B remains unchanged if our knowledge, on the basis of which the probabilities of A and B were judged, is augmented by the additional knowledge that A has actually taken place.

In the fourth part of his *Ars conjectandi*, Jacob Bernoulli throws the bridge from the subjective to the objective conception of probability by means of his "law of large numbers." According to this objective interpretation the probability calculus serves to establish regularities expressed in the mean values of many similar events rather than in the individual event. Beside the strictly valid causal laws we thus have regularities of a statistical nature. Daniel Bernoulli employed the calculus in order to lay the foundation of kinetic gas theory. The objective significance of the probability calculus comes to light not only in physics, but also in a number of biological disciplines (genetics, biometrics), in the modern insurance business, as well as in other applications (economic and social problems, quality control of mass production, etc.). From the subjective point of view described above it is difficult to understand. The question of its justification has given rise to a rich epistemological literature. A scheme which may serve to illustrate all probability problems is the drawing from an urn containing m white and n black balls. By assuming the drawing of any one of the $m + n$ balls as equally possible, the probability of obtaining a white ball is found to be $\dfrac{m}{m + n}$. In the sequence of real events, this probability is supposed to manifest itself, roughly speaking, as the *relative frequency* of drawings of white balls among all drawings, when a large number of drawings are made. (In order to restore the same conditions, it is understood that the drawn ball is always replaced in the urn before the next drawing.) Bernoulli's theorem is as follows: let p be the probability of an event E (say the drawing of a white ball from our urn), ϵ a given arbitrarily small positive number (say $\frac{1}{100}$); if in a series of N trials, in which E may or may not take place (drawings), N' turn out to be favorable, then the relative frequency N'/N *most probably* deviates from p by less than ϵ, provided the number of trials N is sufficiently large. In fact, Bernoulli computes the probability $p(\epsilon, N)$ that N'/N does not lie between $p - \epsilon$ and $p + \epsilon$ as a function of ϵ and N and proves by a suitable estimate that $p(\epsilon, N)$ for fixed ϵ converges to 0 as N increases indefinitely (it converges to a definite limit between 0 and 1 provided

$\epsilon \sqrt{N}$ tends to a limit different from 0). In his calculation the individual trials are treated as statistically independent events. This theorem belongs to pure mathematics. It acquires a relation to reality only by the fact that the occurrence of an event is considered practically certain if its probability deviates from absolute certainty by, say, less than one millionth: we are able to predict with near certainty that in a series of $N \geqq 200,000$ trials the relative frequency will deviate from the *a priori* probability by less than $\frac{1}{100}$. With still greater certainty we can predict that in a large number of such trial series of length N those whose deviation is more than $\frac{1}{100}$ form a small part, say, less than $\frac{1}{1000}$ of all trial series. But there always is a residue of uncertainty, the inexactness of the statement being an essential characteristic.

[There are important cases in which the equiprobability of different results may be derived either from the process leading to the result or, by means of the principle of sufficient reason, from the *symmetry* of the situation. Thus in the bipartition of a large number of entities (e.g. the maturation division of cells in an organism, cf. Appendix B), complementary parts will appear with the same frequency, and consequently their occurrence is to be expected with the same probability. An example of symmetry is the throwing of dice. Once it is admitted that a probability appertains to each of the six faces of a die, then it can only be the same for each, provided the die is really homogeneous. This is an inference similar to that by which Archimedes established the law of equilibrium for a lever with equal arms. The cases in which we have reason to consider such symmetries as exactly or approximately valid are none too rare. If the observed frequencies are in accordance with this assumption, we are satisfied; if not, we are driven to look for the cause of the asymmetry. Thus, if we have a die for which the ace occurs noticeably more often than the six, we shall perhaps bore a hole into the die and find that it is loaded.]

In this example it should still be possible, if the mass distribution within the die is known, to ascertain the probabilities of the six faces by an exact physical analysis. In most cases occurring in everyday life, however, the probability cannot be found *a priori* at all, as the classical definition demands, but only *a posteriori* on the basis of observed relative frequencies. The probability of male births is still a good example — even with our present state of biological knowledge of the mechanism of sex determination. But if the births in a certain region are collected together in their temporal succession and divided up arbitrarily into series, the latter show all the properties which

197

probability theory predicts for series of drawings from an urn containing boys and girls in a certain ratio $m:n$. If in the case of temporally very extended series there are systematic deviations or displacements, then a slow continuous change in the circumstances determining the probability, in particular in the genetic composition of the population, is to be inferred.

But if it is true that probability cannot be determined except by counting of frequencies, then it seems reasonable to base the objective foundation of probability theory directly on the trial series and define the probability p as relative frequency; or, since this is a little too primitive and would not lead to any predictions, as the *limit* of the relative frequency when the number of trials increases indefinitely. Unfortunately, this introduces the impossible fiction of an infinite number of trials having actually been conducted. Moreover, one thereby transcends the content of the probability statement, inasmuch as the agreement between relative frequency and probability p is predicted for such a trial series with absolute certainty rather than with 'a probability approaching certainty indefinitely.' It is asserted that every series of trials conducted under the same conditions will lead to the same frequency value p. In order that the mathematical rules of the probability calculus hold, the trial series would have to comply with certain requirements demanding something like 'order in the large, disorder in the small;' their exact formulation, however, affords very serious difficulties. But does not this 'objective foundation' of statistics, that springs from the epistemological position of strict empiricism, simply conceal the *a priori* probability behind the dogmatic formula of a fictitious frequency limit which is tied to the nonsensical idea of an infinite sequence of trials? As long as one believes in strict causality, statistics must find its proper foundation in a reduction to strict law. If, however, there should be a 'primary probability' for the individual atomic events that cannot be reduced to causal laws — and such seems to be the case according to the most recent development of physics — then we seem to be forced to introduce into the natural laws as an original factor either that probability itself or some quantity connected with it; and the classical definition would be limited to those special cases of symmetry in which the principle of sufficient reason provides adequate guidance.

⸢In 1 gram of hydrogen there are approximately $N = 3 \cdot 10^{23}$ hydrogen molecules swarming about. At a certain pressure the gas will fill out a cubical container C of a certain volume 1. We ask for the probability that an individual molecule will be present in a domain D of the container. The corresponding frequency is here ascertained not by repeating the experiment an innumerable number of times with the same molecule but by 'playing dice' simultaneously with a large number N of molecules. In extending Bernoulli's theorem of large numbers to this situation we make two implicit assumptions: firstly that the forces acting between the molecules may be disregarded, and secondly that the locations of different molecules are statistically independent of one another in the same sense as the repeated drawings from an urn. In this new form, statistics is applied to mechanical systems consisting of many particles of the same kind. Since there is no preference for any particular location, the probability that at a certain instant an individual molecule is in some given subdomain D of the cube is assumed to be equal to the volume of D. Within C we now mentally demarcate a smaller cube c, say of volume $\frac{1}{1000}$. We ask for the probability that at a certain moment the density of the gas in c (i.e. the mass of the molecules present in c divided by the volume $\frac{1}{1000}$ of c) deviates from the over-all value 1 by more than 0.01%. By means of Bernoulli's theorem we find that this probability is given by a fraction with numerator 1 and a denominator consisting of no less than a million digits. We therefore expect that in a state of 'thermodynamic equilibrium' the gas is distributed, macroscopically speaking, with equal density over the container. Noticeable deviations (for instance, the spontaneously occurring 'miracle' that the entire gas collects in a corner of the container) are possible but extraordinarily improbable. If there is initially an uneven distribution, the motion of the molecules will quickly effect the transition to a state of equidistribution, provided there are no exterior forces acting on the system. For the same reason, if coffee and milk are poured into a cup, stirring will soon produce a liquid of uniform color. }

It will now be understandable that most of the physical concepts, especially those concerning matter with its atomic structure (e.g. the density of a gas), are not exact but statistical, that is, they represent mean values affected with a certain degree of indeterminacy. Similarly most of the usual physical 'laws,' especially those concerning matter, must not be construed as strictly valid laws of nature but as statistical regularities. Statics (which treats of the laws valid for thermodynamic equilibrium) and dynamics (which treats of the laws regulating the transition from a disturbed state to a state of equilibrium) are supplemented by a theory of fluctuations, which investigates the fluctuations about the statistical mean values in equilibrium and the physical effects connected therewith. The spontaneous density fluctuations of the air, for instance, are responsible for the diffraction of the sunlight and thus for producing that diffuse daylight which makes the sky appear blue rather than black. However minute they are individually, they still have an observable global effect. In fact, from the color and intensity of the sky's blue it is possible to compute the Avogadro number of gas molecules per volume unit at normal pressure and temperature, and the result checks with those obtained by other methods. The study of such fluctuations that are accompanied by observable cumulative effects has proved to be very fruitful for the determination of the atomic constants and furnishes one of the strongest supports for the molecular theory.

The justification of statistical physics evidently derives from the fact that the hidden complicated molecular processes bear no direct relation to our perceptions. The latter depend on certain mean values, and statistics teaches us how to determine these. Our consciousness does not reflect the molecular chaos of the phenomena but exerts an integrating function with respect to both space and time, from which results the apparent homogeneity and continuity of the phenomena. The statistical mean values may obey relations of dependence indicative of after-effects of the entire history of a system (hysteresis, persisting dispositions, 'memory'), although the exact natural laws are differential equations which connect only infinitely near space-time points with one another. Such phenomena, though not entirely foreign to inorganic nature, occur frequently in the organic world. They cannot, however, as is sometimes done, be cited as proof that organic processes are incapable of 'mechanical' explanation.

In spite of all these applications the physicists at first adhered to the view that the probability calculus is merely a short cut for the derivation of certain consequences of observational significance from the exact causal laws. Consequently the attempt had to be made to derive from them the probability measure on which the statistical analysis is based. This is tantamount to a causal analysis of chance. Chance appears to prevail whenever 'little causes lead to big effects.'[14] A minute deviation in the direction of two projectiles — a deviation perhaps which is entirely beyond our control — may, once the projectiles have traversed a large distance, lead to fatefully different results. Uncontrollable are the circumstances of the act of procreation upon which the fertilization of the egg by a spermatozoon of this or that genetic constitution depends, and thus sex determination, in particular, is left to 'chance.'

〔A simple example may serve as an illustration: the shooting at a circular target which is divided into several sectors and is rotating fast and uniformly. The cause is the instant t of pulling the trigger,

---

[14] "Kleine Ursachen, grosse Wirkungen," a well known German proverb.

the effect the number y of the sector which is hit. y is a discontinuous function of t, $y = f(t)$. The circumstances essential for the chance character of the result are the following: (i) Let Δt be the range within which t can be determined if the greatest possible subjective care is taken; then $f(t)$ assumes all its values during the interval Δt, that is, the intervals on the t-axis in which $f(t)$ remains constant are small in comparison to Δt. (ii) If, in addition, an arbitrary probability distribution on the t-axis is assumed, by setting the probability of pulling the trigger during the infinitely small interval from t to $t + dt$ equal to $\varphi(t)dt$, then the probability of the various y can be mathematically computed. These probabilities are almost independent of the assumed distribution function $\varphi(t)$, as long as $\varphi(t)$ has a fairly regular behavior, i.e. changes little during any interval of length Δt. This analysis, which can be applied to the throwing of dice, reduces the measure of probability (in the domain of y) to the possibility of comparing probabilities as to their *order of magnitude* (in the domain of t).]

So far we had still to rely on such vague terms as 'almost independent' or 'fairly regular.' One is tempted to proceed to a strict formulation by letting the rotational velocity of the target converge to infinity. This idea may be explained by means of the example of the gas which consists of molecules enclosed in a cubical container of volume 1. We assume that the individual punctiform molecule moves in the interior of the container with uniform velocity along a straight line, and that it bounces off the walls according to the usual law of reflection. The initial velocity of the molecule may be given by its three components v_1, v_2, v_3 with respect to the three axes of the cube. We expect that, no matter what the initial velocity, an infinitely extended observation time t will yield that 'disorder' on which statistics is based, and hence we carry out the limiting process $t \to \infty$. For any partial domain D of the container, we understand by the dwelling time t_D of the molecule the total duration of all time intervals during the observation time t during which the molecule is within the domain D. By the relative dwelling time $v(D)$ we understand the limit of the proper fraction t_D/t for infinite observation time t. It is then in fact possible to prove that the relative dwelling time of the molecule equals the volume of D, provided one excepts initial velocities whose components v_i satisfy a linear relation $a_1v_1 + a_2v_2 + a_3v_3 = 0$ with rational coefficients a_i. To be sure, such initial velocities are everywhere dense in the three-dimensional velocity space, but their set is of a 'vanishing measure' as compared to the rest. (It is related to the whole space as the set of rational numbers to the totality of all real numbers.) The appearance of such exceptions corresponds to the fact that Bernoulli's theorem of large numbers equates relative frequency and probability only with a probability which approaches 1 indefinitely. An analogous statement can be proved for an aggregate of N punctiform molecules, each of which follows a path in accordance with the law stated above; but again under the same restriction, namely that no homogeneous linear relation with rational coefficients holds between the $3N$ components of the N initial velocities. In this sense there is statistical independence among the N molecules. And it turns out that, if the 'rational' initial states are excluded, the fraction t_D/t of the observation time t during which the gas deviates noticeably from the state of uniform density converges toward an extremely small number as $t \rightarrow \infty$. By thus interpreting probability statements as statements about dwelling times, the objective significance of the probability calculus within the framework of a world dominated by exact laws would seem to be explained.

[It lies in the nature of things that statements about what happens in the limit for an indefinitely extended observation time t may depend on whether a certain number is rational or irrational — however much this is at variance with the approximate character of all physical measurements and the nature of the continuum. The value of this trend of investigation is questioned by the remark that disturbances can never be entirely eliminated but that the validity of thermodynamics is evidently not affected thereby — or only in the positive sense that the advent of thermodynamic equilibrium is accelerated by the disturbances. Incidentally, in more complicated cases than the one chosen as an example, the analysis hinges on a certain crucial hypothesis known as the ergodic hypothesis, the proof of which long resisted the efforts of the mathematicians. Ironically, when the proof was at last accomplished to the extent that could reasonably be expected, the hypothesis had been deprived of most of its physical significance by quantum mechanics.]

At any rate, in the actual conduct of physical research, statistics today plays at least as important a part as the strict law. Attempts to reduce one to the other have gradually fallen back behind the independent building up of statistical thermodynamics. Of the two laws which are of universal significance for all physical phenomena, the law of conservation of energy and the law of continuously increasing entropy, one is the prototype of a strict law, the other of a statistical law. The latter, in fact, states that heat flows from the warmer to the colder body, that coffee and milk get mixed but not unmixed by stirring, that, in general, an improbable state in the course of time changes into that 'most probable' state from which the overwhelming majority of possible states differs but little; or, otherwise expressed, that order gradually changes to disorder. Boltzmann was actually able to define entropy by the expression $k \cdot \log D$, where D is a measure of disorder and k ($= 3.2983 \cdot 10^{-24}$ cal./°C) a universal constant of nature. When we rely on the methods which have developed from cogent motives we cannot help recognizing the statistical concepts, besides those appertaining to strict laws, as truly original. Perhaps space and time, as the above example of gas molecules locked up in a container suggests, merely have the function of a probability field for physics. According to the new quantum mechanics the situation seems to be such that exact laws of the kind familiar from field physics determine certain probabilities related to the atom. But they determine the observed events in space and time only in the manner in which *a priori* probabilities determine statistical mean values — with an unavoidable uncertainty factor. Evidently it is physics only which can provide us with the ultimate enlightenment about the real meaning of the probability calculus. In former times it was argued that its necessity is based on the impossibility of completely isolating a physical system. Statistics thus was supposed to take into account summarily the influence of the whole infinite universe on the nearly isolated partial system. But more important, perhaps, than the openness toward the infinitely distant fringes of the world is the inner infinitude toward the atoms. The philosophers are impatient people. As a scientist one gains the impression that something reasonable about causality, law, and statistics can again be said only after the riddle of the quanta has been solved,

[Appendices B and C will describe how and to what extent the situation has cleared since 1926 through the development of quantum theory. The 'primary' atomic probability appears as the square of the absolute value of a certain (complex-valued) field quantity ψ. The 'secondary' statistics of quantum-thermodynamics, on the other hand, rests on a simple enumeration of quantum states, and thus the difficulty of an intrinsic probability measure disappears from the theory.]

C. TIME'S ARROW. In a sweeping way and in conformity with the experience in all branches of physics, the law of entropy accounts for 'time's arrow,' for the different roles of past and future, the irreversibility of natural processes. Yet there is a serious difficulty. For the derivation of the law of entropy from the statistics of molecular motion is based exclusively on elementary laws, such as the laws of impact for molecules, which are reversible, that is, which are invariant under a change from t to $-t$. This reversibility is displayed very impressively by the heavenly motions, where disturbing influences are entirely or almost entirely eliminated. Their harmony seems to triumph over the law of universal decay that prevails on earth. A close examination of the statistical derivation of the entropy law will show that indeed an improbable state, one that is far from thermodynamic equilibrium, will be followed with overwhelming probability by a state closer to equilibrium after a small time interval Δt, but also that Δt seconds earlier it will have been preceded, with the same overwhelming probability, by a state closer to equilibrium. In fact this prediction is confirmed by all fluctuation phenomena: they disclose no irreversibility.

[Thus Smoluchowski (*Vorträge über die kinetische Theorie der Materie und der Elektrizität*, 1914) arrives at the conclusion: "If we continued our observation for an immeasurably long time, all processes would appear to be reversible." Or Boltzmann (*Populäre Vorträge*, p. 362): "The laws of probability calculus imply that, if only we imagine the world to be large enough, there will always occur here and there regions of the dimension of the visible sky with a highly improbable state distribution." The circumstance that we happen to witness such an event might be explained by the fact that the possibility of life is bound to an exceptional state of this kind.

Yet this can hardly be considered the last word in the matter. If two previously isolated systems come into contact and begin to influence each other we consider it extremely improbable that they are at that moment in thermic equilibrium with each other. Reversal of the time direction turns the process of joining two systems into a process of separation, and in this case we make the opposite judgment: if a system, which has been isolated against exterior disturbances over a long period of time, is separated into two parts, we consider it extremely probable that the two parts at the moment of separation are in thermic equilibrium with each other. Thus the idea inherent in causality that that which is earlier is the determining reason for what follows, and not vice versa, impresses on our probability judgment a distinguished direction of time.]

It was said above that at the instants $t + \Delta t$ and $t - \Delta t$ the deviation y of a system from the state of equilibrium will, with overwhelming probability, turn out to be smaller than at the time t, provided y is noticeably different from 0 at t; in other words, y falls away rooflike on both sides. But from such roofs, each of which has two slopes, we cannot compose the picture of a unique succession of states. Such a picture can emerge only if at every time point we decide in favor of the forward slope of the roof, or else if at every time point we decide in favor of the backward slope of the roof. We do the former when we pursue, for instance, the process of heat conduction or of diffusion. In other words, as the state S' prevailing at a future moment we predict that state which *results* from the present (improbable) state S with overwhelming probability (and which is more probable than S). As a previous state S' we assume one *from which* the present state S follows with overwhelming probability, — even though this past state S' turns out to be still less probable than the present one. According to the probability judgment directed toward the future, we expect the

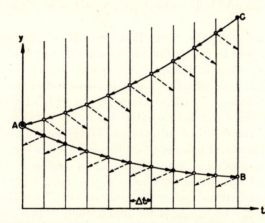

Figure 6. *AB* forward directed and *CA* backward directed probability judgments. In either case the opposite slopes (dotted lines) are ignored.

earth to have cooled off further after a few million years. The same probability judgment directed toward the past would state that also a few million years ago the earth was that much cooler. Instead we assume that it was in that warmer state from which the present one results by virtue of the forward directed probability judgment. The 'miracle' is blamed on the past, and the larger the parts of the universe which are being viewed, the further into the past can it be pushed. Our judgment thus proceeds as if the system with which we are dealing had been *created before our time*. The word 'creation' suggests a metaphysical or even theological interpretation, but this should not prevent us from recognizing the state of affairs which is most aptly expressed by this word.

[An arbitrarily given temperature distribution in a body at an instant $t = 0$ changes in the course of time according to the differential equation of heat conduction. In this way we are able to compute the distribution for any future moment $t \geqq 0$, but in most cases the solution cannot be traced back a single step into the past $t < 0$. Here the act of creation necessarily takes place immediately prior to the present moment $t = 0$. It may consist in the body having been heated here and cooled there by contact with other bodies from which it becomes separated at the moment $t = 0$.

It must of course be admitted that a limited system, isolated over an infinitely long period, has its chance fluctuations which are reversible. But a system which is isolated now behaves as though it might always stay isolated, but not as though it had always been isolated. Therefore the study of reversible fluctuations, so to speak, has only local importance and tells us nothing about the world in the large. }

We thus arrive at the conviction that the distinctiveness of the direction from past to future finds its expression not in the elementary laws but in the probability judgment. It is true that within the framework of classical field physics it was possible to set down a formula (the Liénard-Wiechert formula) according to which the motion of an electron determines the field excited by it, by retaining of the two components of the solution only the one corresponding to the *retarded* potential. This would be a necessary consequence of the field laws only if we could imagine that *prior* to a certain moment the world was empty (with all charges neutralized and the field $= 0$) and that the field begins to radiate only after the separation and creation of charges; it will then never become extinct again even if the charges should later be neutralized. But the classical theory of electrons must today be replaced by quantum theory. According to the basic conception of Niels Bohr, the jump of an electron in the atom sets off a diverging rather than a converging electromagnetic spherical wave, the frequency of which is determined by the energy loss of the atom. And there is no doubt that the non-commutability of emission and absorption must here be understood on the basis of the same principles as in the case of the entropy law.

[After the Einstein equations of gravitation (supplemented by the cosmological term) had yielded as their solution an expanding universe, it was natural to relate the distinguished time direction with this expansion. A massless world, according to Section 16, is represented by de Sitter's hyperboloid. A single star describes a geodetic world line on it, and the light cones which open toward the future and issue from the points of that line fill out the domain of influence D of the star. This domain covers only one half of the hyperboloid, it spreads out fan-like toward the infinite future. D is at the same time the domain of influence of every one of a three-parametric sheaf of geodetic lines. It is plausible to assume that only this half D corresponds to the real world and that those geodetic lines represent the world lines of the stars (provided their mutual gravitational attraction is disregarded). In other words, all stars have a common domain of influence. It is this picture of a community of origin for all stars that leads to an explanation of the red-shift of the spectral lines of the spiral nebulas and yields a value of the order of magnitude of 10^{27} cm. for the world radius. The expansion here appears as a natural consequence of the distinction made in favor of the light cone opening

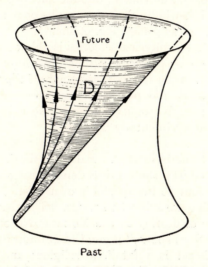

Figure 7. De Sitter's hyperboloid with domain of influence D and world lines of stars.

toward the future. In truth the world is not void of mass; hence de Sitter's solution will have to be replaced by a set of possible solutions, among which it itself as well as Einstein's static solution are special cases. The majority of these solutions are such as to ascribe to the universe a punctiform beginning in time. It has been speculated that its creation was due to a vehement radioactive explosion, of which the fossil of cosmic rays still bears evidence (Lemaître). }

Summarizing we may, with Eddington, describe the situation briefly as follows: "We have swept away the anti-chance from the field of our current physical problems, but we have not got rid of it. When some of us are so misguided as to try to get back milliards of years into the past we find the sweepings piled up like a high wall, forming a boundary — a beginning of time — which we cannot climb over" (*New Pathways in Science*, p. 60).

The contrast of *right and left*, however much it occupies mythical thinking, does not pose as basic a problem to scientific thought as that of past and future. There can be no doubt that all natural laws are invariant with respect to an interchange of right and left. Yet the fact must be noted that the clockwise and anti-clockwise isomers of an optically active compound often occur in nature in very different abundance (e.g. the occurrence of dextro- but not of levo-tartaric acid in fermenting grapes), and that in organic nature the sense of a screw (right-winding or left-winding) sometimes belongs among the inheritable characteristics of a species (e.g. of shells).[15]

None of our present physical theories is able to give an account of the essential difference between *positive and negative electricity*. Is this merely due to the 'accident' that in nature the negative electrons happen to outnumber the positive ones, and the positive protons outnumber their negative equivalents (whose existence has not yet even been proved experimentally)?

D. FREEDOM, PURPOSIVENESS. Fate and destiny, chance, fortune and misfortune, free will, these words denote categories by which man has always been wont to interpret his own existence. With Homer, Gods and men alike are subject to *moira*. The Germanic people personified the power of destiny in the Norns ("No one lives to see the evening, once the Norn spoke," *Edda*, Old Hamdir Song). Islam with its concept of Kismet is a fatalistic religion; and in the sphere of Christian theology Calvin placed the fate of man unconditionally in the hand of God (selection by grace, predestination). The first consistent modern theory of determinism in which natural law appears as the binding force is due to Hobbes. Descartes, on the other hand, clung to the freedom of will, and he had to do so if the self-certainty of thinking guarantees truth as demanded by the principles of his philosophy. He could not comprehend how this is compatible with predestination, which follows necessarily from the omnipotence of God.

By exhibiting the freedom of will in the theoretical acts of affirma-

[15] A by-product of the fermentation of grapes that has the same chemical composition as tartaric acid but is optically inactive, namely racemic acid, was decomposed into dextro- and levo-tartaric acid in a famous experiment carried out by Pasteur in 1848. The human body contains the dextro-rotatory form of glucose and the levo-rotatory form of fructose. That *homo sapiens* contains a screw turning the same way in all individuals is proved in a rather horrid fashion by the fact that man contracts a metabolic disease called phenylketonuria leading to amentia when a certain quantity of levo-phenylalanine is added to his food whilst the dextro form has no such disastrous effect.

tion and denial, Descartes succeeds in bringing out with great clarity the decisive point in the problem of freedom: when with clear reason I judge that $2 + 2 = 4$, this actual judgment does not make itself within me through blind natural causality (a view which would eliminate thought as something which I mean). Rather, a purely spiritual something, namely the circumstance that $2 + 2$ really equals 4, exercises a determining power over my judgment. The issue here is not whether the factors responsible for my action partially lie in me as an existing natural being rather than outside of me. Such freedom is derided by Kant (*Critique of Practical Reason*) as being no better "than that of a mechanical turnspit which, once it has been wound, likewise continues to perform its motion." Nor is it a question of groundless, blind or arbitrary decisions. We are confronted, rather, with the fact that the realm of Being is not closed with respect to its determining factors, but that in the ego, where Meaning and Being are merged in indissoluble union, it is open toward Meaning. However, if this conviction is not erroneous, that open place, a limit of the objective, should manifest itself within nature and its science. Since this was not the case in natural science as it has developed since Galileo from compelling motives and with the claim of embracing all of nature, science became to the modern mind the power which shook the naive belief in the independence of the ego. All evidence indicates that living creatures, apparently endowed with the faculty of will, are not exempt from the exact natural laws. I, too, can impart a momentum to my body only by pushing off from other bodies which together absorb the opposite momentum.

[Descartes tried to save the situation by ascribing to the soul, which cannot alter the total momentum, some kind of directing or 'switching' function. Two quotations from Leibniz may be given here. In an essay on freedom (*Lettres et opuscules inédits de Leibniz*, ed. Foucher de Careil, Paris 1854, p. 178 et seq.) he states "that there may, or even must, be truths which no analysis can reduce to the identical truths or to the principle of contradiction, which, on the contrary, require an infinite series of reasons for their support; a series which is transparent to God only. And this is the essence of all that one considers free and accidental." Further, in his *Monadology* (*Philosophische Schriften*, VI, pp. 607–623; Section 79): "The souls act according to the laws of final causes through appetences, means, and ends. The bodies act according to the laws of efficient causes or motions. And these two realms, of final and efficient causes, are in mutual harmony." Among the clearest formulations of determinism is that by Laplace: "Une intelligence qui pour un instant donné, connaîtrait toutes les forces dont la nature est animée, et la situation respective des êtres qui la composent, si d'ailleurs elle était assez vaste pour soumettre ces données a l'analyse, embrasserait dans la même formule, les mouvements des plus grands corps de l'univers et ceux du plus léger atome: rien ne serait incertain pour elle, et l'avenir comme le passé, serait présent a ses yeux. L'esprit humain offre dans la perfection qu'il a su donner à l'astronomie, une faible esquisse de cette intelligence" (*Essai philosophique sur les probabilités*, second ed., 1814, pp. 3–4). Kant, according to the scientific situation of his time, agrees with this view as far as the world of space-time phenomena is concerned, and he tries, by distinguishing between the phenomenal and the intelligible world, to give a transcendental solution of the conflict between natural causality and freedom of will. His solution, however, can hardly be carried through consistently[16] and even remained obscure to himself to such a degree that he was unable to understand the changes in the character of a person. Any reasonable interpretation of Kant's attempt in the framework of Laplacean physics seems to require the existence of the individual from eternity to eternity, either in the form of a Leibnizian monad or, with the Indians and Schopenhauer, by virtue of metempsychosis. For then the individual is met by every cross-section $t = $ const. and consequently is not completely determined by a state of the world which does not 'cross' its existence. }

The above antinomy, in its most cutting form, concerns the relation of Knowing and Being. Let us assume, with Laplace and field physics, that the state of the world at a given moment, i.e. in a three-dimensional cross-section $t = $ const., determines the course of future events according to known, strictly valid, mathematical laws. It was then thought correct to infer that I can calculate the future from what I know (or can know) at a world point O, from that part of the world which at O is open to perception. *This antinomy existed formerly, but has been dissolved by relativity theory.* For the known is separated from the unknown, not by the cross-section $t = $ const. through O, but by the backward light cone. And it is a mathematical fact that, according to the field laws, the distribution of the state quantities in that cone does *not* uniquely determine the rest of the world. Only immediately *after* an action can I know all the causal premises of my action. Furthermore, it must be remembered that a complete knowledge of the present is possible only on the basis of its consequences which reach to the end of time.

[16] Compare F. Medicus, *Die Freiheit des Willens und ihre Grenzen*, 1926.

However, if we take the problem of determination as one concerning reality alone and not the relation of knowledge and reality, and if we insist on the possibility of free action in this real world, then we must demand that the content of the forward cone issuing from O shall not be completely determined by the remaining part of the world. This would contradict the pure field physics. But its ideal, like the Laplacean, lies in ruins today. From all that has been said it will be clear how little contemporary physics, based as it is half on laws and half on statistics, can pose as a champion of determinism. In quantum physics the elementary processes are not determined by strictly causal laws. If thus the world appears to us today to be much less fettered by 'inviolable laws' than at Laplace's time, it must also be emphasized that the only really consistent form of determinism which maintains the unconditional necessity of everything that happens, has never found a support in physics. Even in the Laplacean world there is an 'open place,' which can be located arbitrarily in one or the other cross-section $t =$ const.

In our natural understanding of living beings *concepts of a teleological origin* play a part beside causal law and statistics. The hand is there to grasp, the eye to see. My body is my real existence in the world. No theory of life that disregards these simple facts can succeed. An abundance of the most minute mutual adaptations of all parts is necessary in order to enable the sense organs to fulfill their task of being portals for the sensory perception of the external world. It is all too cheap to declare emphatically, as has been done by some positivistically inclined scientists, "that it is only we who read purpose into the life of an organism" or "that purposes in life are man-made, not found." (Nor is much light thrown on the relation between causality and finality by a remark such as that 'I would soon enough feel compelled to abandon the teleological view regarding my car in favor of causal analysis if, while driving to visit a friend, I have motor trouble on the way.') The multitude of adaptations to organic and inorganic environment and of functional adjustments must first be recognized as such before the attempt can be made (as was done by Darwinism) to explain them causally. The functions of the organs, the preservation of the individual, and the maintenance of life, in the particular forms once evolved, by propagation beyond the death of the individual, none of these are man-made purposes. One aspect of the purposiveness of organisms may be seen in the fact that the accomplishment of certain constant or nearly constant end effects of a morphological or functional kind is guaranteed in them, even in the face of far-reaching disturbances of the external conditions or bodily damages (regeneration). Purpose and freedom appear as two aspects of the same thing, when free purposive action is opposed to lawful necessary events. Of myself I know that I am open toward a purely spiritual world of images. Here lies the origin of my free insight and of my concern for truth, as well as of my free action and my responsibility. But I am at the same time possessed of a human body and therefore a living creature profoundly akin to all other living beings. It is thus natural to search for a seed of freedom and idea in all creatures.

Order, organization, is the characteristic of life. Hence the impression that life, in its evolution on earth, resists the plunge into the abyss of 'thermal death,' which the entropy law imposes upon inorganic matter. Bergson has coined the grandiose word 'élan vital' for this resisting power. Even a crystal, as it grows, creates order in the substance which it seizes. When water crystallizes into snow or ice it gives off heat to its environment, that is, the entropy of the environment, its 'disorder,' increases. Thus, in toto, the entropy law remains intact. Similarly an organism which grows up from the fertilized egg by progressive differentiation has the capacity to create order—at the expense of the environment, to be sure, whose entropy, or disorder, increases correspondingly. Photosynthesis in the green leaves of plants is accompanied by absorption and 'disorganization' of the incident light. Amino-acids and purine bases carried by the blood stream are synthesized into complex proteins and nucleo-proteins in the cells, and self-duplication of genes and chromosomes takes place in cell-division—while heat is released to the environment and substances of lower organization (water, carbon dioxide, urea, etc.) are discharged.

The opinion, generally held until Berzelius' time, that organic compounds may come into being only by a 'vital force' peculiar to living matter, had to be given up when Wöhler in 1828 produced urea from purely mineral material (ammonium cyanate). It is a fact that most organic compounds are much more complex than anything encountered in inorganic chemistry. A typical protein molecule consists of something like 6,000 atoms. This complexity probably accounts for some of the most characteristic features of living matter. Yet it is a complication by degree only; for E. Fischer (1901) and his successors (M. Bergmann and others) have succeeded in synthesizing a great number of polypeptides, chains of amino-acids of considerable length, which from a systematic standpoint are nothing but the simplest proteins. Hence if there is a difference in principle between life and death it is certainly not to be sought in the chemical constitution of the material substrate.

212

[This much for the facts. The physicist, with Leibniz, will not refrain from pointing out that the strict reversible natural laws, such as the law of planetary motion, are indifferent with respect to causality and finality. The law of light refraction in a medium of spatially variable coefficient of refraction (= velocity of propagation) may either be formulated differentially in the usual fashion, in which case the path of a light ray issuing from a given point A in a given direction, or arriving at a given point B from a given direction, is uniquely determined by it through stepwise integration. On the other hand, it may be formulated, with Fermat, as stating that the light, in order to get from A to B, follows that path which will require the least time. This looks as though nature accomplishes its ends in the most economical way possible. Such principles of variation, in particular Hamilton's action principle, indeed play an important part in mechanics and physics. But mathematically they are equivalent to differential laws, which establish connections only among what is infinitely closely adjacent. To see in them the expression of a purposive economy of nature is unwarranted for the further reason that Hamilton's quantity of action does not necessarily take on the smallest value possible under the given conditions; it merely takes on an extreme value, i.e. it remains unchanged under an arbitrary infinitely small alteration of the independent state quantities.

With the introduction of the concepts of probability and entropy the symmetry of causality and finality which holds for the reversible natural laws vanishes altogether. Within the theoretical construction these concepts evidently furnish an adequate expression for the metaphysical idea of causation, according to which the earlier is the determining reason of the later.

In the scheme of Newtonian physics, which was still valid for Laplace, only the initial state is free, e.g. the order of the planets in the solar system. This, to Newton, affords enough play for the activity of purposiveness, for "the wisdom and skill of a powerful ever-living agent." He is convinced that "blind fate could never make all the planets move one and the same way in orbs concentric" (later Kant and Laplace tried to give a scientific explanation of this feature). "Such a wonderful uniformity in the planetary system must be allowed the effect of choice. And so must the uniformity in the bodies of animals." (Conclusion of Newton's *Opticks*.)]

In Kant's view, purposiveness is a mere regulative principle for judging natural beings. Once we know that the eye serves to see we feel urged to find out what internal organization enables the eye by virtue of the functional laws that hold for it as well as for other physical objects, to perform this task. Kepler, on the other hand, stated: "It is not the construction of the eye which determines the peculiar ability of the mind to grasp quantitative relations, but conversely, this basic qualification of thought requires the corresponding construction of the eye." A statement like this has its place within a metaphysical interpretation of the world, such as has been attempted by Fichte, under the teleological viewpoint that the world should become conscious of itself. For the idealist this *telos* of the world is posited by its essence itself.

[Schelling and, more recently, Driesch as well as the biological school of Holists, substituted the concept of the *whole* for purposiveness. The wholeness conception in biology has been represented with a variety of nuances by Bertalanffy, Haldane, A. Meyer, Alverdes, v. Uexküll, Woltereck and others. Within psychology, gestalt psychology exhibits a similar tendency. Driesch's 'proofs' that the organic processes are incapable of 'mechanical explanation' will hardly be recognized as stringent by any physicist, since the argument is based on an altogether too limited notion of mechanical explanation of nature. Even the atomic physical processes have very little similarity with the gross macroscopic action of a machine. Every atom is already a whole of quite definite structure; its organization is the foundation for possible organizations and structures of the utmost complexity. Incidentally, more recent research on chromosomes has exhibited such physical structures preserving their genetic constitution through all cell divisions as Driesch had declared to be unthinkable. There is no reason to see why the theoretical symbolic construction should come to a halt before the facts of life and of psyche. It may well be that the sciences concerned have not as yet reached the required level. But that this limitation is neither fundamental nor permanent is already shown by psychoanalysis, in my opinion. The fact that in nature "all is woven into one whole," that space, matter, gravitation, the forces arising from the electromagnetic field, the animate and inanimate are all indissolubly connected, strongly supports the belief in the unity of nature and hence in the unity of scientific method. There are no reasons to distrust it.

Biology is certainly not applied physics or chemistry but an independent science, not only with its own subject domain but also with laws of its own; the latter may be characterized, in contrast to the general laws of matter treated in physics, as laws of specific complication (N. Hartmann). A striking testimony is the chromosome theory of inheritance, which accounts for a vast array of facts. But the great progress in biology since the turn of the century has been achieved by the same methods of generalizing induction, of comparison and experiment, which also govern physics. The fact that in the present state of biology comparative methods still prevail over experimental ones to a larger extent than in physics is not a matter of principle. During the last few decades this situation has already changed considerably. However essential the emphasis on planning and wholeness may be for the characterization and understanding of biological forms and functions, it must still be pointed out, with Kant, that this *poses* rather than *solves* a problem. The solution must be sought in the same methodical manner which we apply to wholeness concepts in physics. "The innermost of nature is penetrated by observation and analysis of phenomena, and one cannot know how far these will lead in time" (Kant). True, there are limits to cognizable necessity, and we already encounter them in the inorganic domain, in fact as it seems, in every atom. If inorganic matter behaves in such manner as follows from statistics provided the causally independent atomic processes are treated as being also statistically independent, then in the organic domain the power of life in a living whole might be theoretically representable as a non-spatial factor by statistical correlations between those elementary processes. It must be said, however, that so far the statistical-thermodynamical laws of nature have proved to be valid even within living organisms. At present we possess only comparatively primitive criteria of life which are unsuitable for precise formulation and exact research. All these questions as to the essence of life and the possibility of spontaneous generation are premature and must rest until the day when the laws of life will be known to us to a much wider extent.

The body-soul problem belongs here too. I do not believe that insurmountable difficulties will be encountered in any unprejudiced attempt to subject the entire reality, which undoubtedly is of a psycho-physical nature, to theoretical construction — provided the soul is interpreted merely as the aggregate of the real psychic acts in an individual. It is an altogether too mechanical conception of causality which views the mutual effects of body and soul as being so paradoxical that one would rather resort, like Descartes, to the occasionalistic intervention of God or, like Leibniz, to a harmony instituted at the beginning of time.}

The real riddle, if I am not mistaken, lies in the double position of the ego: it is not merely an existing individual which carries out real psychic acts but also 'vision,' a self-penetrating light (sense-giving consciousness, knowledge, image, or however you may call it); as an individual capable of positing reality, its vision open to reason; "a

215

force into which an eye has been put," as Fichte says, or "an organization turned toward two worlds at once," in the words of Schelling ("Erster Entwurf eines Systems der Naturphilosophie," *Werke*, Cotta 1858, III, p. 148). But this secret, by its very nature, lies beyond the cognitive means of natural science.

[In conclusion I want to emphasize once more that it has not been my intention to write a history of philosophical thought within the natural sciences. This would require much more comprehensive historical studies, such as have been made for instance by Lasswitz for his *Geschichte der Atomistik* or by Cassirer for his work *Das Erkenntnisproblem in der Philosophie und Wissenschaft der neueren Zeit*. Primarily interested in mathematical research, I am wanting, in both time and love, for such work.

The more I look into the philosophical literature the more I am impressed with the general agreement regarding the most essential insights of natural philosophy as it is found among all those who approach the problems seriously and with a free and independent mind rather than in the light of traditional schemes — or if not agreement then at least a common direction in their development. Whether one talks about space in the language of phenomenology like Husserl or 'physiologically' like Helmholtz is less important, in view of their substantial concordance, than it appears to the 'standpoint philosophers' who swear by set formulas.

Exact natural science, if not the most important, is the most distinctive feature of our culture in comparison to other cultures. Philosophy has the task to understand this feature in its peculiarity and singularity. The ideas collected here, which have their firm foundation in the first, mathematical part, should be looked upon as endeavors toward this end — although I cannot but admit that the task is at present far from being completely accomplished.]

REFERENCES

D. HUME, *Treatise on Human Nature*.
—— *Enquiry Concerning Human Understanding*.
I. KANT, *Kritik der Urteilskraft*.
E. KOENIG, *Die Entwicklung des Kausalproblems*, 1888–90.
E. WENTSCHER, *Geschichte des Kausalproblems in der neueren Philosophie*, 1921.
J. S. MILL, *A System of Logic, Ratiocinative and Inductive*, many editions.
H. POINCARÉ, *La valeur de la science*, Paris, 1905. (See also H. WEBER's remarks on causality in the German edition, Leipzig, 1910.)
E. MACH, *Analyse der Empfindungen*, third edition, 1902, Chap. V.
F. EXNER, *Vorlesungen über die physikalischen Grundlagen der Naturwissenschaften*, 1919.
P. FRANK, *Le principe de causalité et ses limites*, Paris, 1937.

M. Planck, *Wege zur physikalischen Erkenntnis, Reden und Vorträge,* second edition, Leipzig 1934.

N. Bohr, *Atomic Theory and the Description of Nature,* Cambridge, 1934.
—— Kausalität und Komplementarität, *Erkenntnis* 14 (1937), p. 293.

L. Rosenfeld, L'évolution de l'idée de causalité, *Mem. Soc. Roy. Sci. Lièges,* 4e série, VI (1942), pp. 59–86.

E. Cassirer, *Determinismus und Indeterminismus in der modernen Physik,* Göteborg, 1936.

N. K. Brahma, *Causality and Science,* London, 1939.

A. D'Abro, *The Decline of Mechanism,* New York, 1939.

F. Gonseth, *Déterminisme et libre arbitre,* Neuchâtel, 1944.

P. S. de Laplace, *Théorie analytique des probabilités,* third edition, Paris, 1820.

G. Boole, *An Investigation of the Laws of Thought,* London, 1854.

J. von Kries, *Die Prinzipien der Wahrscheinlichkeitsrechnung,* Tübingen, 1886.

J. Venn, *The Logic of Chances,* third edition, London, 1888.

M. von Smoluchowski, Über den Begriff des Zufalls und den Ursprung der Wahrscheinlichkeitsgesetze in der Physik, *Die Naturwissenschaften,* 6 (1918), pp. 253 ff.

P. and T. Ehrenfest, Begriffliche Grundlagen der statistischen Auffassung in der Mechanik, *Enzyklopädie der mathematischen Wissenschaften,* IV (Mechanik), Art. 32.

J. M. Keynes, *A Treatise on Probability,* London, 1929.

R. von Mises, *Probability, Statistics and Truth,* New York, 1939 (German edition, Vienna, 1928).

E. Nagel, Principles of the Theory of Probability, *Internat. Encycl. Unif. Sci.,* I, No. 6, Chicago, 1939.

A Symposium on Probability, I, II, III, in: *Philosophy and Phenomenological Research,* 5, pp. 449–532; 6, pp. 11–86, 590–622 (1945).

D'Arcy W. Thompson, *On Growth and Form,* Cambridge, 1917.

W. Köhler, *Die physischen Gestalten in Ruhe und im stationären Zustand,* Erlangen, Philosophische Akademie, 1924.

H. Friedmann, *Die Welt der Formen, System eines morphologischen Idealismus,* Munich, 1930.

C. R. Darwin, *The Origin of Species,* London, 1859.

H. Driesch, *Analytische Theorie der organischen Entwicklung,* Leipzig 1894.
—— *Philosophie des Organischen,* Leipzig, 1909.
—— Metaphysik der Natur, in *Handbuch der Philosophie,* edited by Baeumler and Schröter, Vol. 2, Munich, 1927.

J. Schaxel, *Grundzüge der Theorienbildung in der Biologie,* Jena, 1919.

P. Jensen, *Organische Zweckmässigkeit, Entwicklung und Vererbung,* Jena, 1907; *Über den chemischen Unterschied zwischen dem lebendigen und toten Organismus,* Munich and Wiesbaden, 1921.

J. von Uexküll, Die Rolle des Subjekts in der Biologie, *Naturwissenschaften,* 19 (1931), p. 385.

H. Bergson, *L'Évolution créatrice,* Paris, many editions.

J. Huxley, *Evolution, The Modern Synthesis,* London, 1942.

C. H. Waddington, *An Introduction to Modern Genetics,* London, 1939.

J. Reinke, *Das dynamische Weltbild,* Leipzig, 1926.

F. Medicus, *Die Freiheit des Willens und ihre Grenzen,* Tübingen, 1926.

E. Cassirer, *Das Erkenntnisproblem in der Philosophie und Wissenschaft der neueren Zeit,* 3 Vols., Berlin, 1911–20.
—— *Philosophie der symbolischen Formen, Vol. 3, Phänomenologie der Erkenntnis,* Berlin, 1929.

B. Bavink, *Ergebnisse und Probleme der Naturwissenschaften, Eine Einführung in die heutige Naturphilosophie,* eighth edition, Bern, 1945.

M. Schlick, Naturphilosophie, in M. Dessoir, *Lehrbuch der Philosophie,* Vol. 2, Berlin, 1925;
—— *Gesammelte Aufsätze,* Vienna, 1938.

M. Hartmann, *Philosophie der Naturwissenschaften*, Berlin, 1937.

A. N. Whitehead, *An Enquiry concerning the Principles of Natural Knowledge*, Cambridge, 1919.

M. R. Cohen, *Reason and Nature, An Essay on the Meaning of Scientific Method*, New York, 1931.

A. S. Eddington, *The Philosophy of Physical Science*, Cambridge and New York, 1939.

D. de Santillana and E. Zilsel, The Development of Rationalism and Empiricism, in *Internat. Encycl. Unif. Sci.*, II, No. 8, Chicago, 1941.

Appendices

The Structure of Mathematics

1. The aim of Hilbert's "Beweistheorie" was, as he declared, "die Grundlagenfragen einfürallemal aus der Welt zu schaffen." In 1926 there was reason for the optimistic expectation that by a few years' sustained effort he and his collaborators would succeed in establishing consistency for the formal equivalent of our classical mathematics. The first steps had been inspiring and promising indeed. But such bright hopes were dashed by a discovery in 1931 due to Kurt Gödel, which questioned the whole program. Since then the prevailing attitude has been one of resignation. The ultimate foundations and the ultimate meaning of mathematics remain an open problem; we do not know in what direction it will find its solution, nor even whether a final objective answer can be expected at all. "Mathematizing" may well be a creative activity of man, like music, the products of which not only in form but also in substance are conditioned by the decisions of history and therefore defy complete objective rationalization. The undecisive outcome of Hilbert's bold enterprise cannot fail to affect the philosophical interpretation. Yet I find little to change in what I said about it in this book in 1926, although I should probably now set my words a little more cautiously.

Gödel showed that in Hilbert's formalism, in fact in any formal system M that is not too narrow, two strange things happen: (1) One can point out arithmetic propositions Φ of comparatively elementary nature that are evidently true yet cannot be deduced within the formalism. (2) The formula Ω that expresses the consistency of M is itself not deducible within M. More precisely, a deduction of Φ or Ω within the formalism M would lead straight to a contradiction in M, i.e. to a deduction in M of the formula $\sim (1 = 1)$. From the first fact one learns that the fields of propositions accessible to insight on the one hand and to deduction on the other overlap, neither of the two being contained in the other. Symbolic mathematics, while in some directions going far beyond what is capable of verification based on evidence, in other directions accomplishes less. Although the idea of a transcendental world existing and complete in itself is the guiding principle in building up our formalism, that formalism at any fixed stage has the character of incompleteness, inasmuch as there will always be problems, even problems of a simple arithmetical nature, that can be formulated within the formalism and decided by insight, but not decided by deduction within the formalism. We are not surprised that a concrete chunk of nature, taken in its isolated phenomenal existence, challenges our analysis by its inexhaustibility and incompleteness; it is for the sake of completeness, as we have seen, that physics projects what is given onto the background of the possible. But it is surprising that a construct created by mind itself, the sequence of integers, the simplest and most diaphanous thing for the constructive mind, assumes a similar aspect of obscurity and deficiency when viewed from the axiomatic angle.

Gödel's second theorem is even more disquieting, for it confronts us with this alternative: either the reasoning by which consistency of the formalism is established must contain some argument that has no formal counterpart within the system, i.e. we have not succeeded in completely formalizing the procedure of mathematical induction; or the idea of a strictly 'finitistic' proof of consistency must be given up altogether. As a matter of experience, mathematical induction in whatever form it has so far been used by mathematicians in their actual research falls under the scheme adopted as an axiomatic rule in Hilbert's formalism. Thus the hopes for a finitistic proof of consistency have become dim indeed. G. Gentzen's ingenious proof of consistency for arithmetic (1936) is not finitistic in Hilbert's sense; the price of a substantially lower standard of evidence is exacted from him, and he is forced to accept as evident a type of inductive reasoning that penetrates into Cantor's "second class of ordinal numbers." Thus the boundary line of what is intuitively trustworthy has once more become vague. After this Pyrrhic victory nobody had the courage to carry arms into the field of analysis; yet it is here that the ultimate test for Hilbert's conception would lie.

Gödel's construction is closely connected with the *logical paradoxes* that had provoked so many discussions among the ancient philosophers[1] and after a long period of oblivion have again during the last fifty years become a ferment in the development of our thoughts about

[1] The Socratic philosophers of the school of Megara, Euclides, Eubulides, etc., revelled in paradoxes of this sort, which clearly belong in a class different from that of the Eleatic paradoxes of motion as formulated by Zeno. Aristotle dedicates to them a whole book, *De Sophisticis Elenchis;* the Stoic Chrysippus dealt extensively with them (see the list of titles of Chrysippus's logical treatises in *Diogenes Laertius*, VII, 189–198); under the Roman Empire they formed part of the regular school curriculum in dialectics. The medieval scholastic development culminates in Paulus Venetus (died 1428) (*Logica Magna*, Venetiis 1499, fol., in particular *De Insolubilibus*, 192r. B et seq.). Typical for the attitude of most modern philosophers is C. Prantl's contemptuous remark in his classical *Geschichte der Logik im Abendlande:* "Lappalien, wie die Mehrzahl der Fangschlüsse sind, wird die wahre Logik überhaupt gar nicht berücksichtigen" (I, p. 95).

the foundations of mathematics. For this reason and because of the deep insight it affords into the structure of mathematics and thereby of all theoretical science, I shall attempt to describe Gödel's discovery with as much precision as is possible without becoming too technical.

2. The symbolism outlined in Section 10 will be adopted with the following modifications (formalism H). In the manner indicated loc. cit. we differentiate between factual and numerical formulas. The rule

$$(1) \qquad (b = c) \rightarrow (A(b) \rightarrow A(c))$$

is adopted for factual formulas A only; for numerical formulas A it is to be replaced by

$$(b = c) \rightarrow (A(b) = A(c)).$$

Closed numerical and factual formulas will be called formal numbers and formal propositions respectively. Since we wish to restrict ourselves to arithmetic, we drop the symbol ε and the transfinite set-theoretic rule (I) on p. 58. As our variables will then stand for 'arbitrary natural numbers' only (and not for sets of such numbers and the like), the operator N and the two rules concerning it,

$$N1; \quad Nb \rightarrow N(\sigma b),$$

are to be discarded. We let the sequence of natural numbers start with 0 instead of 1 and replace the symbol σ by $'$. From the outset we add the axioms that contain the inductive definition of addition and multiplication of numbers. Hence the specific arithmetic rules (in which b and c can be any formal numbers) now read

$$('b = 'c) \rightarrow (b = c).$$
$$\sim ('b = 0).$$
$$b + 0 = b. \quad b + 'c = '(b + c).$$
$$b \cdot 0 = 0. \quad b \cdot 'c = (b \cdot c) + b.$$

The principle of mathematical induction is introduced in the form

$$\Pi_x(A(x) \rightarrow A('x)) \rightarrow (A(0) \rightarrow A(b))$$

in which $A(x)$ may be any x-predicate ($=$ factual formula containing no free variable except possibly x) and b any formal number.

The axioms

$$(2) \qquad A(b) \rightarrow \Sigma_x A(x), \quad \Sigma_x A(x) \rightarrow A(\varepsilon_x A(x))$$

are of course adopted for *any* variable x occurring in our symbolism.

In the main text the axiomatic rules of the elementary calculus of propositions have not been enumerated explicitly. We mention the following three because we shall need them later:

221

1. $(b \to c) \to ((a \to b) \to (a \to c))$.
2. $(a \to \sim a) \to \sim a$.
3. $(a \to b) \to (\sim b \to \sim a)$

(a, b, c, any formal propositions). If in the game of deduction two formulas of the type $a \to b$ and $b \to c$ have arisen, we may use Rule 1 and the syllogism to pass on from them to the formula $a \to c$. In the same manner we can pass from a formula $a \to \sim a$ to $\sim a$, and from the implication $a \to b$ to the 'inverted' implication $\sim b \to \sim a$.

In our metamathematical reasoning we make use of the (actual) numbers

(3) $$0, \quad '0, \quad ''0, \quad '''0, \ldots$$

(sometimes referred to by the customary abbreviations 0, 1, 2, 3, . . .) and of a non-axiomatic theory of numbers which is based on the possibility of passing from any given number ν to the next $'\nu$. Hence we introduce functions $\mathfrak{f}(\nu)$ and properties $\mathfrak{F}(\nu)$ by recursive definitions that enable us actually to compute the values of $\mathfrak{f}(\nu)$ for $\nu = 0, 1, 2, 3,$ . . . one after the other, or to decide in the same manner for each $\nu = 0, 1, 2, 3, \ldots$, whether or not $\mathfrak{F}(\nu)$ holds. For instance we introduce a function $\mathfrak{f}$ by the recursive definition

$$\mathfrak{f}(0) = 0, \quad \mathfrak{f}('\nu) = ''\mathfrak{f}(\nu),$$

from which we compute

$$\mathfrak{f}(0) = 0, \quad \mathfrak{f}(1) = 2, \quad \mathfrak{f}(2) = 4, \quad \mathfrak{f}(3) = 6, \ldots.$$

Or we introduce the complementary properties $\mathfrak{F}$, $\overline{\overline{\mathfrak{F}}}$ of *even* and *odd* by a procedure common to all armies the world over, that of 'counting off by two's': '0 is even; an even ν is succeeded by an odd $'\nu$ and an odd ν by an even $'\nu$.' Recursive properties turn up in such complementary pairs $\mathfrak{F}$ and $\overline{\overline{\mathfrak{F}}}$.

The intuitively defined function $\mathfrak{f}(\nu)$ must be distinguished from the numerical formula $x + x$, which for the moment may be indicated by the abbreviation $f(x)$. If μ is the value of $\mathfrak{f}(\nu)$ for a given number ν, then the formula $f(\nu) = \mu$ is *deducible* in the formalism; e.g. because $\mathfrak{f}(2) = 4$, the formula $''0 + ''0 = ''''0$ is deducible. In order to prove this in general one has to deduce the formula $\Pi_x(('x + 'x) = ''(x + x))$, abbreviated $\Pi_x(f('x) = ''f(x))$, which yields the formula $f('a) = ''f(a)$ not merely for an actual but for any formal number a. In the same manner the x-predicate $F(x) \equiv \Sigma_y(x = (y + y))$ is a formal equivalent of the property $\mathfrak{F}$ of being even. Indeed, if $\mathfrak{F}(\nu)$ holds for a given number ν (if ν is even) then the formula $F(\nu)$ is deducible; but if $\overline{\overline{\mathfrak{F}}}(\nu)$ holds (if ν is odd) then the formula $\sim F(\nu)$ is deducible. Our formalism must be wide enough to allow representation of recursively defined functions and relations by numerical and factual formulas to the extent that our construction will require. Incidentally any recursive property $\mathfrak{F}(\nu)$ can be written in the form of an equation $\mathfrak{f}(\nu) = 0$ where $\mathfrak{f}$ is a recursively defined function.

We are now in a position to describe more precisely what Gödel's first theorem consists in. He constructs inductively a definite property $\mathfrak{F}$ which has an x-predicate F as its formal equivalent, such that no actual number ν has the property $\mathfrak{F}$, while the formula $\sim \Sigma_x F(x)$ (which symbolizes the statement 'No number has the property $\mathfrak{F}$') is not deducible in the formalism. Since $\overline{\overline{\mathfrak{F}}}(\nu)$ holds, $\sim F(\nu)$ is deducible. Use the abbreviation a for $\epsilon_x F(x)$. Considering that $\Sigma_x F(x) \rightarrow F(a)$ or $\sim F(a) \rightarrow \sim \Sigma_x F(x)$, we face this situation: $\sim F(\nu)$ is deducible whenever ν is an actual number, namely one of the formulas (3); however $\sim F(a)$ is not deducible for a certain formal number a explicitly constructed.

3. 'Paradoxy' enters into Gödel's construction in the form of the "diagonal process" by which Cantor proved that the continuum is not denumerable. Writing a real number in the interval $\langle 01 \rangle$ as a dual fraction, Cantor replaces that number by an infinite sequence R of 0's and 1's,

$$R = r_1 r_2 \ldots r_\nu \ldots ; \qquad (r_\nu = 0 \text{ or } 1).$$

The assumption that all such 'dual fractions' can themselves be arranged in a sequence $R^{(1)}, R^{(2)}, \ldots$,

$$R^{(\mu)} = r_1^{(\mu)} r_2^{(\mu)} r_3^{(\mu)} \ldots \qquad (\mu = 1, 2, 3, \ldots)$$

leads to a contradiction. Indeed, write them in this order, one below the other, and then go along the diagonal transmuting every 0 into a 1, every 1 into a 0; you obtain a sequence $Q = q_1 q_2 \ldots$ that differs from $R^{(1)}$ in the first place, $q_1 \neq r_1^{(1)}$, from $R^{(2)}$ in the second place, $q_2 \neq r_2^{(2)}$, etc., and therefore coincides with none of the R's in the sequence. This contradiction proves the impossibility of enumerating all dual fractions R.

On interpreting 0, 1 as the two truth values true and false, a sequence like $R = r_1 r_2 \ldots$ becomes equivalent to the *predicate* $R(x)$ that holds for a number $x = \nu$ if $r_\nu = 0$, and does not hold if $r_\nu = 1$. Our entire matrix $r_\nu^{(\mu)}$ of 0's and 1's becomes equivalent to a binary relation $S(y, x)$ that obtains for the numbers $x = \nu$, $y = \mu$ if $r_\nu^{(\mu)} = 0$ and does not obtain if $r_\nu^{(\mu)} = 1$. The predicate $R^{(\mu)}(x)$ coincides with $S(\mu, x)$. Cantor's construction amounts to forming the predicate $\sim S(x, x)$. Should it coincide with one of the predicates $R^{(\mu)}(x)$, say with $R^{(\gamma)}(x) = S(\gamma, x)$, then $S(\gamma, x)$ would hold for a number x whenever $\sim S(x, x)$ holds, and *vice versa*. For $x = \gamma$ this leads to the contradiction that $\sim S(\gamma, \gamma)$ implies $S(\gamma, \gamma)$ and *vice versa*. This of course is no paradox, but simply an indirect proof for the non-denumerability of the continuum of dual fractions (or of the predicates of natural numbers).

It becomes, however, a paradox, as was first pointed out by Richard in 1905, when one holds the view that any real number is definable by a (written or printed) English text. A text is a succession of elementary signs, and there are no more than about 50 different such signs, namely the letters of the alphabet (we can do without capitals!), the digits from 0 to 9, and the signs of punctuation, including the lacuna of one em and the brackets. Call the number of signs of which a text is composed its length. As there is only a limited number of sequences of signs of given length we can enumerate all texts, writing down first the sequences of length 1, then the sequences of length 2, etc. It is a mechanization of this method on which the 'Lullian art' was based (Ramon Lull, died 1315) and which Swift's "projector in speculative learning" at the Grand Academy of Lagado (*Gulliver's Travels*, visit to Laputa) practices. Thus whatever is definable, in particular every dual fraction, will find its place in a file arranged according to the numbers 1, 2, 3, ...

In this form the paradox cannot be discussed mathematically because it refers to the meaning of sentences in the English language, which is of course a somewhat vague affair. But there are better means for definition than language. At the very beginning of this book, in Section 1, we spoke of the logical structure of propositions and propositional functions and saw that properties of numbers are defined by iterated combined application of a few principles of construction, described there on pp. 5–6. Afterwards we learned that we have to forego use of the quantifiers Σ_x and Π_x if we insist that our propositions have an intuitively verifiable meaning; instead a principle of definition by induction is to be added. In a symbolic formalism, however, where the question of meaning and truth is not raised, we may even include something like Hilbert's ϵ_x. But whether we describe x-predicates by a systematized process of recursive definitions (system Δ) or by a formalism M, we can always enumerate the constructible x-predicates (so that this or that among them may be referred to as property No. 17 or property No. 919). We then form the binary relation with two variables $S(z, x)$, 'x has the property No. z.' Richard's paradox is inevitable provided (i) the propositions of our system are decidable and (ii) the relation S itself is constructible. For Δ the assumption (i) is fulfilled, and we therefore conclude that S is not constructible. It was essentially in this form that I discussed and solved Richard's paradox in my book, *Das Kontinuum*,

1918. Gödel however discovered that in a formalism M of sufficient latitude S is constructible, and he therefore concludes that not all of the formulas of M are decidable.

4. From these heuristic arguments let us now turn to an exact description of Gödel's procedure. Before we can enumerate all possible formulas we must label by numbers all the symbols which our formalism H makes use of. In the following table the number κ for each symbol is written under it. Room is left for an *unlimited supply*

$$0, \sim, \rightarrow, \&, \mathbf{v}, =, ', +, \cdot \quad \text{[constants and operators]}$$
$$1, \ 6, \ 11, 16, 21, 26, 31, 36, 41 \quad \text{[numbers of the form } 5n + 1]$$

$x, y, z, u, \cdots$ [variables] 2, 7, 12, 17, $\cdots$ [numbers $5n + 2$]	$\Sigma_x, \Sigma_y, \Sigma_z, \Sigma_u, \cdots$ 3, 8, 13, 18, $\cdots$ [$5n + 3$]
$\Pi_x, \Pi_y, \Pi_z, \Pi_u, \cdots$ 4, 9, 14, 19, $\cdots$ [$5n + 4$]	$\epsilon_x, \epsilon_y, \epsilon_z, \epsilon_u, \cdots$ 5, 10, 15, 20, $\cdots$ [$5n + 5$]

of variables. (This essential feature of the formalism would have forced us anyhow to abandon letters as symbols of variables and to use letters with numbers as indices, like x_7, instead.) The numbers 46, 51, . . . , are kept in reserve for constants and operators that may be introduced later when an extension of the formalism is considered. The first two variables x, y, with the numbers 2 and 7, will play a special role.

A formula is a 'tree' of symbols where each symbol is immediately followed by at most two 'descendents.' In accordance with the Platonic diagram, p. 53, we can therefore mark the several places of the tree as 'numerals' $\mathfrak{i}$, i.e. as finite sequences of 0's and 1's that start with a 1. The head bears the mark 1, and the marking is continued throughout the tree according to the rule

$$\begin{array}{ccc} \mathfrak{i} & & \mathfrak{i} \\ | & & / \ \backslash \\ \mathfrak{i}0 & & \mathfrak{i}0 \quad \mathfrak{i}1 \end{array} \qquad \text{(for one and two descendents respectively)}.$$

In the Platonic fashion we interpret a 'numeral'

$$\mathfrak{i} = r_1 \cdots r_h \quad (r_1 = 1; r_2, \cdots, r_h = 0 \text{ or } 1)$$

as the dyadic representation of the number

$$i = r_1 2^{h-1} + r_2 2^{h-2} + \cdots + r_h.$$

Thus the description of the formula consists in giving for each place mark i of its tree the label κ_i of the symbol occupying that place.

225

Take for instance the simple formula

(4) $\qquad \sim (('0 + x) = y) \quad$ (or abbreviated $1 + x \neq y$).

The skeleton of the tree is printed beside it, its places marked the first time by the numerals i and a second time by the corresponding numbers i in the usual decimal writing. On looking up the labels κ of

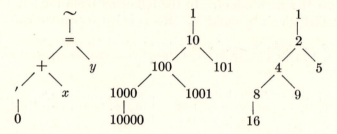

the symbols $0, \sim, =, ', +, x, y$ in our table we find for this formula (4) the following list of κ's by which it is completely characterized:

$$\kappa_1 = 6, \quad \kappa_2 = 26, \quad \kappa_4 = 36, \quad \kappa_5 = 7, \quad \kappa_8 = 31, \quad \kappa_9 = 2, \quad \kappa_{16} = 1.$$

Let now $\pi_1, \pi_2, \ldots$ be the prime numbers in their natural order,

$$\pi_1 = 2, \quad \pi_2 = 3, \quad \pi_3 = 5, \quad \pi_4 = 7, \cdots.$$

(According to Euclid the series is infinite.) To the formula tree in which the place i is occupied by the symbol labeled κ_i we ascribe the number (characteristic)

(5) $\qquad \nu = \Pi_i \pi_i{}^{\kappa_i},$

the product extending over the several place marks i of the tree. For instance the characteristic of the formula (4) is

$$\pi_1^6 \pi_2^{26} \pi_4^{36} \pi_5^7 \pi_8^{31} \pi_9^2 \pi_{16}^1 \quad \text{or} \quad 2^6 \cdot 3^{26} \cdot 7^{36} \cdot 11^7 \cdot 19^{31} \cdot 23^2 \cdot 53$$

(which is quite some number!). Because the factorization of a number into primes is unique, distinct formulas have distinct characteristics, and for any given number ν we can uniquely determine (i) whether it corresponds to a formula and (ii) if so what that formula is.

The pattern P of a *deduction*, when read in the direction from conclusion to premisses is a tree of (closed factual) formulas: each member c of the tree has either no direct descendent, and then it is an axiom; or it has two, a minor and a major, and if the minor is b, then the major is $b \rightarrow c$. The formula at the head is the one *deduced* by the pattern. Having characterized every conceivable formula by a number ν we can again apply the principle on which the formation (5) of the characteristic of a tree of symbols is based, and thus determine the characteristic number μ of a deduction P by the product $\Pi_i \pi_i{}^{\nu_i}$ where ν_i is the number of the formula occupying the place marked i in our tree P and the product extends over all the place marks i of P.

Simple as this method of arithmetizing a formalism is, it leads to an insight of considerable philosophical interest — namely that the natural numbers with their arithmetic constitute a field so wide that any theory (once it is completely formalized) can be mapped into it. This amazing power of number, which Pythagoras and Plato recognized more or less clearly, and which Swift made fun of, was utilized by Gödel for the purpose of the metamathematical study of a given mathematical formalism.

If ν is a number, written out explicitly as one of the formulas (3), and μ is the number of an x-predicate $F(x)$, then we denote by $\mathfrak{s}(\mu, \nu)$ the number of the formula $F(\nu)$; if, however, μ is not the number of an x-predicate, we set $\mathfrak{s}(\mu, \nu) = 0$. The formula $F(\nu)$ of course arises from $F(x)$ by substituting the number ν for x wherever x occurs free in F. This function $\mathfrak{s}$ of two arbitrary numbers μ, ν can be defined by induction. It is a fact that in the formalism H we can construct a numerical formula $s(z, x)$ with the two free variables x, z that is the formal counterpart of $\mathfrak{s}$ in the sense previously described: if μ and ν are any two actual numbers and $\mathfrak{s}(\mu, \nu)$ equals κ then the formula $s(\mu, \nu) = \kappa$ is deducible. $\mathfrak{t}$ and $t(z, y)$ have the corresponding significance for y-predicates: λ being the number of any given y-predicate $F(y)$, $\mathfrak{t}(\lambda, \nu) = \kappa$ is the number of $F(\nu)$, and the equation $t(\lambda, \nu) = \kappa$ is deducible under these circumstances.

If μ, ν are any two numbers we can decide whether ν is the characteristic of a formal proposition a and μ the characteristic of a pattern P of deduction the head formula of which is a. Let $\mathfrak{D}(\mu, \nu)$ (read: the pattern No. μ is the deduction of formula No. ν) denote this relation between two arbitrary numbers μ and ν, and $\overline{\mathfrak{D}}(\mu, \nu)$ the complementary relation. $\mathfrak{D}$ and $\overline{\mathfrak{D}}$ can be defined by recursion. The formalism H allows representation of $\mathfrak{D}$ by a formula $D(y, x)$ with two free variables x, y: whenever μ and ν are numbers for which $\mathfrak{D}(\mu, \nu)$ holds, the formula $D(\mu, \nu)$ is deducible; whenever $\overline{\mathfrak{D}}(\mu, \nu)$ holds, $\sim D(\mu, \nu)$ is deducible.

Construction of the two formulas $s(z, x)$ and $D(y, x)$ is a minutely detailed and somewhat laborious but not a particularly difficult task.

The Cantor-Richard paradox depends on an operator S performing substitution. If we had a formula $S(z, x)$ with two free variables x and z such that for any formal number a and any x-predicate $F(x)$ the formulas

$$(6) \qquad F(a) \rightarrow S(\varphi, a) \quad S(\varphi, a) \rightarrow F(a)$$

227

were deducible, φ denoting the number of the formula $F(x)$, then the system would certainly be contradictory. The s here constructed cannot serve this purpose, for the simple reason that $s(\mu,\ \nu)$ is a numerical rather than a factual formula. But it is obvious how to proceed: $S(\mu,\ \nu)$ ought to stand for the statement that the formula with the number $s(\mu,\ \nu)$ is true. Interpreting here truth as deducibility, one arrives at this definition of S in our formalism:

$$S(z,\ x) \equiv \Sigma_y\, D(y,\ s(z,\ x)).$$

It is a task for further inquiry to ascertain the extent to which formulas like (6) are deducible for this S.

After this intermezzo we resume our systematic exposition. Form the x-predicate $\sim S(x,\ x)$, i.e.

$$\sim \Sigma_y\, D(y,\ s(xx))$$

and compute its characteristic γ (a number of fantastic magnitude!). Substitute in this formula of characteristic γ the number γ for x, thus obtaining the formula

(7) $$\sim \Sigma_y\, D(y,\ s(\gamma\gamma))$$

with the number $\mathfrak{s}(\gamma,\gamma) = \beta$. Of (7) it may therefore be said that *it states its own untruth.*

Here the classical paradox of the *pseudomenos* comes to mind.[2] Its driest form is the sentence: 'This statement (the statement which I now make) is false.' Like language, our formalism enables us to

---

[2] Eubulides is the inventor of this paradox, according to Diogenes Laertius's testimony (*D.L.*, II, 108). Aristotle describes it and gives his own solution in *Soph. Elench.*, *25*, 180a, 35 ff. Of ancient formulations which have come down to us I mention Cicero, *Academica*, II, 29: "Si te mentiri dicis idque verum dicis, mentiris an verum dicis?"; and Alexander Aphrodisiensis (about A.D. 200) *ad Soph. Elench.*, Aldina f. 54 r. [M. Wallies, *Comm. in Arist. Graeca*, Vol. II, pars III, Berlin, 1898, p. 171, l. 18], ἀλλὰ μὴν ὁ λέγων "ἐγὼ ψεύδομαι" ἅμα καὶ ψεύδεται καὶ ἀληθεύει. Athenaeus, *Deipnosophists*, IX, 401e, gives a distichon commemorating the poet Philitas of Cos (Theocritus's master) who allegedly was killed by his vain attempts to solve the paradox. The pseudomenos is the chief topic of at least seven of Chrysippus's logical treatises. As governor of the island Barataria, Sancho Panza is faced with a problem of the pseudomenos type, and his Solomonic wisdom finds a drastic solution (Cervantes, *Don Quixote*, II, 51). For a detailed history of the pseudomenos cf. Alexander Rüstow, *Der Lügner*, Leipzig, 1910. The similar but less pointed form of the "Epimenides" belongs to the Christian era. Sending Titus to preach the gospel to the Cretans, Paul warns him (*Ep. to Titus*, I, 12): "One of themselves, even a prophet of their own, said:

κρῆτες ἀεὶ ψεῦσται, κακὰ θηριά, γαστέρες ἀργαί.
The Cretans are always liars, evil beasts, slow bellies,"

and sensing no paradox, Paul adds: "This witness is true." Early patristic tradition (Clemens Alexandrinus, *Stromata*, I, 59, ed. Stählin, Leipzig, 1906, II, p. 37) identifies the "prophet" with Epimenides (cf. Diels, *Fragmente der Vorsokratiker*, fourth state the truth of a sentence (here a formula) by a sentence (formula); indeed, for any closed formula a we can form $\Sigma_y D(y, a)$. Beyond this the paradox of the pseudomenos depends on the demonstrative words 'this,' 'I,' 'now,' by which the meaning of the sentence explodes ("cassatio" in the terminology of Paulus Venetus). Here this paradoxical character is brought about by Cantor's diagonal process. But no explosion takes place: γ and β are definite numbers the computation of which is described with perfect clarity and unambiguity.

$\{F(y) \equiv D(y, s(\gamma\gamma))$ is the formal counterpart of $\mathfrak{F}(\mu) = \mathfrak{D}(\mu, \beta)$. Indeed if μ is any number for which $\mathfrak{D}(\mu, \beta)$ holds then $D(\mu, \beta)$ is deducible. Moreover $\beta = s(\gamma\gamma)$ is deducible. Hence the following instance of the axiomatic rule (1),

$$(\beta = s(\gamma\gamma)) \rightarrow \{D(\mu, \beta) \rightarrow D(\mu, s(\gamma\gamma))\},$$

yields in two syllogistic steps the formula $D(\mu, s(\gamma\gamma))$. For similar reasons $\sim D(\mu, s(\gamma\gamma))$ is deducible if $\overline{\mathfrak{D}}(\mu, \beta)$ holds. As previously announced, we show two things: (i) the formula $\sim \Sigma_y F(y)$ is not deducible; (ii) and yet if μ is any given number then $\overline{\overline{\mathfrak{F}}}(\mu)$ holds. The proof depends on an assumption and a fact: (A) the formalism is consistent; (B) it is wide enough to provide us with formal counterparts s and D of the function $\overline{s}(\mu, \nu)$ and the relation $\mathfrak{D}(\mu, \nu)$.

The proof for (i) is indirect. Suppose we had deduced the formula (7) whose characteristic number is β. We should then have a definite number μ such that $\mathfrak{D}(\mu, \beta)$ holds. Therefore $D(\mu, \beta)$ and $D(\mu, s(\gamma\gamma))$ are deducible. But

$$D(\mu, s(\gamma\gamma)) \rightarrow \Sigma_y D(y, s(\gamma\gamma))$$

according to the general axiomatic rule $A(b) \rightarrow \Sigma_y A(y)$. Hence we have deduced the formula $\Sigma_y D(y, s(\gamma\gamma))$ as well as (7), and that constitutes an actual contradiction.

(ii) If μ is any number, $\mathfrak{D}(\mu, \beta)$ will not hold. For if it did we would have an actual deduction of the formula (7) whose characteristic

---

ed., II, Berlin, 1922, Epimenides fr. 1, pp. 188–189), and it is not unlikely that someone like Clement or Jerome who had been educated in 'pagan' dialectics linked the pseudomenos to Paul's disparagement of the Cretans. However, no earlier reference than Angelus Politianus (1454–1494), Ep. ad Manutium (*Opera omnia*, Basel, 1553, p. 91), seems to be known. Phocylides (around 530 B.C.) is mentioned by Strabo, *Geography*, 10.487, as the author of the following statement that resembles the lying Cretan but blunts the sharp edge of paradox: "The Lerians are bad people; not merely this one or that one, but all, except Proclees; and Proclees is a Lerian." A variation substituting the inhabitants of Chios for the Lerians is ascribed, we do not know with what authenticity, to Demodocus in *Anth. Pal.*, 11.235.

number is β — contrary to what we have proved under (i). Hence $\overline{\mathfrak{D}}(\mu, \beta)$ holds.

Thus Gödel's first theorem is proved. The second results from it by the observation that the reasoning by which we proved the first theorem can be transmuted into a formal deduction. Exact formulation of the theorem must make use of an abbreviation 'neg' that arithmetizes the negation $\sim$. If ν is the number of a factual formula a then $\sim a$ has a certain number $\mathfrak{neg}(\nu)$; if ν is not the number of a factual formula we set $\mathfrak{neg}(\nu) = 0$. The arithmetical function $\mathfrak{neg}$ thus defined has its formal counterpart neg. Consistency may now be expressed by the following formula $\Omega(b)$ involving an arbitrary formal number b:

$$\Sigma_y D(y, b) \rightarrow \sim \Sigma_y D(y, \text{neg } b).$$

We then maintain: *If the formalism is actually consistent, i.e. if the formula* $\sim (0 = 0)$ *cannot be deduced, then we are able to point out a definite instance of the formula of consistency* $\Omega(b)$ *that is not deducible.*

From our argument it should also be clear what that particular instance is. Let φ be the number of the y-predicate $D(y, s(\gamma\gamma))$ and a an abbreviation for $\epsilon_y D(y, s(\gamma\gamma))$. Then we choose for b the formula $t(\varphi, a)$. This is not the place to go through the various steps of the argument. It boils down to a description of a definite process by which a (hypothetic) deduction of the specific instance $\Omega(b)$ of the *formula of consistency* may be transformed into a deduction of $\sim (0 = 0)$, i.e. of *actual inconsistency.* }

5. Since Gödel has left us little hope that a formalism wide enough to encompass classical mathematics will be supported by a proof of consistency, the axiomatic systems developed before Hilbert without such ambitious dreams gain renewed interest. Here the logical terms 'not' $\sim$, 'if then' $\rightarrow$, 'there is' Σ_x, etc., were still understood in their meaning, and deduction took place by that sort of transcendental logic on which one is used to rely in geometry and analysis, including the free use of 'there exists' and 'all' with reference to the objects of the axiomatic system. If the formalism of symbolic logic was employed, this was done merely for the sake of conciseness. Intending moreover to follow Dedekind and Frege in founding arithmetic on set theory, one introduced no arithmetical axioms. The whole interest was concentrated on the basic set-theoretic relation $x \, \varepsilon \, y$, 'x is a member of the set y.' The assumption that to any well-defined property $F(x)$ there corresponds a set φ such that $x \, \varepsilon \, \varphi$ if and only if x has the property F,

(8) $\qquad F(x) \rightarrow (x \, \varepsilon \, \varphi), \quad (x \, \varepsilon \, \varphi) \rightarrow F(x)$ for all x,

leads to a formal contradiction (cf. p. 58). Indeed on writing $S(y, x)$ instead of $x \,\varepsilon\, y$, the formulas (8) become identical with (6), except for the fact that now the set φ is not supposed to be, or to be representable by, an actual number. We are familiar with the construction by which the contradiction arises: Take $\sim (x \,\varepsilon\, x)$ for $F(x)$ and substitute φ for x. This is Russell's paradox of the "set of all things that are not members of themselves."

Hence some limitation must be imposed on the conversion of properties into sets, a limitation strict enough to block Russell's paradox but affording as much latitude for mathematics as possible. The attitude is frankly pragmatic; one cures the visible symptoms but neither diagnoses nor attacks the underlying disease. The foremost example of this kind of axiomatic approach is Zermelo's system (1908), later improved by Fraenkel, von Neumann, Bernays and others; the theory of numbers, classical analysis, and even Cantor's general set-theory can be based on it. Zermelo's axioms deal with but one category of objects called elements or sets, and one basic relation $x \,\varepsilon\, y$. The fundamental idea is this: instead of using properties for the definition of sets one uses them only to cut out a subset from a given set. Hence his axiom of selection: "Given a well defined property B and an element a, there is an element b such that $x \,\varepsilon\, b$ if and only if x is a member of a and at the same time has the property B."

The notion of a well-defined property which enters into it is somewhat vague. But we know that we can make it precise by constructing properties by iterated combined application of some elementary constructive processes. Instead of saying that x has the property A, let us say that x is a member of the class A, $x \,\varepsilon\, A$. We thus distinguish between elements or sets on the one hand, classes on the other, and formulate the axioms in terms of two undefined categories of objects, elements and classes. Since we postulate that two elements a, b are identical in case $x \,\varepsilon\, a$ implies $x \,\varepsilon\, b$ and vice versa, and since each element a is associated with the class A of all elements x satisfying the condition $x \,\varepsilon\, a$, we are justified in identifying a with that class A. Then *every element is a class*, and the axioms deal with one undefined fundamental relation $x \,\varepsilon\, Y$, 'the element x is member of the class Y,' which has absorbed Zermelo's relation $x \,\varepsilon\, y$ between elements. The principles for the construction of properties are replaced by corresponding axioms for classes; e.g. given two classes A and B, there exists a class C such that $x \,\varepsilon\, C$ whenever $(x \,\varepsilon\, A) \,\mathbf{v}\, (x \,\varepsilon\, B)$ and vice versa.

Since the axiom of selection can only generate smaller sets out of a given set, we need some vehicle that carries us in the opposite direction. Therefore two axioms are added guaranteeing the existence of the set of all subsets of a given set and the join of a given set of sets. It is essential that they be limited to sets = elements, and do not apply to classes.

With the introduction of classes, the axioms assume the same self-sustaining character as, for instance, the axioms of geometry; no longer do such general notions as 'any well-defined property' penetrate into the axiomatic system from the outside. A complete table of axioms for this system, Z as we shall call it, is to be found on the first pages of Gödel's monograph, *Consistency of the Continuum Hypothesis*. Such classes as are too 'big' are here excluded from admission to the club of the decent sets, and in this way the disaster of antinomies is averted. Even before the turn of the century Cantor himself had moved in the same direction by distinguishing 'consistent classes' = sets, and inconsistent classes.

The system Z seems to be the most adequate basis for what is actually done in present day mathematics. In particular the 'existential' Dedekind-Frege theory of numbers can be derived from it (Zermelo), and Gödel was able to show (*op. cit.*) that Zermelo's far-reaching axiom of choice in a very sharp form is consistent with the other axioms of Z (provided they are consistent!).

A different idea, incorporated in the transfinite set-theoretic rule **(I)** proposed on p. 58, underlies Bertrand Russell's theory of types. Sets of numbers are objects of a higher type than the numbers themselves: In **(I)** the variable x is limited to the realm of (natural) numbers while the set y coresponding to the given property $A(x)$ is not, or is at least not assumed to be, a number. Russell therefore admits an infinite series of types: the first type consists of the primary objects, e.g. the numbers, the second of the properties of numbers, the third of the properties of properties of numbers, etc. These types have to be kept separate, and the relation $x_i \, \varepsilon \, x_{i+1}$ connects an arbitrary element x_i of type i with one, x_{i+1}, of the next higher type $i + 1$. Thereby the formation of $\sim (x \, \varepsilon \, x)$, which led to Russell's paradox, is prevented. It is advisable to treat binary, ternary, . . . relations along with the properties. A relation $R(x_1, x_2, \ldots, x_n)$ with n variables of given types $\tau_1, \tau_2, \ldots, \tau_n$ is of a 'higher' type

$$\tau = \{\tau_1, \tau_2, \cdot \cdot \cdot, \tau_n\}$$

uniquely determined by the types τ_i. We draw the diagram of a family tree for τ in which $\tau_1, \ldots, \tau_n$ appear as the immediate descendents of τ. They in turn have their own descendents, and thus any type is depicted by a tree that in all its branches ends with the ground type (or one of the ground types if there are several). An axiomatic system U based on this hierarchy of types is described in outline in

H. Weyl, "Mathematics and Logic" (*American Math. Monthly*, 53, 1946, pp. 6–7).

Since 'there is' and 'all' are applied without hesitation to objects of all types, we must assume that the objects of any given type form an existential category, in the sense that the question whether there exists an element in that category of such and such a property A always has a meaning and that there either exists such an element or every element of the category has the complementary property $\sim A$. Elementary geometry is a field in which the primary objects (the points, straight lines, and planes) are considered as *given* and as constituting existential categories, while the properties and relations between the primary objects are *constructed* from a few undefined basic relations. Our description of axiomatics in Section 4 has tacitly assumed this standard situation. However in intuitive number theory both the relations and the primary objects are constructed, not given; while in a phenomenology of nature one will have to deal not only with categories of objects, as 'bodies' or 'events,' but also with whole categories of properties which are prior to all construction, e.g. with the continuum of color qualities. The system U, if taken realistically as a description of a world that is, goes much farther in the latter direction; in this world there are types upon types of objects, related by ε, but existing in themselves, individually and in their totality. We are as remote as possible from Dedekind's thesis that numbers are free creations of the constructive intellect; nor is mathematics based here on logic alone (the reduction of mathematics to logic had been the lodestar in the Russell-Whitehead enterprise of *Principia Mathematica*), but on a sort of logician's paradise, a universe endowed with an "ultimate furniture" of rather complex structure and governed by quite a number of sweeping axioms of closure. The motives are clear, but would any realistically-minded man dare say he believes in this transcendental world? Zermelo's system Z, which is of simpler structure and even greater power, taxes the strength of our faith to no lesser degree. (Since the very structure of U may not be described without resort to the intuitive concept of iteration, it seems foolish to base a theory of natural numbers in Dedekind-Frege style on this system U — while in Z this is a perfectly legitimate enterprise.)

It is no serious problem to formalize the systems Z and U completely by adding axioms of elementary and transcendental logic.

Strange was the way in which Russell arrived at his conception. He saw clearly that if properties are constructed then the prescription 'Form the property that attaches to a number if there is a property such that . . . ' as a rule of construction involves a vicious circle. He was thus driven to recognize different levels of properties of numbers, a property of level $l + 1$ being defined in terms of the totality of properties of level l.   But afraid of the radical consequences of this critical insight he at once reduced everything to the ground level by his axiom of reducibility, and the idea of a ramified hierarchy of types and levels was nipped in the bud.   While he saw that the differentiation of the several types alone is sufficient to stop the known antinomies, he hardly realized to the full how completely he had abandoned the road of analysis and construction in favor of the existential-axiomatic standpoint.

For any attempt to get at the real source of the antinomies Russell's analysis of the levels is more important than the subsequent 'betrayal' by the axiom of reduction.   And yet it is more or less an historic accident that the lever was first applied at this point.   The deepest root of the trouble lies elsewhere: a field of possibilities open into infinity has been mistaken for a closed realm of things existing in themselves.   As Brouwer pointed out, this is a fallacy, the Fall and Original Sin of set-theory, even if no paradoxes result from it.

It was for this reason that I put comparatively little emphasis on the paradoxes in the account given in the main part of this book.   It is not surprising that Gödel, who derived from them the leading idea for his shattering discovery, judges differently.   In a recent appraisal of Russell's contribution to mathematical logic he says that the paradoxes reveal "the amazing fact that our logical intuitions are self-contradictory."   I confess that in this respect I remain steadfastly on the side of Brouwer, who blames the paradoxes not on some transcendental logical intuition which deceives us but on an error inadvertently committed in the passage from finite to infinite sets.

Whereas Russell's axiomatic universe U avoids the paradoxes by the hierarchy of types among sets, and Zermelo escapes them by their limitation in size, Gödel hints at a theory in which "every concept is significant everywhere except for certain 'singular points' . . . so that the paradoxes would appear as something analogous to dividing by zero.   Our logical intuitions would then remain correct up to certain minor corrections, i.e. they could then be considered to give an essentially correct only somewhat blurred picture of the real state of affairs."   On the ground of all his experience Gödel makes a strong plea for the realistic standpoint where classes are conceived as real objects, namely as "pluralities of things," or as structures consisting of such pluralities, and he adds, "It seems to me that the assumption of such objects (classes or concepts) is quite as legitimate as the assumption of physical bodies and there is quite as much reason to believe in their existence.   They are in the same sense necessary to obtain a satisfactory system of mathematics as physical bodies are necessary for a satisfactory theory of our sense perceptions, and in both cases it is impossible to interpret the propositions one wants to assert about these entities as propositions about the 'data.' Logic and mathematics (just as physics) are built upon axioms with a real content which cannot be 'explained away.'" He adds the warning, "Many symptoms show only too clearly however that the primitive concepts need further clarification."

It is impossible to discuss realism in logic without drawing in the empirical sciences. Then consistency appears as that part of concordance which can be settled independently of the empirical physical facts (cf. Section 17, p. 122). We can never be sure whether concordance, however complete for the moment, will still survive when our observations expand and become more accurate. Gödel seems to suggest that we are no better off as far as consistency is concerned. No Hilbert will be able to assure us of consistency forever; we must be content if a simple axiomatic system of mathematics has met the test of our elaborate mathematical experiments so far. It will be early enough to change the foundations when, at a later stage, discrepancies appear. That is a position against which I cannot find much to say. But what are the guiding principles of our axiomatic construction? Gödel, with his basic trust in transcendental logic, likes to think that our logical optics is only slightly out of focus and hopes that after some minor correction of it we shall see *sharp*, and then everybody will agree that we see *right*. But he who does not share this trust will be disturbed by the high degree of arbitrariness involved in a system like Z, or even in Hilbert's system. How much more convincing and closer to facts are the heuristic arguments and the subsequent systematic constructions in Einstein's general relativity theory, or the Heisenberg-Schrödinger quantum mechanics. A truly realistic mathematics should be conceived, in line with physics, as a branch of the theoretical construction of the one real world, and should adopt the same sober and cautious attitude toward hypothetic extensions of its foundations as is exhibited by physics.

REFERENCES

B. RUSSELL, Mathematical Logic as based on the Theory of Types, *Am. Jour. Math.*, *30* (1908), pp. 222–262.

A. N. WHITEHEAD and B. RUSSELL, *Principia Mathematica*, 3 Vols., Cambridge, 1910–13; second ed., 1925–27.

E. ZERMELO, Untersuchungen über die Grundlagen der Mengenlehre, *Mathematische Annalen*, *65* (1908), pp. 261–281.

K. GÖDEL, Über formal unentscheidbare Sätze der Principia Mathematica und verwandter Systeme, *Monatsh. Math. Phys.*, *38* (1931), pp. 173–198.

—— *Consistency of the Continuum Hypothesis*, Annals of Mathematics Studies No. 3, Princeton, University Press, 1940.

—— Russell's Mathematical Logic, in *The Philosophy of Bertrand Russell*, Northwestern University 1944, 125–153.

G. GENTZEN, *Mathematische Annalen, 112* (1936), 493–565; Die gegenwärtige Lage in der mathematischen Grundlagenforschung, in *Forschungen zur Logik und zur Grundlegung der exakten Wissenschaften*, her. von H. Scholz, Neue Folge, Heft 4, Leipzig, 1938.

CARL PRANTL, *Geschichte der Logik im Abendlande*, 4 Vols., Leipzig, 1855–1870. (Reprint 1927).

Ars Combinatoria

MARSCHALLIN (*looking into her hand-mirror*):
Wie kann das wirklich sein,
dass ich die kleine Resi war
und dass ich auch einmal die alte Frau sein werd'
. . . .
Wie macht denn das der liebe Gott?
Wo ich doch immer die gleiche bin.
Und wenn er's schon so machen muss,
warum lässt er mich denn zuschaun dabei
mit gar so klarem Sinn! Warum versteckt er's nicht vor mir?
Das alles ist geheim, so viel geheim . . .
　　　　　H. VON HOFMANNSTHAL, *Der Rosenkavalier*, Act I.

---

1. Perhaps the philosophically most relevant feature of modern science is the emergence of abstract symbolic structures as the hard core of objectivity behind — as Eddington puts it — the colorful tale of the subjective storyteller mind.   In Appendix A we have discussed the structure of mathematics as such.   The present appendix deals with some of the simplest structures imaginable, the combinatorics of aggregates and complexes.   It is gratifying that this primitive piece of symbolic mathematics is so closely related to the philosophically important problems of individuation and probability, and that it accounts for some of the most fundamental phenomena in inorganic and organic nature.   The same structural viewpoint will govern our account of the foundations of quantum mechanics in Appendix C. In a widely different field J. von Neumann's and O. Morgenstern's recent attempt to found economics on a theory of games is characteristic of the same trend.   The network of nerves joining the brain with the sense organs is a subject that by its very nature invites combinatorial investigation.   Modern computing machines translate our insight into the combinatorial structure of mathematics into practice by mechanical and electronic devices.

It is in view of this general situation that we are now going to insert a few auxiliary combinatorial considerations of an elementary nature concerning aggregates of individuals.   The reader should be warned beforehand that in their application to genetics the lines are drawn somewhat more sharply than the circumstances warrant.   In the progress of science such elementary structures as roughly correspond to obvious facts are often later recognized as founded on structures of a deeper level, and in this reduction the limits of their validity are

237

revealed.   This hierarchy of structures will be illustrated in Appendix D by the theory of chemical valence.

An aggregate of white, red, and green balls may contain several white balls.   Generally speaking, in a given aggregate there may occur several individuals, or *elements*, of the same *kind* (e.g. several white balls) or, as we shall also say, the same *entity* (e.g. the entity white ball) may occur in several *copies*.   One has to distinguish between *quale* and *quid*, between equal (= of the same kind) and identical. To the question of individuation thus arising, Leibniz gave an *a priori* answer by his *principium identitatis indiscernibilium*.   Physics has recently arrived at a precise and compelling empirical solution as far as the ultimate elementary particles, especially the photons and electrons, are concerned.   Closely related is the question of the conservation of identity in time; the identical 'I' of my inner experience is the philosophically most significant instance.[1]   Our decision as to what is to be considered as equal or different influences the counting of 'different' cases on which the determination of probabilities is based, and thus the problem of individuation touches the roots of the calculus of probability.   It is through the combinatorial theory of aggregates that these things find their exact mathematical interpretation, and there is hardly another branch of knowledge where the relationship of idea and mathematics presents itself in a more transparent form.

The simplest combinatorial process is the *partition* of a set S of n elements into two complementary subsets $S_1 + S_2$.   For the sake of identification and recording, we attach arbitrarily chosen distinct marks p to the elements.   Only such relations and statements have objective significance as are not affected by any change in the choice of the labels p; this is the principle of relativity.   Its abstract formulation reveals its triviality.   An 'individual' subset S_1 is characterized by stating for each element, marked p, whether it is a member of S_1, $p \, \varepsilon \, S_1$, or of the complementary subset S_2.   As the construction of S_1 thus depends on the decision of n alternatives ($p \, \varepsilon \, S_1$ or $p \, \varepsilon \, S_2$ for each of the n elements p) there are 2^n individually distinct possible subsets (including the vacuous null set as well as the total set S). However, this number is reduced to $n + 1$ if the n elements are considered as indiscernible.   For then a subset S_1 is completely characterized by the number n_1 of its elements, a number that is capable of the $n + 1$ values 0, 1, $\cdots$, n; and the partition $S = S_1 + S_2$ is characterized by the decomposition $n = n_1 + n_2$ of n into a sum of

---

[1] The riddle of the identical ego, that is an onlooker at what is done to him and by him, is movingly expressed by the above lines from the *Rosenkavalier*.

two terms (n_1, n_2 being the numbers of elements in S_1, S_2 respectively.)[2]
One will ask how many individually distinct partitions $S = S_1 + S_2$
lead to the same 'visible' partition as characterized by the decomposition $n = n_1 + n_2$. The answer is $\dfrac{n!}{n_1!\, n_2!}$. Consequently the
total number 2^n of all individual partitions must equal the sum

$$\sum \frac{n!}{n_1!\, n_2!}$$

extending over the $n + 1$ different decompositions $n = n_1 + n_2$:

$$(1) \qquad 2^n = \frac{n!}{0!\, n!} + \frac{n!}{1!\,(n-1)!} + \frac{n!}{2!\,(n-2)!} + \cdots + \frac{n!}{n!\, 0!}.$$

This simplest case affords but little interest. Moving closer to
reality, let us now assume that there is a certain respect in which
elements may be equal ($\sim$) or different.[3] Balls may be white, red,
or green; electrons may be in this or that position; animals in a zoo
may be mammals or fish or birds or reptiles; atoms in a molecule may
be H, He, Li, . . . atoms. The universal expression for such 'equality in kind' is by means of a binary relation $a \sim b$ satisfying the
axioms of equivalence: $a \sim a$; if $a \sim b$ then $b \sim a$; if $a \sim b$, $b \sim c$
then $a \sim c$. Various words are in use to indicate equivalence, $a \sim b$,
of two arbitrary elements a, b under a given equivalence relation $\sim$:
a and b are said to be the same *kind* or *nature*, they are said to belong
to the same *class*, or to be in the same *state*. An *aggregate* S is a set of
elements each of which is in a definite state; hence the term aggregate
is used in the sense of 'set of elements with equivalence relation.' Let
us assume that an element is capable of k distinct states $C_1, \ldots C_k$.
A definite *individual state* of the aggregate S is then given if it is
known, for each of the n marks p, to which of the k classes the element
marked p belongs. Thus there are k^n possible *individual states* of S.
If, however, no artificial differences between elements are introduced
by their labels p and merely the intrinsic differences of state are made
use of, then the aggregate is completely characterized by assigning to
each class C_i ($i = 1, \cdots, k$) the number n_i of elements of S that
belong to C_i. These numbers, the sum of which equals n, describe
what may conveniently be called the *visible or effective state* of the
system S. Each individual state of the system is connected with an

---

[2] n_1 and n_2 are understood to be natural numbers ranging over 0, 1, 2, ...
[3] A symbol for negation is no longer needed. It is therefore hoped that no
confusion will result when from now on we use the sign $\sim$ for equivalence.

effective state, and any two individual states are connected with the same effective state if and only if one may be carried into the other by a permutation of the labels; here the principle of relativity finds expression in the postulate of invariance with respect to the group of all permutations. The number of different effective states equals that of the 'ordered' decompositions $n = n_1 + n_2 + \cdots + n_k$ of n into k summands n_i, a number for which one readily finds the value

$$(2) \qquad \frac{(n+1) \cdots (n+k-1)}{1 \cdots (k-1)} = \frac{(n+k-1)!}{n! \, (k-1)!}.$$

Nor is it difficult to ascertain how many distinct individual states are connected with the same visible state and thus to explain the discrepancy between the two numbers k^n and (2), just as the equation (1) explains the discrepancy between the value 2^n on the left and the number $n + 1$ of terms in the sum on the right.

The number of individually distinct possible partitions of S into two complementary subsets S_1, S_2, or the number of individually distinct sub-aggregates S_1, has been found to be 2^n; but since the elements are now discernible according to their 'kind,' an effective sub-aggregate S_1 is fixed by assigning to each class C_i the number $n_i^{(1)}$ of elements with which that class is represented in S_1. Since $n_i^{(1)}$ is capable of the $n_i + 1$ values $0, 1, \cdots, n_i$ there are

$$(3) \qquad (n_1 + 1) \cdots (n_k + 1)$$

different possible effective partitions $S = S_1 + S_2$. The number (3) is therefore of necessity smaller than or equal to

$$2^n \qquad (n = n_1 + \cdots + n_k).$$

The maximum 2^n is attained if all n_i have the value 0 or 1, i.e. if no two elements of S are ever found in the same class. Indeed, that being the case, the elements of S may be completely characterized by the classes to which they belong (by their state or their 'nature') and there is no need then for an artificial differentiation by labels. In this case we speak of a *monomial* aggregate.

The process inverse to the partition of an aggregate S into two complementary sub-aggregates S_1, S_2 is the *union* of two given (disjoint) aggregates S_1, S_2 into a whole $S = S_1 + S_2$. The combinatorial theory of aggregates and of the mutually inverse operations of partition and union finds a particularly important application in *genetics*. The development of two organisms may run a different course, owing to 'external circumstances,' even if they are of the same genetic constitution (have the same germ plasm or are of the same genotype, in Weismann's and Johannsen's terminologies). This duality of constitution and environment, 'nature and nurture,' is basic for our interpretation of the facts of inheritance. It may be called an *a priori* conception like the somewhat similar duality of inertia and force in mechanics. Roughly, the environmental factors are characterized as being external to the organism, (relatively) variable and controllable, in contrast to the internal, given and (relatively) stable constitution. Constitution is often inferred and thus of hypothetical rather than manifest nature, as for instance the atoms that constitute a chemical compound. What belongs to the social environment of an individual may be a constitutive characteristic of the society in which he lives. As in the case of other fundamental conceptions, the precise meaning for each field gradually unfolds with a theory of the relevant phenomena: on the basis of a somewhat vague but natural interpretation one discovers certain laws that surprise by their exact form and are welded together into a theory; by holding on to these simple laws and interpreting the ever increasing array of detailed facts in the light of the theory one succeeds in making the original conception more and more precise. In this sense there is an overwhelming amount of empirical evidence in biology for the distinction of nature and nurture, although it never becomes a perfectly sharp one.

By breeding experiments one has succeeded in dissolving the genetic constitution into an aggregate of individual *genes* or 'points,' much as chemistry dissolves a molecule into an aggregate of atoms. And as an individual atom may be in one of the various states (may be one of the various 'entities atom') indicated by the chemical symbols H, He, Li, . . . , so are the genetic points capable of different discrete states called alleles. In the act of fertilization (syngamy) two aggregates S_1, S_2, the 'gametes' (sperm and egg), are united into a 'zygote' or germ cell $S = S_1 + S_2$. A gamete is produced by an organism, S_1 by Ω_1, S_2 by Ω_2 (Ω_2 is not necessarily distinct from Ω_1, both may be the same self-fertilizing plant). All body cells of the organism Ω that develops from the germ cell S are, notwithstanding their functional differentiation, as far as their genetic constitution is concerned, replicas of its zygote S. Part of the body cells at a certain stage of the life-cycle undergo the inverse process of partition into two complementary subaggregates (maturation division or meiosis);[4] the organism Ω is therefore capable of producing as many constitutionally distinct gametes as there exist effective different sub-aggregates of S. This interplay of syngamy and meiosis, union and partition, explains

---

[4] The actual process (a two-step process accompanied by the longitudinal splitting of each chromosome into two chromatids) is slightly more complicated than this its combinatorial result.

the essential features of heredity: *constancy prevails in so far as the sum of two aggregates is uniquely determined by both parts, variability prevails inasmuch as partition of an aggregate into two complementary parts may be performed in various ways.*

An organism produces the gametes S' contained as sub-aggregates in its zygote S with certain relative frequencies (probabilities) $\gamma = \gamma(S')$. The probabilities will be influenced by external circumstances, in particular by temperature, and are thus, in contrast to the discrete aggregates, capable of continuous variation. But it is evident *a priori* that the complementary gamete S'' must occur with the same frequency as S', $\gamma(S') = \gamma(S'')$. (Even if, as is the case for a ♀ organism, one of the two complementary parts S', S'' is eliminated after maturation division by degenerating into the polar body, one will hold on to the assumption $\gamma(S') = \gamma(S'')$ for the *a priori* probabilities.) It is plausible that the probability of syngamy between a gamete S' produced by Ω and a gamete S'_* produced by Ω_* is the same for the various kinds of gametes S' and S'_* that are produced by the two organisms Ω, Ω_*. The probability that the pair of parents Ω, Ω_* beget a child with the zygote $S' + S'_*$ is therefore presumed to be $\gamma(S') \cdot \gamma_*(S'_*)$.

Returning to the abstract theory, let us pass to a discussion of *temporal changes of state* of a given aggregate S. As long as elements are capable of discrete states only, we are forced to dissolve time also into a succession of discrete moments, $t = \cdots, -2, -1, 0, 1, 2, \cdots$. Transition of the system from its state at the time t into its state at time $t + 1$ will then be a jump-like mutation. With the n elements individualized by their labels p, the changing state of affairs will be described by giving the state $C(p; t)$ of the element p at the time t as a function of p and t. This 'individual' description, by means of the function $C(p; t)$, is to be supplemented by the principle of relativity according to which the association between the individuals and their identification marks p is a matter of arbitrary choice; but it is an association for all time, and once established it is not to be tampered with. If, on the other hand, at each moment attention is given to the visible state only, then the numbers $n_1(t), \ldots, n_k(t)$ in their dependence on t contain the complete picture — however incomplete this information is from the 'individualistic' standpoint. For now we are told only how many elements, namely $n_i(t)$, are found in the state C_i at any time t, but no clues are available whereby to follow up the identity of the n individuals through time; we do not know, nor is it proper to ask, whether an element that is now in the state, say C_5, was a moment before in the state C_2 or C_6. The world is created, as it were, anew at every moment, no bond of identity

242

joins the beings present at this moment with those encountered in the next. This is a philosophical attitude towards the changing world taken by the early Islamic philosophers, the Mutakallimûn. This non-individualizing description is applicable even if the total number $n_1(t) + \cdots + n_k(t) = n(t)$ of elements does not remain constant in time.

Wherever in reality identification of the same being at different times is carried out, it is of necessity based on the observable state. For a continuous flow of time and a continuous manifold of states, the underlying principle is by and large to be formulated as follows: suppose there exists at a time t but one individual in a certain state C appreciably different from the states of all other individuals; if afterwards, especially if shortly afterwards, at a time t', one and only one individual is encountered in a state C' deviating but little from C, or 'typically similar' to C, then the presumption is justified that one is dealing with the *same* individual at both moments t and t'. Instead of t and t' one may have a whole sequence of moments $t, t', t'' \ldots$. Think of following a wave moving over the surface of the water! Even in recognizing people, we are dependent on such means (the famous scenes of recognition in world literature, from the Odyssey on, come to mind) as long as the inner certitude of the identity of one's own ego and communications based thereon ("I am the same man who once met you then and there") are left out of play.

2. We saw that in speaking of different kinds or states or classes, reference is made to an underlying notion of equivalence. It is a frequent occurrence that classes of elements break up into sub-classes; we prefer to speak of genus and class, rather than of class and subclass.[5] Every class C belongs to a definite genus $G = [C]$, and an individual by being a member of the class C belongs also to the genus $[C]$. Thus the animals of a zoo are divided into mammals, fish, birds, etc., and the mammals again into monkeys, lions, tigers, etc. States may coincide in a certain character; this character then corresponds to a genus, and the state to a class, of elements. The division into genera and classes is based on a coarser and on a finer notion of equivalence: $a \sim b$ and $a \approx b$, where $a \sim b$ implies $a \approx b$. In different fields of knowledge this graded division appears under different terminological disguises. The aggregates of genetics are an example in point. The genes correspond to the genera, the alleles to the classes; a gene may have two or several alleles. The fact that an element p of the aggregate belongs to the class C and thereby to the

---

[5] This is the terminology used in number theory; a genus of quadratic forms is wider than a class. Biological taxonomy with its graded hierarchy of kingdom, class, order, family, genus, species, variety, favors the opposite usage.

genus $[C]$ is here expressed by the words 'the point p is occupied by the allele C of the gene $[C]$.'

I mention here a few of the names given in genetics to the basic notions of the combinatorial theory of aggregates, and describe the special circumstances 'normally' prevailing in procreation. An individual aggregate S is known if for every one of its points p the class C_p is known to which p belongs; p then belongs also to the corresponding genus $[C_p]$. Two individual aggregates, S and S^*, are of the same *constitution*, $S \approx S^*$, if the labels p employed for the points of the first aggregate can be mapped in a one-to-one way upon the labels p^* employed for the points of S^*, $p \rightleftarrows p^*$, such that homologous points p and p^* in the two aggregates always belong to the same class (isomorphic mapping). According to the principle of relativity, aggregates of the same constitution are to be considered as indiscernible. Under given external circumstances the zygote S completely determines the phenotype, the visible development of an organism; the phenotype is necessarily the same for zygotes of the same constitution. An effective aggregate S is described by assigning to each class C the number n_C of the points of S in C; the number n_G of points in a genus G then equals the sum Σn_C extending over those classes C for which $[C] = G$. Individual aggregates are connected with the same effective aggregate if and only if they are isomorphic, i.e. of the same constitution. Two individual aggregates S and S^* are said to be of the same *species* σ if, with regard to a suitable one-to-one mapping $p \rightleftarrows p^*$, homologous points p and p^* in S and S^* always belong to the same *genus*.[6] Coincidence of the numbers n_G and n_G^* for all possible genera G is the necessary and sufficient condition for this to be the case; the numbers n_G therefore contain a complete description of the species σ of an aggregate. An aggregate S was called monomial if different points of S never belong to the same class; it is called *haploid* if different points of S never belong to the same genus, i.e. if for each genus G the number n_G equals 0 or 1. The corresponding species then deserves the name haploid. If S contains two points of different classes but of the same genus, then S is said to be heterozygous (*hybrid*); if it contains two points of different classes but of the same given genus G, it is heterozygous with respect to G. Union of two aggregates S_1, S_2 into

---

[6] This natural but purely combinatorial concept is related to but not identical with the meaning of the word species current in biology. There is no doubt that the latter, in spite of the difficulty of giving a precise definition, corresponds to a fundamental fact. As an example indicative of the wide gap between the two notions I quote Dobzhansky's 'dynamic' definition (*Philosophy of Science, 2,* 1935, pp. 344–355): "Species is that stage of evolutionary process at which the once actually or potentially interbreeding array of forms becomes segregated in two or more separate arrays which are physiologically incapable of interbreeding."

a whole $S = S_1 + S_2$ and the inverse process of partition of S into
S_1, S_2 may be called *balanced* in case the parts are of the same species.
This is what normally happens in syngamy and meiosis. *Under
balanced syngamy and meiosis, species remain constant throughout the
sequence of generations.* Indeed, let S, S^* be two gametes of the same
species σ that have united to form the zygote $S + S^*$; if the latter
splits by balanced meiosis into S_1, S_2, then S_1, S_2 are necessarily of
the same species σ as S, S^*. This remains true even if mutations are
admitted by which a point may change its class but not its genus (point
mutations). In particular, if the game of 'balanced' reproduction
starts with two gametes of the same haploid species σ, then only haploid
gametes of that species (and diploid zygotes) will turn up in the successive generations; this is the most common case that Gregor Mendel
was dealing with. Assuming the zygote of a self-fertilizing organism
Ω to be non-hybrid, all direct and indirect descendants of Ω will have
the same genotype as Ω. Differences of phenotype in such a 'pure
line,' if they occur, must be due to different external circumstances,
and thus the invariable genetic constitution is most clearly separated
from the variable environmental factors (W. Johannsen's experiments
with beans, 1903).

3. In physics one aims at making division into classes so fine that
no further refinement is possible; in other words, one aims at a *complete*
description of state. Two individuals in the same 'complete state'
are indiscernible by any intrinsic characters — although they may not
be the same thing. Classical mechanics takes the state of a point of
given mass (and charge) to be completely described by position and
velocity, because by taking this view it reaches agreement with the
principle of causality, which asserts that the (complete) state of a mass
point at one moment determines its state at all times. The simplest
example of a mass point is the linear oscillator; it oscillates on a
definite line (thus requiring a space of one dimension only) and has a
definite frequency ν (= number of oscillations per 2π seconds).[7] The
possible states of an oscillator as specified by position and velocity
form a two-dimensional continuous manifold. According to quantum
mechanics, however, it is capable of a discrete variety of different
states only, specified by a number n assuming the values 0, 1, 2, . . . ;
In the state n the oscillator has the energy $n \cdot h\nu$ where $h = 1.042
\times 10^{-27}$ erg $\times$ sec is Planck's action quantum (the number now
usually designated by a crossed $\hbar$). Radiation in a room, the walls of
which are perfect mirrors ('Hohlraum'), is equivalent to a superposi-

[7] It is unfortunate that in English the word 'frequency' is used in two entirely
different senses — for the number of occurrences in a statistical ensemble (German
'Häufigkeit') and for the number of oscillations.

tion of harmonic oscillations, each marked by an index α and having a definite frequency ν_α. Hence the Hohlraum radiation may be considered as an aggregate of linear oscillators endowed with certain proper frequencies ν_α. (The static part of the electromagnetic field is here disregarded.) According to the quantum mechanics of the individual oscillator, the complete state of our field of radiation therefore assigns an integer n_α to each oscillator α; in this state the oscillator α has the energy $n_\alpha \cdot h\nu_\alpha$, and the sum $\Sigma_\alpha n_\alpha \cdot h\nu_\alpha$ extending over all oscillators is the total energy. In the language of photons one expresses this by saying that n_α photons in the state α and of energy $h\nu_\alpha$ are present. In the language of oscillators the index α specifies the individual oscillator and the integer n_α its state; whilst, in the language of photons, α designates the state of a photon and n_α the number of photons in that state. After translation into the photon language, radiation appears as a gas of photons.

Since the possible complete states of an individual form a discrete manifold in quantum mechanics, application of statistics consists here in a mere counting of states. Once the question of complete description of states is solved, all probabilities are evaluated by simple enumerations, and the problem of comparing probabilities in a continuous 'phase space' by measurement does not arise. Since photons come into being and disappear, are emitted and absorbed, they are individuals without identity. No specification beyond what was previously termed the effective state of an aggregate is therefore possible. Hence the state of a photon gas is known when for each possible state α of a photon the number n_α of photons in that state is given (Bose-Einstein statistics of radiation).

While one need not penetrate deep into the nature of light before it reveals, by such phenomena as diffraction, interference, etc., its undulatory character, its corpuscular features are more concealed. For the electrons the opposite is true; they openly show their corpuscular nature by hitting here or there, whereas their undulatory features were discovered by the experimentalists only simultaneously with the development of quantum theory. Yet it is clear that matter, like radiation, is to be represented by a wave field, the laws of which will form a counterpart to Maxwell's equations of the electromagnetic field (de Broglie, Schrödinger, Dirac). Once this has been accomplished for the individual electron, the same considerations take place as applied above to the Hohlraum radiation; a gas of electrons is described by giving for each state α the number of electrons n_α that exist in this state and possess the corresponding energy $h\nu_\alpha$. With a gas of innumerable free electrons we contrast the shell of the few electrons that are tied to a positively charged atomic nucleus, and together with that nucleus constitute an *atom*. The ideas of discrete energy levels and of the photon have scored their most brilliant success in their application to the latter situation; for they lead straight to Bohr's frequency rule, according to which the energy $h\nu$ gained by an electron jumping from a higher to a lower energy level in the atom is emitted as a photon of frequency ν. This rule gives the clue for the explanation of the vast array of very accurate observations accumulated by the spectroscopists concerning the emission of spectral lines by radiating atoms and molecules. But full agreement is reached only after adding the assumption that no two electrons are ever found in the same complete state (Pauli's exclusion principle). This is the decisive fact for an understanding of the so-called periodic system of chemical elements. The quantum theory of chemical bonds rests on the same principle (cf. Appendix D). Once deduced from the spectroscopical facts, the principle could be applied to such free electrons as take care of electric conduction in metals or knock about in the interior of stars; and here too the results were found to be in accordance with experience. The upshot of it all is that the electrons satisfy Leibniz's *principium identitatis indiscernibilium*, or that the electronic gas is a "monomial aggregate" (Fermi-Dirac statistics). In a profound and precise sense physics corroborates the Mutakallimûn; neither to the photon nor to the (positive and negative) electron can one ascribe individuality. As to the Leibniz-Pauli exclusion principle, it is found to hold for electrons but not for photons.

4. The aggregates considered so far have been without *structure*. But the aggregate of atoms in a molecule possesses a structure characterized in a schematic way by Kekulé's valence strokes. It is to be presumed that the aggregate of gene points that constitutes a gamete or zygote likewise is not without a structure. Experience has taught that this structure is based on a simple binary relation of 'neighborhood' between points. We say that two neighboring points are 'joined.' With a name borrowed from topology, an aggregate endowed with this kind of structure may be called a *complex*.[8] Two complexes K, K^* are of the same $\begin{Bmatrix} \text{constitution} \\ \text{species} \end{Bmatrix}$ if a one-to-one correspondence $p \rightleftarrows p^*$ can be established between the points p and p^* of K and K^* respectively such that (i) two homologous points always belong to one and the same $\begin{Bmatrix} \text{class} \\ \text{genus} \end{Bmatrix}$ and (ii) p^*, q^* are neighbors in K^* if and only if p and q are neighbors in K. Complexes of the same constitution (isomorphic complexes) are to be held indiscernible. A

---

[8] The complexes of topology consist of elements without quality, whereas our elements possess different qualities in so far as they belong to different classes.

complex K consists of the two *separate* parts $K_1 + K_2$ provided no point of K_1 is neighbor to a point of K_2; K is *connected* if no decomposition into two separate parts is possible (except the trivial one in which one part is vacuous and the other the whole K). In a unique manner any complex may be decomposed into separate connected components. According to the combined experiences of genetics and cytology these components are to be identified with the chromosomes in the nucleus of a cell, and we shall therefore call them by this name. A connected complex that decomposes into two separate parts after removal of any one of its joins is said to be a *tree*. (Incidentally the trees used in Appendix A to depict formulas and demonstrations are of this kind.) Under given external circumstances the complex of points that constitutes the zygote of an organism Ω determines the phenotype [of Ω; or more precisely, the phenotype is the same for two isomorphic zygotes. This implies that generally speaking the phenotype not only depends on the *aggregate* K but also on the structure of the *complex* K; the structural influence is known under the name of position-effect.

In carrying out union and partition of complexes no joins must be severed nor new joins be established. If this were the whole story chromosomes would behave as indivisible wholes and there would be no way of distinguishing between different genes in the same chromosome. Under these circumstances Mendel's rule of independent assortment would hold, asserting that the probabilities γ for the various constitutionally different gametes produced by a definite organism are all alike. Whereas Mendel is right in that two points in two different chromosomes are independent, it has been found that points in the same chromosome are not absolutely but only more or less tightly linked together. This phenomenon of linkage has been studied with paramount success by T. H. Morgan and his school for the fruit fly *Drosophila melanogaster*, and has resulted in detailed gene maps from which quantitative information can be drawn about the probabilities γ. Morgan has explained linkage by the process of *crossing-over*. Suppose a zygote $K + K^*$ has been formed by balanced syngamy from two gametes K, K^* of the same species σ that are related to each other by the isomorphic mapping $p \rightleftarrows p^*$. Let a, b be a pair of neighboring points in K and a^*, b^* the homologous pair in K^*. Then a^*, b^* will be neighbors in K^*; the points a, b will lie in one chromosome K_0 of K and a^*, b^* in the homologous chromosome K_0^* of K^*. Crossing-over consists in breaking the joins ab and a^*b^* and joining instead a with b^* and b with a^*. If K_0 is a tree, then this process carries the disconnected pair (K_0, K_0^*) into a pair $(\bar{K}_0, \bar{K}_0^*)$ of chromosomes that are isomorphic to K_0. Points that before crossing-over were in the same chromosome K_0 may now be separated, one belonging to $\bar{K}_0$, the other to $\bar{K}_0^*$. Pairs of homologous chromosomes in the nucleus of a cell are seen to put themselves in a position for such an operation of crossing-over immediately before meiosis takes place; they extend side by side, each point in one chromosome opposite the homologous point in the other (synapsis). If afterwards balanced meiosis occurs, the new gametes $\bar{K}$, $\bar{K}^*$ will be of the same species σ K, K^*. *Linkage between two points a, b of a chromosome will be the looser the more ways there exist to separate them by crossing-over.*

Complexes may undergo two sorts of *mutations.* Besides the point mutations in which the joins are not tampered with while the points p change their classes C_p (without changing their genera), we have structure mutations that leave the state of the points undisturbed but alter the joins.[9] The operation described above as crossing-over may be performed with *any* four distinct points a, b, a^*, b^* (and may then be called 'switching-over'). A simple break and this process of switching-over seem to play the role of elementary operations for structure mutations. Mutations are rare events, in contrast to crossing-over for which an opportunity is provided by synapsis before every meiosis.

The simplest connected complexes are the rod $a_1 — a_2 — \cdots$ $— a_h$ (in which consecutive points a_i are connected by the joins $—$) and the ring. With few exceptions the chromosomes seem to be rods (T. H. Morgan's *law of linear arrangement*). However, switching-over when occurring in one rod (not between two rods) may produce a rod plus a ring (or reproduce the rod with an inverted section). A complex consisting of separate rods and rings will preserve this character under any breaks and switchings-over.

A chromosome has a centromere. If a structural change produces chromosomes with no or two centromeres then they are usually left behind when the cell divides, and thus *deficiencies* result. There are also several ways in which the whole chromosome outfit of a cell nucleus, or an individual chromosome, or part of a chromosome, may be *duplicated*.

Here we have attempted to develop the formal scheme of genetics in such general form as to comprise all more or less irregular occurrences. Nowhere in this scheme was it necessary so far to speak of *sex;* but of course the fact cannot be ignored forever that syngamy between two gametes takes place only if one is a sperm, the other an

[9] On the basis of the position-effect, R. Goldschmidt has recently challenged the entire conception of gene and the distinction between point and structure mutations.

egg. This is a polarity (gamete sexuality) that has nothing to do with genes.[10] On the other hand, whether an individual organism is a sperm-producing male or an egg-producing female (zygote sexuality) is determined like all its other 'visible characters' by the genotype of its zygote — in conjunction with the external circumstances influencing development. Experience shows that it is not a single gene in the zygote, but a balance between many genes, that determines the sex. The sex chromosome (where it is distinguishable from the autosomes) merely tips the scales. This explains the phenomenon of intersexes and modifies the common belief that sees in sex the outstanding example of a non-quantitative, an either-or, character.

⟨5. Our remarks about entropy and statistics in Section 23B of the main text and in the following Appendix C on quantum physics will be made clearer if we say at this place, in parenthesis as it were, a few words about the foundations of *statistical thermodynamics*. Here quantum theory has introduced a decisive simplification. Indeed, in quantum physics a system Σ is capable of no more than a discrete series of (complete) states with definite energy levels

$$U_i \qquad (i = 0, 1, 2, \ldots).$$

In view of the conservation law for energy, let us distribute a large number N of systems Σ at random over its possible states i, yet so that the total energy of the N systems has a preassigned value $N \cdot A$ (A = average energy of the individual system). One finds that in the overwhelming majority of all distributions the relative frequencies N_i/N with which the several states i are represented is, in the limit for $N \to \infty$, proportional to $e^{-\alpha U_i}$. Here α denotes a constant that is to be determined in terms of the given average energy A. We therefore define the *canonic distribution of parameter* α by assigning the relative probability $w_i = e^{-\alpha U_i}$ to the state i. (Relative probabilities need not satisfy the normalizing condition $\Sigma_i w_i = 1$.) Any quantity Z_i dependent on the state i will then have the mean value

$$\langle Z \rangle_\alpha = \sum_i Z_i \cdot e^{-\alpha U_i} \Big/ \sum_i e^{-\alpha U_i},$$

and it is this value that we ascribe to the quantity in 'thermic equilib-

---

[10] Denote by $\Omega_{\alpha\beta}$ an organism arising from syngamy of a sperm of genotype α and an egg of genotype β. The fact that there are cases when, even under equal circumstances, the reciprocal cross $\Omega_{\beta\alpha}$ differs in appearance from $\Omega_{\alpha\beta}$ is evidence that interpretation of organic development in terms of genes alone will not always suffice. Besides the genes in the chromosomes of the cell nuclei, other hereditary agents influencing the development have to be assumed in the cytoplasm. This problem however is still far from a satisfactory solution.

rium.' The parameter α is connected with the given mean energy A by the equation $\langle U \rangle_\alpha = A$. The systems occurring in nature are capable of states with arbitrarily high energy values; consequently α must be positive, and the standard distribution which assigns to every state i the same probability $w_i{}^0 = 1$ may only approximately be realized as long as the energy A stays finite (namely for large values of A and correspondingly low values of α). The reciprocal number α^{-1} has the dimension of energy and may for the moment be called statistical temperature.

Because of the conservation law of energy, the canonical distribution is stationary in time. The possible states of a system Σ consisting of two parts Σ, Σ' is characterized by the pairs (i, k) formed from any state i of the system Σ and any state k of the system Σ'. Let U_i, U'_k designate the energy of Σ in the state i and of Σ' in the state k respectively; then the energy of Σ in the state (i, k) equals $U_i + U'_k$, provided no interaction takes place between the two parts. For the probability of the state (i, k) in thermic equilibrium we obtain

$$\mathbf{w}_{ik} = e^{-\alpha(U_i + U_k')} = e^{-\alpha U_i} \cdot e^{-\alpha U_k'} \; (= w_i \cdot w'_k).$$

This means three things: (1) thermic equilibrium of the whole implies thermic equilibrium for the parts; (2) the law of statistical independence prevails for the combination of the parts, $\mathbf{w}_{ik} = w_i \cdot w'_k$; (3) the parameter α has the same value for both parts, with which the value for the total system also agrees. On account of this third point statistical temperature shares with ordinary observable temperature (a more-or-less rather than a quantitative character of bodies) the decisive property that bodies in contact level their temperatures.

An ideal gas, i.e. an aggregate of n particles the states of which are completely described by position and velocity and the interaction of which is negligible, will occupy a definite volume V under a definite pressure p. Application of classical physics and the canonical distribution to such a gas yields the value pV/n for its statistical temperature. Hence if T designates the (absolute) temperature read from a gas thermometer in contact with the system Σ and filled with an ideal gas, then the statistical temperature turns out to be kT where k is a universal factor of proportionality (Boltzmann's constant) that must be added in order to reduce the scale of temperature to the customary Celsius degrees ($100°C$ = difference of the boiling and freezing points of water under pressure of one atmosphere).[11] The

---

[11] The mass $M = n \cdot \mu$ of the gas is proportional to the number n of particles. $v = V/M$ denoting the specific volume, the Gay-Lussac laws are obtained in the usual form $pv = RT$ with a constant $R = k/\mu$ characteristic for the gas. Consequently k is of atomistic smallness, and for different gases the product of R and the 'molecular weight' has the same value.

entire theory of thermic equilibrium thus boils down to this one principle holding in quantum as well as classical physics: the canonical distribution w arises from the standard distribution w^0 by means of the equation

$$w = w^0 \cdot e^{-U/kT}$$

where U denotes the energy (variable from state to state) and T the fixed temperature of the system (or of the heat bath in which the system is immersed). }

REFERENCES

C. H. WADDINGTON, *An Introduction to Modern Genetics*, London, 1939.

E. SCHRÖDINGER, *Statistical Thermodynamics*, Cambridge, 1946.

R. GOLDSCHMIDT, Position Effect and the Theory of the Corpuscular Gene, *Experientia*, 2 (1946), pp. 197–203, 250–256.

J. H. WOODGER, *The Axiomatic Method in Biology*, Cambridge, 1937.

J. VON NEUMANN and O. MORGENSTERN, *Theory of Games and Economic Behavior*, second ed., Princeton, 1947.

Quantum Physics and Causality

1. Modern quantum theory has done away with strict causal determination for the elementary atomic processes. It does not deny strict laws altogether, but the quantities with which they deal regulate the observable phenomena only statistically. Quantum theory is incompatible with the idea that a strictly causal theory of unknown content stands behind it — in the manner in which it may be true that strictly causal motion of individual particles stands behind the statistical-thermodynamical regularities of a gas consisting of many particles. What is thinkable for the laws of a collective is demonstrably impossible for the elementary quantum laws. The uncertainty of the outcome of an atomic experiment is not such as to be gradually reducible to zero by increasing knowledge of the determining factors. The reasons for the passage from classical to quantum physics are no less compelling than those for the relinquishment of absolute space and time by relativity theory; the success, if measured by the empirical facts made intelligible, is incomparably greater. True, a final stage has not yet been attained; certain serious difficulties remain unsolved. But whatever the future may bring, the road will not lead back to the old classical scheme.

One of the most poignant revelations of light's corpuscular nature is the photoelectric effect: a metal plate when irradiated with ultraviolet or X-rays releases electrons. Observation shows that strangely enough their energy is determined by the *color* of the incident radiation, namely equal to or less than h times its frequency. One thus arrives at the conception that light of frequency v gets absorbed in discontinuous quanta (photons) of energy hv (Einstein 1905). This energy is used for the emission of an electron (whose kinetic energy may be short of hv on account of the work used in the liberation of the electron). The intensity of the radiation does not determine the energy of the individual electron, but the number of electrons released per time unit. The process sets in at once even if the radiation is so weak that it would take hours before the accumulated influx of field energy into the region of an atom reaches the amount hv necessary for the ejection of an electron. What a continuous field theory would describe as *presence of a fraction* of the radiating energy hv must in fact be interpreted as *small probability* for the presence of a whole photon of that energy. The process inverse to the photoelectric effect is the transformation of primary electrons into secondary X-rays in a tube whose anode stops the electrons. Since the stopping may take place in several steps, a continuous spectrum for the X-rays is to be expected with a sharp edge at the frequency $\nu = eV/h$ (where $-e$ is the charge of the electron and V the voltage of the tube). Experience has confirmed these relations first predicted by Einstein, including the numerical value for h that has to agree with Planck's constant derived from the thermodynamical laws of Hohlraum radiation.

The problem of reconciling the conceptions of light wave and photon is perhaps best illustrated by polarization. Let a plane monochromatic light wave that propagates in a definite direction $\vec{a}$ be linearly polarized; the direction of polarization, represented by a vector $\vec{s}$ of length 1, is perpendicular to $\vec{a}$. Choose an arbitrary point O as origin and draw a 'cross' G consisting of two mutually perpendicular axes 1 and 2 through O that are perpendicular to $\vec{a}$. The cross as well as the vector $\vec{s}$ lies in the plane perpendicular to $\vec{a}$. Assume that the light ray passes through a Nicol prism of orientation G; it splits into two parts 1 and 2, the part 1 being linearly polarized in the direction 1, the part 2 in the direction 2. The relative intensities of the two partial rays with respect to the total ray are given by the squares (of the lengths of the projection of $\vec{s}$ upon the axes 1 and 2, or) of the components s_1, s_2 of the vector $\vec{s}$ with respect to the cross G. On assuming the light ray to consist of photons we are forced to conclude that photons of two distinct 'characters' 1 and 2 (white balls and black balls as it were) occur with the respective probabilities s_1^2 and s_2^2. (By Pythagoras's theorem, $s_1^2 + s_2^2 = 1$, the probability for a photon to be 'white or black' equals unity, as it should.) The characters 1 and 2 are relative to G. The photons of both kinds are separated by the Nicol prism, that operates like a sieve catching 1 and letting 2 pass. One should therefore expect the ray that has passed the Nicol prism and is polarized in the direction 2 to be more homogeneous than the original ray. Such however is not the case, as the polarized plane monochromatic light wave represents the highest degree of homogeneity obtainable for light. The partial ray polarized in the direction 2, when sent through a second Nicol prism of a different orientation G', will again split up according to the rule of intensity just described. Something similar to polarization happens when a ray of silver atoms is acted upon by a non-homogeneous magnetic field. Using Cartesian coordinates x, y, z, let us assume that the field strength depends on x only. A silver atom is a small magnetic dipole with a vectorial magnetic momentum $\vec{m}$. The field should split the ray into various parts according to the various values of the

x-component m_x of $\vec{m}$. Since only two partial rays (of opposite curvature) are observed, one must conclude that this component is capable of two values only, $+\mu$ and $-\mu$ (μ = 'magneton'). This must hold whatever direction the x-axis has. A vector, however, whose components in every possible direction are either $+\mu$ or $-\mu$ is geometric nonsense! The impossibility of simultaneously ascribing to the photon or silver atom the several characters that correspond to the various orientations G of the "sieve" lies clearly in the nature of things and is not due to human limitations. The sorting by a sieve of orientation G is destroyed by a subsequent sifting of orientation G'. But for the particles passed by the sieve G one may ask what probability they have to pass the sieve G'; this probability can be computed *a priori* in terms of the orientation of G' with respect to G.

By a prism as Newton used it, or by a grating, light is decomposed according to its various monochromatic constituents. A 'sieve' permitting separation not only of two but of several 'sizes of grain' may, therefore, be called a 'grating.' The photoelectric effect sifts photons according to the place where they hit. For good reasons Eddington says in his beautiful book *New Pathways in Science* (p. 267): "In Einstein's theory of relativity the observer is a man who sets out in quest of truth armed with a measuring-rod. In quantum theory he sets out armed with a sieve." By a certain operation M the mammals are sifted from the other animals in a zoo. Let F be the corresponding operation for fish. The iterated operation MM yields no other result than the simple M, whereas the catch of MF, that is of the operation M followed by F, is zero. In view of the equations

$$MM = M, \quad FF = F, \quad MF = FM = 0,$$

M and F are called mutually orthogonal idempotent operators. A grating in classical physics is nothing but a classification of the states that a given physical system is capable of. Here the states are considered as the elements of an aggregate. Assuming the number of states to be finite, let $E_1, \ldots, E_r$ denote the several classes. The operation sifting the states of class E_i from the rest may also be designated by E_i. The grating $G = \{E_1, \ldots, E_r\}$ may be refined by dividing each class into subclasses. There is a finest division in which each class contains only one member. Two gratings, the division into the classes E_i and the division into the classes E'_k, may be superposed. The operation $E_i E'_k$ sorts out the members common to the two classes E_i and E'_k; commutativity prevails, $E_i E'_k = E'_k E_i$, and the total aggregate is divided up into the classes $E_i E'_k$ (some of which may be vacuous). After associating distinct numbers α_i with the several classes

255

E_i one may speak of an *observable* (or state quantity) A that assumes the value α_i when the state of the system belongs to the class E_i.

This classical scheme is now to be confronted with that of quantum physics as adumbrated by the typical example of polarization. The aggregate of n states has to be replaced by an n-dimensional Euclidean vector space **I**. Given a linear subspace **E** of **I**, any vector $\vec{x}$ may be orthogonally projected upon **E**; the projection is a vector $\overrightarrow{Ex}$, and the operation E of projection is idempotent, $EE = E$. Two linear subspaces E_1, E_2 are said to be orthogonal if each vector in the one is orthogonal to each vector in the other. We may then form their sum $\mathbf{E} = \mathbf{E}_1 + \mathbf{E}_2$ consisting of all sums of a vector $\vec{x_1}$ in $\mathbf{E}_1$ and a vector $\vec{x_2}$ in $\mathbf{E}_2$. The decomposition of a vector $\vec{x}$ of **E** into these two summands $\vec{x_1}$ and $\vec{x_2}$ is unique and effected by the orthogonal projections E_1, E_2. In this way a three-dimensional vector space, for instance, splits into a horizontal plane and a vertical line. The situation is hardly more complicated when we deal with more than two mutually orthogonal subspaces $\mathbf{E}_1$, $\mathbf{E}_2$, . . . $\mathbf{E}_r$. If their sum is the total space, then every vector $\vec{x}$ splits into r component vectors lying in the several subspaces $\mathbf{E}_i$, according to the formula

$$\vec{x} = E_1\vec{x} + \cdots + E_r\vec{x},$$

and the total space is said to be split into $\mathbf{E}_1 + \cdots + \mathbf{E}_r$. The projections E_i ($i = 1, \cdots, r$) are idempotent and mutually orthogonal. Let $|\vec{x}|$ designate the length of a vector $\vec{x}$.

By a direction in **I**, or by a vector $\vec{x}$ laid off in that direction, quantum physics represents the *wave state* of the physical system under investigation (be it a simple particle, or an aggregate of many, or even of an indeterminate number of particles). A *grating* $G = \{E_1, \cdots, E_r\}$ is a splitting of the total vector space into mutually orthogonal subspaces $\mathbf{E}_1 + \cdots + \mathbf{E}_r$. We speak of a *character i*, more precisely $(G; i)$, corresponding to any one of these subspaces $\mathbf{E}_i$. If the system is in the wave state $\vec{x}$, then its probability of having the character i equals

(1) $$p_i = |E_i\vec{x}|^2/|\vec{x}|^2.$$

Pythagoras's equation,

$$|E_1\vec{x}|^2 + \cdots + |E_r\vec{x}|^2 = |\vec{x}|^2,$$

states that the sum of these probabilities p_i equals unity. No grating can furnish more than n distinct characters; thus our model, like the classical model it is compared with, corresponds to a situation where the number of character values is limited. That is not so in nature, but no essential features of quantum mechanics are lost by using the finite-dimensional model. It is clear wherein the *refinement* of a grating G consists: indeed every $\mathbf{E}_i$ may again be split into mutually orthogonal subspaces. A *finest* grating consists of n mutually orthogonal axes and is thus identical with a 'cross' ($=$ Cartesian coordinate system). Let us from now on refer to the characters $(G; 1)$, . . . , $(G; r)$ as the r *quantum states* defined by the grating G and call them *complete* quantum states if G is a finest grating.

I cannot refrain from pointing out that, without thinking of quantum physics, I used this very model at the end of Section 17 to illustrate the relationship between object, observer, and observed phenomenon. Its decisive difference in comparison to the classical model is the fact that gratings in vector space defy superposition. Suppose $\mathbf{I}$ has been split into orthogonal subspaces in two manners,

$$\mathbf{I} = \mathbf{E}_1 + \cdots + \mathbf{E}_r \quad \text{and} \quad \mathbf{I} = \mathbf{E}'_1 + \cdots + \mathbf{E}'_s.$$

It is of course possible to split any vector $\overrightarrow{x}$ of $\mathbf{E}_i$ into its components $\mathbf{E}'_k \overrightarrow{x}$ ($k = 1, \cdots, s$); but they will in general no longer lie in $\mathbf{E}_i$. Only if the operator E_i commutes with the operators $E'_1, E'_2, \ldots$, will this be so. Thus combination of two gratings presupposes commutability of the r operators E_i with the s operators E'_k, and, that condition satisfied, the order of combination, whether G is followed by G' or G' by G, is irrelevant. The strange thing about the quantum-physical sieves is exactly this feature, that two such sieves may, and usually will, be non-commutative because of their 'discordant orientations.' Characters i and k referring to two non-commutative gratings are incompatible, and in a situation where i is determined, k is not. In this sense, position and momentum of a particle are incompatible. If Δx, Δp are the uncertainties of the x-coordinate of a particle and of the x-component of its momentum respectively, then the product $\Delta x \cdot \Delta p$ necessarily exceeds h. This *principle of indeterminacy* due to Heisenberg embodies the idea of complementarity in a precise form. If color and shape of a body were such incompatible characters, it would make sense to ask whether a body is green, and it would also make sense to ask whether it is round; but the question "Is it green *and* round?" would make no sense. Here as in the classical model distinct numbers α_i may be assigned to the several quantum states $(G; i)$ and thereby an observable A be defined that is capable of the

257

r values $\alpha_1, \ldots, \alpha_r$ and assumes the value α_i with the probability p_i, (1), provided the system is in the wave state $\vec{x}$.[1]

Decomposition by a grating G of the vector space **I** into the several subspaces $\mathbf{E}_i$ is in itself a purely ideal process. The actual application of the grating to a physical system, however, throws the system from the wave state $\vec{x}$ into one of the wave states $E_1\vec{x}, \ldots, E_r\vec{x}$. Which wave state cannot be foretold; only the relative probabilities $|E_i\vec{x}|^2$ of these r events are predetermined. In this sense every measurement or observation implies an encroachment on the phenomenon, with results of no more than statistical predictability. At no time have experimental physicists closed their eyes to the fact that every measurement is coupled with a reaction of the measuring instrument on the object under investigation. As long as the hypothesis seemed admissible that the instrument could be made infinitely more sensitive than the object, this involved no difficulty of principle. But what if the object itself is of atomic refinement, which cannot be surpassed by any instrument? Then the very idea of facts prevailing independently of observation becomes dubious.

2. Let us for a moment return to the classical model, and use it to depict the temporal succession of events. The simplifying hypothesis of a finite number of states forces us to operate within a discontinuous time. The dynamical law will then assert that from one moment t to the next $t + 1$ the n states $1, \cdots, n$ undergo a certain permutation s, the same at every moment $t = \cdots -2, -1, 0, 1, 2, \cdots$. If this permutation s is of 'order m,' i.e. if one reaches identity after having performed the permutation m times, then the system returns to its initial stage after each period of length m (eternal recurrence). Quantum physics does not force a discontinuous time upon us even if the number of quantum states separable by a grating is universally limited. During the infinitesimal time interval dt the vector space experiences a certain infinitesimal rotation imparting the increment $\vec{dx} = L\vec{x} \cdot dt$ to the arbitrary vector $\vec{x}$. This dynamical law

$$(2) \qquad \vec{dx}/dt = L\vec{x}$$

(in which the operation L is independent of t and $\vec{x}$) is expressed in terms of Cartesian coordinates x_i by equations of the form

$$dx_i/dt = \sum_j l_{ij}\, x_j(t), \qquad (i, j = 1, \cdots, n),$$

with given constant antisymmetric coefficients l_{ij} $(l_{ji} = -l_{ij})$. The

---

[1] According to their definition such quantities can be added and multiplied provided they belong to commuting gratings.

salient point is that the wave state $\overrightarrow{x}$ varies according to a strict causal law; its mathematical simplicity is gratifying. A grating $G = \{E_1, \ldots, E_r\}$ and the corresponding quantum states $(G; 1), \ldots, (G; r)$ are *stationary* if the subspaces $\mathbf{E}_i$ are invariant in time, i.e. if the linear operators E_i commute with the linear operator L.

What in Appendix B has been called state of a particle or of an aggregate is now to be more precisely interpreted as *quantum state*. A circumstance that may have caused some misgivings there, the relativity of the notion of photon with respect to the Hohlraum and its proper frequencies, now appears as a special instance of a general phenomenon: the distinction of quantum states is relative to a grating.

Measurement means application of a sieve or grating. One must not imagine the wave state as something given independently of such measurements. In fact the monochromatic polarized light ray that is sent through the Nicol prism had itself been sorted out by a grating from natural light of unknown quality. This is in accordance with the fundamental fact that only the *relative* position of one Cartesian coordinate system with respect to another may be characterized in objective terms. Given the gratings $G = \{E_1, \cdots, E_r\}$ and $G' = \{E'_1, \cdots, E'_s\}$, we are, however, entitled to ask questions of this type: 'If the first grating shows our particle to be in the quantum state $(G; i)$, between what limits does the probability lie that a test by the second grating G' finds it in the quantum state $(G'; k)$?' In geometric terms this amounts to the following question: (I) 'Between which limits does the quotient $|E'_k \overrightarrow{x}|^2/|\overrightarrow{x}|^2$ lie if $\overrightarrow{x}$ varies freely over the space $\mathbf{E}_i$?' Should time elapse between application of the first and the second grating, then the change of the wave state between the two moments as determined by the dynamical law (2) has to be taken into account.

A system is never completely isolated from its surroundings, and its wave state is therefore subject to perpetual disturbances. This is the reason why the secondary statistics of thermodynamics is to be superimposed upon the primary statistics dealing with a given wave state and its reaction to a grating. In Euclidean space there is an *a priori* probability for the random distribution of vectors of length 1 according to which regions of equal area on the unit sphere are of equal probability. This 'standard distribution' assigns to the r quantum states $i = 1, \cdots, r$ defined by a grating $G = \{E_1, \cdots, E_r\}$ the probabilities n_i/n where n_i is the dimensionality of $\mathbf{E}_i$, and in particular equal probabilities to the n quantum states defined by a *complete* grating. The actual probability distribution (Gibbs ensemble) need by no means coincide with this standard distribution. At the end of the previous section it has been described how the particular canonical distribution for a system embedded in a heat bath of known temperature proceeds from the standard distribution. A general question somewhat different from the one considered above now arises, namely: (II) 'Suppose a grating G and the statistical distribution of wave states is given, what probabilities result therefrom for the several quantum states $(G; i)$?' A grating and a statistical ensemble, rather than two gratings, are here compared with each other. To find the answer one has to average the probability p_i, (1), which depends on $\vec{x}$, according to the given statistical distribution of vectors $\vec{x}$ over the unit sphere. In the same manner one may ascertain the average probability that a particle in the quantum state $(G; i)$ is encountered in the quantum state $(G'; k)$ if tested by another grating G'. Whether question (I) or (II) is posed, the interest will always be focused, especially when we are concerned with systems consisting of numerous particles, on such events as can be foretold with overwhelming probability. In splitting a light ray by a Nicol prism the fate of the individual photon is unpredictable. Predictable however are the relative intensities of the two partial rays with an accuracy that increases with the number of photons.

The description here given must be corrected throughout in one point: the coordinates x_j in the underlying n-dimensional vector space are not real but arbitrary *complex* numbers and as such have an absolute value $|\vec{x}|$ and a phase. The square of the length of the vector is expressed in terms of a Cartesian coordinate system by the sum of the squares of the absolute values of the coordinates. The simplest of all dynamical laws (2) in such a complex space is of the form

$$(3) \qquad\qquad d\vec{x}/dt = i\nu\vec{x}, \qquad (i = \sqrt{-1}).$$

Here ν is a real constant. The wave state $\vec{x}$ then carries out a simple oscillation of frequency ν,

$$\vec{x} = \vec{x}_0\{\cos(\nu t) + i\sin(\nu t)\} \qquad (\vec{x}_0 = \text{const.}),$$

and hence the energy has the definite constant value $h\nu$ (Planck's law). But whatever the dynamical law (2), the space can always be broken up into a number of mutually orthogonal subspaces $\mathbf{E}_j$ $(j = 1, \cdots, r)$ such that an equation (3) with a definite frequency $\nu = \nu_j$ holds in $\mathbf{E}_j$. The grating $G = \{E_1, \cdots, E_r\}$ thus obtained is stationary and effects a sifting with respect to different frequencies ν_j and corresponding energy levels $U_j = h\nu_j$. Thermodynamics is based on this G.

Any vector $\vec{x}$ in E_j satisfies the equation $L\vec{x} = i\nu \cdot \vec{x}$ $(\nu = \nu_j)$, and this fact is expressed in mathematical language by saying that $\vec{x}$ is an eigenvector of the operation L with the eigenvalue $i\nu$. The operator $H = \frac{h}{i} L$, called energy, has the same eigenvectors, but the corresponding eigenvalues are the energy levels $h\nu$. The general equation (2) now reads $\frac{h}{i} \frac{d\vec{x}}{dt} = H\vec{x}$ (Schrödinger's equation).

The 'physical process' undisturbed by observation is represented by a mathematical formalism without intuitive (anschauliche) interpretation; only the concrete experiment, the measurement by means of a grating, can be described in intuitive terms. This contrast of physical process and measurement has its analogue in the contrast of formalism and meaningful thinking in Hilbert's system of mathematics. As it is possible to formalize an intuitive mathematical argument, so is it true that measurement by a grating G may be interpreted as a physical process. In doing so one has to extend the original system Σ to a system Σ^* by inclusion of the grating G. But as soon as we want to learn something about Σ^* that can be told in concrete terms, then the undisturbed course of events as ruled by the dynamical law (2) must again be disrupted by subjecting Σ^* to the test of a grating outside Σ^*.

3. Given two systems Σ, Σ', their union $\mathbf{\Sigma} = \Sigma + \Sigma'$ is capable of all states (i, k) consisting of a combination of an arbitrary state i of Σ and an arbitrary state k of Σ'. That is the prescription for combination given by classical physics. Quantum physics agrees provided state means quantum state; a (finest) grating for Σ together with a (finest) grating for Σ' yields a (finest) grating for $\mathbf{\Sigma}$. Assuming therefore that a wave state of the first system is represented by a vector $\vec{x} = (x_1, \cdots, x_m)$ in an m-dimensional Euclidean space S referred to a Cartesian coordinate system, and that the generic vector $\vec{y} = (y_1, \cdots, y_n)$ of an n-dimensional space S' has the same significance for the second system, we conclude that the wave state of the united system is represented by a vector

$$\vec{z} = (z_{11}, \cdots, z_{m1}, z_{12}, \cdots, z_{m2}, \cdots, z_{1n}, \cdots, z_{mn})$$

in an mn-dimensional 'product space' $\mathbf{S} = S \times S'$. A vector $\vec{x}$ in S and a vector $\vec{y}$ in S' determine the vector $\vec{z} = \vec{x} \times \vec{y}$ with the components

(4) $$z_{ik} = x_i y_k \quad (i = 1, \cdots, m; k = 1, \cdots, n)$$

261

in S. This fixes what rotation in S is induced by two arbitrary rotations of the coordinate systems in S and S'. Since (4) implies $|z_{ik}|^2 = |x_i|^2 |y_k|^2$ one finds the probabilities of the quantum states of the two parts Σ and Σ' to be independent of each other in a wave state of Σ of the special kind $\vec{z} = \vec{x} \times \vec{y}$. But the manifold of the possible wave states of the joint system Σ is much larger than those representable by the combinations $\vec{x} \times \vec{y}$ of arbitrary wave states $\vec{x}$ and $\vec{y}$ of the two parts. In fact *every* vector $\vec{z}$ in the product space represents a possible wave state. In this very radical sense quantum physics supports the doctrine that *the whole is more than the combination of its parts*. In general the probabilities of the quantum states of the whole system cannot be determined from the probabilities of the quantum states of the parts by the product rule of statistical independence. And this is so even when both parts are not in dynamical interaction.

This consideration is of special importance for a pair of two *equal* systems Σ, Σ', e.g. for a pair of electrons. Then the wave states of both parts are represented by vectors in one and the same Euclidean space S, and among the vectors $\vec{z} = (z_{ik})$ of the product space $S \times S$ ('tensors') one may distinguish the antisymmetric ones satisfying the condition $z_{ki} = -z_{ik}$ and the symmetric ones with the property $z_{ki} = z_{ik}$. Once the pair is in an antisymmetric wave state, its wave state will remain antisymmetric; no external influences can alter that because equal particles enter into the law of action in a symmetric fashion. It is therefore to be expected that the wave state of a pair of electrons has a definite symmetry character, that it is either antisymmetric or symmetric. Experience proves the first alternative to be correct. For an antisymmetric vector z_{ik} the equation $|z_{ii}|^2 = 0$ holds, i.e. the probability that both electrons are found in the same complete quantum state i is zero; *the permanent antisymmetry of the wave state thus explains Pauli's exclusion principle*. The statistical independence of the quantum states of two electrons could scarcely be denied in a more radical fashion than by this principle! The hydrogen molecule may be treated, at least in first approximation, as a system of two electrons circling around two fixed nuclei, and it is obvious that the restriction by antisymmetry of the wave state of the electronic pair must be of decisive influence upon the result of the computation of their motion. It leads indeed, as London and Heitler have shown, to a full explanation of the chemical binding of neutral atoms in a molecule. Under the reign of classical physics this had remained an inscrutable conundrum.

The fact of antisymmetry carries over from two to more electrons.

Since statistical independence of several particles is at variance with this law, it is not the same whether again and again an electron of given wave state or a simultaneous shower of many electrons is sent through a grating. A similar remark applies to a shower of photons. Its wave state is to be restricted by the condition of symmetry rather than of antisymmetry. (We know indeed that the exclusion principle does not hold for photons!) In its final form the theory does not require the number of particles to be constant. Not only may photons appear and disappear, but owing to a bold interpretation of Dirac's it also accounts for the process of mutual annihilation of a positive and negative electron under emission of a photon of corresponding energy ("Zerstrahlung") and the inverse process.

4. I summarize those features of quantum physics which seem to me of paramount philosophical significance.

(1) Observation is impossible without an encroachment the effect of which can be predicted only in a statistical sense. Thus new light is thrown on the relationship of subject and object; they are more closely tied together than classical physics had realized. It has been said in Section 20 that quantitative results derived from the observation of reactions of a body with other bodies are ascribed as inherent characters to the body itself, whether or not the reactions are actually carried out. We now see that this 'Euler principle' has very serious limitations. There are obvious analogies to this situation in the domain of psychic self-observation.

(2) Characters referring to two different gratings cannot meaningfully be combined by 'and' or 'or.' Classical logic does not fit in with quantum physics and is to be replaced by a kind of 'quantum logic.'

(3) The principle of causality holds for the temporal change of the wave state, but must be dropped as far as the relation between wave and quantum states is concerned.

(4) The whole is always more, is capable of a much greater variety of wave states, than the combination of the parts. Disjoint parts in an isolated system of fixed wave state are in general not statistically independent even if they do not interact.

(5) The Leibniz-Pauli exclusion principle, according to which no two electrons may be in the same quantum state, is made comprehensible by quantum physics as a consequence of the law of antisymmetry.

(6) There exists a primary probability, as a basic trait of nature itself, that has nothing to do with the observer's knowledge or ignorance. The probabilities $|x_i|^2$ of the individual complete quantum states i are derived from the components x_i of a vector quantity $\vec{x} = (x_1, \cdots, x_n)$ describing the 'wave state.' This seems to me to confirm the opinion expressed in the main text, namely that proba-

bility is connected with certain basic physical quantities and can in general be determined only on the ground of empirical laws governing these quantities.

It must be admitted that the meaning of quantum physics, in spite of all its achievements, is not yet clarified as thoroughly as, for instance, the ideas underlying relativity theory. The relation of reality and observation is the central problem. We seem to need a deeper epistemological analysis of what constitutes an experiment, a measurement, and what sort of language is used to communicate its result. Is it that of classical physics, as Niels Bohr seems to think, or is it the 'natural language,' in which everyone in the conduct of his daily life encounters the world, his fellow men, and himself? The analogy with Hilbert's mathematics, where the practical manipulation of concrete symbols rather than the data of some 'pure consciousness' serves as the essential extra-logical basis, seems to suggest the latter. Does this mean that the development of modern mathematics and physics points in the same direction as the movement we observe in current philosophy, away from an idealistic toward an 'existential' standpoint?

Aside from the riddles of epistemological interpretation, quantum physics is also beset by serious internal difficulties; we do not yet possess a really consistent and complete quantum theory of the interaction between electromagnetic radiation and (negative and positive) electrons, let alone the other elementary particles.

Returning to safer ground, let us add a word about the position of quantum physics towards the problem of *past and future* as discussed in Section 23C. What one wishes to understand is why light is emitted only 'towards the future.' We saw that physics can account for this distinction of the future from the past half of the light cone by keeping merely the retarded part of the potential in the Liénard-Wiechert formula. Quantum theory describes the interaction between the electrons of an atom and the field of Hohlraum radiation as a sequence of individual acts in which a light-quantum is emitted or absorbed under a corresponding energy jump of the atom. The formula for the frequencies of these acts can be interpreted as indicating that the individual act is either spontaneous or enforced. The frequency of the enforced acts is proportional to the density of the radiation, while the spontaneous acts are independent of it. The enforced part is symmetric with respect to past and future. Not so the spontaneous part; there is only spontaneous emission, but no spontaneous absorption. This asymmetry is accounted for by probability arguments of the same sort as led to the law of increasing entropy. Hence the distinction of the future half of the light cone has its roots here in the statistical principles of thermodynamics rather than in any elementary laws.

REFERENCES

P. A. M. DIRAC, *The Principles of Quantum Mechanics*, third ed., Oxford, 1947.

J. VON NEUMANN, *Mathematische Grundlagen der Quantenmechanik*, Berlin, 1932.

H. WEYL, *Gruppentheorie und Quantenmechanik*, second ed., Leipzig, 1931.

G. WENTZEL, *Einführung in die Quantentheorie der Wellenfelder*, Vienna, 1943.

NIELS BOHR, *Atomic Theory and the Description of Nature*, Cambridge, 1934.

—— Kausalität und Komplementarität, *Erkenntnis*, *14* (1937), p. 293.

M. BORN, *Atomic Physics*, transl. by J. DOUGALL, second ed., London & Glasgow, 1937.

—— *Experiment and Theory in Physics*, Cambridge Univ. Press, 1943.

H. REICHENBACH, *Philosophic Foundations of Quantum Mechanics*, Univ. of Calif. Press, 1944.

L. ROSENFELD, L'évolution de l'idée de causalité, *Mém. Soc. Roy. Sci. Lièges*, 4ᵉ sér., VI (1942).

Chemical Valence
and the Hierarchy of Structures

The symbolic structure in terms of which quantum theory explains the atomic phenomena may well be of a primitive and irreducible nature. In contrast, the aggregate of atomic points joined by valence strokes, as which Kekulé depicts a chemical molecule, is only of an intermediary character. Indeed the valence bonds are an abbreviated symbol for the actual quantum-physical forces acting between the atoms, which in themselves are complex dynamical systems. The Kekulé diagram is thus seen to be founded on a more primary structure, that of quantum mechanics. This is one instance of what Hilbert generally described as "Tieferlegung der Fundamente."

The theory of chemical bondage affords such a striking illustration of the hierarchy of structures that I cannot refrain from describing it in a little more detail. The electronic spin and the exclusion principle are those features made responsible by quantum mechanics for chemical valence. *Position* is certainly a character of the electron. If separation according to position were a finest grating then the wave state of an electron would be given by a (complex-valued) function $\psi(P)$ of an argument P ranging over all points in space [the square of the length of this vector being the integral of $|\psi(P)|^2$]; and the wave state of an aggregate of f electrons $1, 2, \ldots, f$ would be an antisymmetric function $\psi(P_1, \ldots, P_f)$ of their positions $P_1, \ldots, P_f$. The exclusion principle is a consequence of antisymmetry. Because of the inner likeness of all electrons no dynamic action is imaginable that would ever carry an antisymmetric ψ over into one that is not. A function of several arguments i, $\psi(i_1, \ldots, i_f)$, is symmetric if it stays unaltered under all $f!$ permutations of its f arguments; it is anti-symmetric if all even permutations leave it unchanged, all odd permutations carry it into $-\psi$. The nature of the argument i does not matter. It may range over a finite number of values $i = 1, 2, \ldots, n$, as we assumed for simplicity's sake in our exposition of quantum mechanics, or range over a whole continuum, as P does. We have previously seen that the distinction between even and odd permutations is the combinatorial basis for the polarity of left and right; we now find that it lies at the root of the periodic system of chemical elements and of a number of decisive traits of the physical world that defy explanation by the notions of classical physics.

Spectroscopic experience has shown that, over and in addition to separation of electrons by position, a splitting into two beams takes place, e.g. under the influence of a magnetic field. We have to conclude that the wave function of a single electron $\psi(P\rho)$ depends on two variables, the continuous variable of position P, and a second variable ρ, called spin, that is capable of two values $+1$ and -1 only. The two components $\psi(P, +1) = \psi_+(P)$ and $\psi(P, -1) = \psi_-(P)$ are relative to a Cartesian frame and, as W. Pauli first recognized, transform according to the spinor representation mentioned in Section 15, when one passes by rotation to another such frame. The wave function of an aggregate of f electrons is an antisymmetric function $\psi(P_1\rho_1, P_2\rho_2, \ldots, P_f\rho_f)$ of f pairs $(P\rho)$.

A third circumstance besides spin and antisymmetry is relevant: with considerable approximation the dynamic influence of the spin may be disregarded. Let us assume that it is strictly nil, i.e. that the dynamical operator H of energy operates only on the positional variables P, not on the spin variables. At first one may think that then one could ignore the spin altogether. That this is not so is due to the condition of antisymmetry with respect to the pairs $(P\rho)$. Let $\eta(P_1, \ldots, P_f)$ be an eigenfunction of the operator H with the eigenvalue $h\nu$, $H\eta = h\nu \cdot \eta$, η thus representing a stationary wave state of energy $h\nu$. Assume η to be antisymmetric in its f arguments P. On taking the existence of the spin into account one obtains a whole linear manifold of wave functions ψ of energy $h\nu$,

$$\psi(P_1\rho_1, \cdots, P_f\rho_f) = \eta(P_1, \cdots, P_f) \cdot \varphi(\rho_1, \cdots, \rho_f).$$

Here the second factor $\varphi(\rho_1 \ldots \rho_f)$ could be any function of the f spin variables ρ; but antisymmetry of ψ requires φ to be symmetric. A symmetric function φ of $\rho_1, \ldots, \rho_f$ assumes a definite value φ_g if a given number g of the arguments ρ are $+1$ and $f - g$ of them are -1; and the function is completely characterized by its $f + 1$ values $\varphi_f, \varphi_{f-1}, \ldots, \varphi_0$. Hence the linear manifold of the symmetric functions φ is $(f + 1)$-dimensional, and the existence of the spin has the effect that the energy level $h\nu$ or 'term' ν acquires the multiplicity $f + 1$. Only when the actually existing weak interaction of the spins is taken into account, this term of multiplicity $f + 1$ splits up into a *multiplet* of $f + 1$ different terms. Consider for a moment what would happen, on the other hand, if η were symmetric. Then φ must be antisymmetric. But an anti-symmetric function vanishes whenever two of its arguments have equal values. Hence if the individual argument is capable of two values only the function will vanish identically provided $f > 2$, and the term ν corresponding to a symmetric η is wiped out, its multiplicity becomes 0. Owing to the low dimensionality 2 of the spin space the possible permutational symmetry characters of functions $\varphi(\rho_1 \ldots \rho_f)$ can be described by one number, the valence v, which is capable of all values $0 \leq v \leq f$ that differ from f by an even number. States in which η is antisymmetric (and hence φ symmetric) are of valence f. A term of valence v has, merely on account of the permutability of electrons, the multiplicity $v + 1$.

We consider a neutral atom as an aggregate of f electrons of charge $-e$ that move in the field of a nucleus of charge $f \cdot e$ fixed at a center O. Non-relativistic mechanics is applied to this model in taking into account only the electrostatic forces between these charges plus the kinetic energy of the electrons. Let $\eta(P_1 \ldots P_f)$ be an antisymmetric eigenfunction of the energy operator H corresponding to the term ν. The atom in this state η has the energy $h\nu$ and the highest possible valence f. (Any permutation of the f points $P_1, \ldots, P_f$ would change η into an eigenfunction for the same term ν; but because of antisymmetry this causes no permutational multiplicity.) It is also true that the effect of any common rotation about O of the points $P_1, \ldots, P_f$ transforms $\eta(P_1 \ldots P_f)$ into an eigenfunction for the same term ν. Hence if we wish to avoid 'rotational' multiplicity of ν we must assume that the function $\eta(P_1 \ldots P_f)$ of the constellation $P_1 \ldots P_f$ of the electrons is invariant with respect to all rotations (central symmetry). Such a stationary wave state is called an S-state in spectroscopy. Thus we assume the atom to have its highest valence f and to be in an S-state. The probability $\mathcal{P}(r)$ to find one of the electrons at a distance greater than r from the center O is determined by η, and it turns out that $\mathcal{P}(r)$ falls off exponentially with increasing r.

After introducing two 'indeterminates' x_+, x_- corresponding to the two values $\rho = +1$ and -1 of the spin, a symmetric function $\varphi(\rho_1 \ldots \rho_f)$ is conveniently represented by the algebraic form of x_+, x_- of degree f,

$$\sum_\rho \varphi(\rho_1 \cdots \rho_f) x_{\rho_1} \cdots x_{\rho_f} = \sum_g \frac{f!}{g!(f-g)!} \varphi_g \, x_+^g \, x_-^{f-g}$$

with the coefficients φ. The sum at the left consists of 2^f terms as each ρ takes on its two values $+$ and $-$ while the range of g at the right side is the sequence $f, f - 1, \ldots, 0$. Considering the indeterminates x_+, x_- as components of a vector x in a plane we submit them to an arbitrary linear transformation

(1) $$x_+ = \alpha x_+' + \beta x_-', \quad x_- = \gamma x_+' + \delta x_-'$$

of modulus $\alpha\delta - \beta\gamma = 1$. A form $F(x, y, \ldots)$ of several indeterminate vectors $x, y, \ldots$ which is of degree f_a in x, f_b in y, $\ldots$, is said to be an *invariant* if $F(x, y, \cdots) = F(x', y', \cdots)$ whenever x and x', y and y', . . . , are connected by the same transformation (1) of modulus 1.

Envisage now a number of neutral atoms $a, b, \ldots$ of $f_a, f_b, \ldots$ electrons with their nuclei fixed at definite points in space $O_a, O_b, \ldots,$ and suppose that each is in a stationary S-state of highest valence, their respective energy levels being $h\nu_a, h\nu_b, \ldots$. This implies that the combined system of these atoms has the energy $h\nu_0$, $\nu_0 = \nu_a + \nu_b + \cdots$, and that its state belongs to a linear manifold Π of $(f_a + 1)(f_b + 1) \cdots$ dimensions. Speaking in this way we have disregarded the mutual interaction of the atoms and thus violated the essential likeness of all $f = f_a + f_b + \cdots$ electrons by assigning f_a of them to the entourage of O_a and letting these f_a electrons interact only among themselves and with the nucleus O_a. We assume the mutual distances r of the nuclei at $O_a, O_b, \ldots$ to be large in comparison to the Bohr radius h^2/me^2. Taking now the interaction between the several atoms into account as a small perturbation one finds that the term ν_0 breaks up by 'permutational resonance' into a number of term systems of the molecule according to the various possible valences $v = f, f - 2,$ . . . . The states of valence v form a linear submanifold Π_v of n_v dimensions that, in the approximation of perturbation theory, as a whole is invariant with respect to the energy operator H and hence stationary. The corresponding n_v terms $\nu = \nu_0 + \Delta\nu$ and individual stationary states of the molecule are to be determined as the eigenvalues and eigenfunctions of H operating in Π_v. Since each of the n_v terms ν of the molecule in a state of valence v has the multiplicity $v + 1$, comparison of dimension leads to the equation

$$(f_a + 1)(f_b + 1) \cdots = \sum_v n_v(v + 1).$$

The shifts $\Delta\nu = V(O_a, O_b \cdots)$ are functions of the constellation $O_a O_b \ldots$ of the nuclei which are found to be of the same type as the probability $\wp(r)$ mentioned above, namely falling off exponentially with increasing distances r. This accounts for the fact that the homopolar bond between neutral atoms is a short range force. (The attraction of two ions of opposite charges at a distance r, the heteropolar bond, is no mystery at all. Its energy follows the Coulomb law $1/r$ and is thus of the long range type.)

Let an indeterminate binary vector $x = (x_+, x_-)$, y, . . . be associated with each of the atoms $a, b, \ldots$ and add one more 'free' vector l. Then Π_v is best described as the linear manifold of all invariants $J(x, y, \ldots, l)$ depending on the indeterminate vectors $x, y, \ldots, l$ with the given degrees $f_a, f_b, \ldots, v$. The details do

269

not matter here. But so much should be clear that the two-dimensional vectors and invariance with respect to linear transformations play a role because, owing to the spin, the state (φ_+, φ_-) of an electron is such a vector. The dimensionality n_v of Π_v is the number of linearly independent invariants (of degrees $f_a, f_b, \ldots$ in the vectors $x, y, \ldots$ and) of degree v in the free vector l.

The simplest invariant, linearly depending on two indeterminate vectors x, y, is the 'bracket factor' $[xy] = x_+y_- - x_-y_+$. Any product of such bracket factors is called a monomial invariant. A monomial invariant is completely described by a diagram in which each of the argument vectors $x, y, \ldots, l$ is represented by a point and each bracket factor like $[xy]$ by a line joining the points x and y. (A bracket factor $[xl]$ involving the free vector l may instead be represented by a stroke issuing from x the other end of which remains free.) The degrees $f_a, f_b, \ldots, v$ of the monomial invariant are the numbers of strokes ending at the respective points $x, y, \ldots, l$. Hence the monomial invariants correspond completely to the Kekulé valence diagrams. We shall therefore call a state described by such an invariant a pure valence state. The first main theorem of the theory of invariants states that every invariant of given degrees is a linear combination of monomial invariants of those degrees.

For a molecule consisting of two atoms x, y of valences a and b, $a \geq b$, we find only one invariant

$$[xy]^d [xl]^{a-d} [yl]^{b-d}$$

for each of the possible molecular valences $v = a + b - 2d$, $d = 0, 1, \ldots, b$. This corresponds exactly to what the valence diagrams would have us expect; d is the number of valence strokes joining the two atoms, and $a - d$, $b - d$ are the numbers of free valence strokes issuing from x and y respectively. For the corresponding term $v_0 + \Delta v$ of the molecule one finds $\Delta v = \lambda \cdot V(r)$ where $V(r)$ is a function of the distance r of the two atoms that does not depend on d, while the form factor $\lambda = (a - d)(b - d) - d$ depends on d but not on r. The function $V(r)$ is difficult to compute, but in the simplest cases turns out to be positive for large r. Assuming this to be generally true one obtains a force of attraction or repulsion according to whether the form factor λ is negative or positive. λ is negative for $d = b$; but since λ varies from $-b$ to ab while d assumes the values $b, b - 1, \ldots 0$, the form factor will be negative only for the strongest binding $d = b$, or possibly for a few of the stronger bindings $d = b$, $b - 1, \cdots$.

The picture changes somewhat when more than two atoms come into play. Then the number n_v of linearly independent invariants is less than that of the possible diagrams with v free valence strokes, owing to the existence of linear relations among the monomial invariants. Moreover the individual stationary states with definite energy levels $v_0 + \Delta v$ do no longer coincide with any of the pure valence states. There are clear indications for this in chemistry. For instance, Kekulé's famous formula for the benzene ring, a regular arrangement of six CH groups, foresees two possibilities whereas the study of ortho-derivatives proves conclusively that there is only one

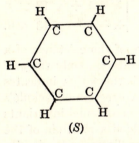

(S)

in nature. The skeleton shown in the valence diagram S has the full hexagonal symmetry which one expects for the benzene ring; but it leaves one valence electron in each C-atom unattached. This conception of a fixed skeleton upon which the variable state of bondage between the remaining valence electrons is superimposed may be an unwarranted simplification, but is useful for a first orientation and

reduces our problem (which in fact involves forty-two electrons) to that of six equal one-electron atoms arranged in a regular hexagon of side r. Here the states of valence 0 form the linear manifold of all invariants depending linearly on six argument vectors 1, 2, 3, 4, 5, 6. The five monomial invariants A, A'; B_1, B_2, B_3 shown below represent a basis for that manifold. (Their diagrams should be superimposed

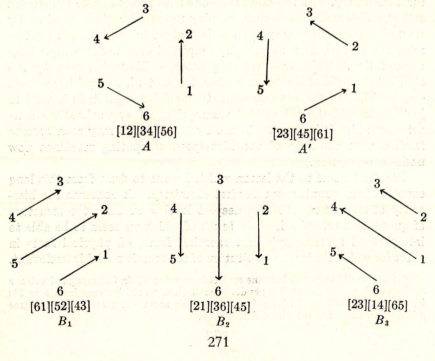

upon the skeleton S; the corresponding five pure valence states are 'in resonance.') The term shifts $\Delta \nu = \lambda \cdot V(r)$ of the various stationary states η differ by the form factor λ; the potential function $V(r)$ of r, however, is a common factor of the short range exponential type. Here is a list of the stationary states with their form factors λ:

$$
\begin{array}{l|l}
A + A' & \lambda = 0 \\
\beta_1 B_1 + \beta_2 B_2 + \beta_3 B_3 \quad (\beta_1 + \beta_2 + \beta_3 = 0) & \lambda = 2 \\
6(A - A') - (1 + \sqrt{13})(B_1 + B_2 + B_3) & \lambda = 1 + \sqrt{13} \\
\eta = 6(A - A') - (1 - \sqrt{13})(B_1 + B_2 + B_3) & \lambda = 1 - \sqrt{13} < 0.
\end{array}
$$

As we know, a negative value of λ is suggestive of the existence of a stable molecule in the corresponding quantum state η. Only the η of the last line satisfies this condition, and thus it is this η that indicates the direction in which the quantum mechanical correction of Kekulé's benzene formula will lie: To the difference of the monomial invariants A, A' depicted by Kekulé's two diagrams is added a multiple of the sum of the three terms represented by the Dewar diagrams B_1, B_2, B_3. (Both $A - A'$ and $B_1 + B_2 + B_3$ change sign under the influence of a rotation of the hexagon by 60° and remain unaltered by reflection in one of the three diagonals.)[1]

The notion of quantum-mechanical resonance between states of (nearly) equal energy levels plays an important role in modern structural chemistry. At the same time one tries to hold on to well-tested and plausible valence schemes, keeping the modifications required by resonance to a minimum; and one is content in most cases to determine the perturbation energies by empirical observation rather than computation. This conservative procedure, illustrated here by the classical example of the benzene ring, has met with surprising success — surprising to the scrupulous mathematician, who finds it hard to justify some of the 'plausible' assumptions of approximative character on which it is based. (A more exact analysis may soon become feasible with the help of the high-speed computing machines now under construction.)

Finally I come to the lesson which I want to draw from this long excursus into quantum-mechanical chemistry. It concerns the hierarchy of structures. On the deepest level α we have the structure of quantum mechanics itself in terms of which we seem to be able to interpret all spectroscopic and chemical facts, all physical facts in short for which the inner constitution of the atomic nuclei is irrelevant.

[1] If one would treat the benzene molecule as a ring of six CH groups of valence 3, the number of resonant independent possibilities would increase from 5 to 34; for the linear manifold of binary invariants of six argument vectors and of degree 3 in each of them has the dimensionality 34.

On the second level β the structure representing a molecule in its various possible states is the linear manifold of binary invariants. This picture has limited truth only. Above all it does not refer to the ready-made molecule but to the aggregate of its atoms with their nuclei fixed at distances large in comparison to the extension of the atoms. Moreover, as to the individual atoms, conditions as simple as possible are assumed with respect to the permutations of its electrons and rotation of their configuration in space. The structures which are used on the third level γ for the interpretation of chemical facts are the valence diagrams. In the light of β the picture γ is correct in one essential respect: all possible states of the molecule (all invariants) are indeed linear combinations of the pure valence states (monomial invariants). But it errs on three other counts: (1) There are not only a few discrete states, such as the pure valence states, but rather a whole linear manifold of wave states; this, of course, is the decisive contrast between classical and quantum mechanics. (2) The linear relations between the monomial invariants are ignored, and therefore too high a value is obtained for the number n_v of independent possibilities. (3) The n_v stationary quantum states coincide in general with none of the pure valence states but are certain linear combinations of them.

Contrary to our exposition, the historical order is that of descent to an ever deeper level, $\gamma \rightarrow \beta \rightarrow \alpha$. A. Kekulé developed his graphical representation of chemical structure in 1859. The intermediate level β was first reached by J. J. Sylvester in 1878[1] (he was later followed

---

[1] Sylvester's paper published in the first volume of the *American Journal of Mathematics*, which he founded at Johns Hopkins, bears the title *On an application of the new atomic theory to the graphical representation of the invariants and covariants of binary quantics*. Of the opening sentences the first is such a characteristic statement of 19th century natural philosophy and the second such a charming piece of Sylvesterian prose that they may be quoted here. "By the *new* Atomic Theory I mean that sublime invention of Kekulé which stands to the *old* in a somewhat similar relation as the Astronomy of Kepler to Ptolemy's, or the System of Nature of Darwin to that of Linnaeus; — like the latter it lies outside of the immediate sphere of energetics, basing its laws on pure relations of form, and like the former as perfected by Newton, these laws admit of exact arithmetical definitions. — Casting about, as I lay awake in bed one night, to discover some means of conveying an intelligible conception of the objects of modern algebra to a mixed society mainly composed of physicists, chemists and biologists, interspersed only with a few mathematicians, to which I stood engaged to give some account of my recent researches in this subject of my predilection, and impressed as I had long been with the feeling of affinity, if not identity of object, between the inquiry into compound radicals and the search for 'Grundformen' or irreducible invariants, I was agreeably surprised to find of a sudden distinctly pictured on my mental retina a chemico-graphical image serving to embody and illustrate the relations of these derived algebraic forms to their primitives and to each other, which would perfectly accomplish the object I had in view, as I will now proceed to explain."

by the German invariant theorist P. Gordan and the Russian chemist W. Alexejeff). However in the absence of a physical interpretation for the addition of invariants and of dynamical laws by which to determine the binding forces and the actual stationary states the chemists stuck to their familiar valence diagrams. We can see today that only such radical departure as that of quantum mechanics could reveal the significance of the picture that Sylvester had stumbled upon as a purely formal, though very appealing, mathematical analogy.

The moral of this story is evident: do not take too literally such preliminary combinatorial schemes as the valence diagrams, useful as they are as a first guide in a seemingly incoherent mass of facts. A picture of reality drawn in a few sharp lines can not be expected to be adequate to the variety of all its shades. Yet even so the draftsman must have the courage to draw the lines firm. There is no doubt that the gene aggregates of genetics with their linkages are structures of no less preliminary character than the valence diagrams of chemistry. The cytological study of cells reveals complicated motions of chromosomes and multiform physical processes whose details are capable of continuous variation and of whose result the discrete genetic diagrams are no more than abbreviated abstracts of limited validity. I should therefore not vouch too much for the adequacy of the primitive combinatorial scheme as depicted in Appendix B, and yet it seemed best to make the picture itself, however limited its value, as definite as possible. (This is a principle which Nicolaus Cusanus stressed in *De docta ignorantia*: if the transcendental is accessible to us only through the medium of images and symbols, let the symbols at least be as distinct and unambiguous as mathematics will permit.)

The facts related in the next Appendix leave little doubt that the laws of inheritance are ultimately based on the same structure as the laws of chemistry: on the structure of quantum mechanics. A structure that could serve to mediate between the genetic diagrams and quantum physics should be one that takes into account the chemical complexity of the carriers of life. Perhaps the simplest combinatorial entity is the group of the $n!$ permutations of n things. This group has a different constitution for each individual number n. The question is whether there are nevertheless some asymptotic uniformities prevailing for large n or for some distinctive class of large n. Mathematics has still little to tell about such problems. One wonders whether a quantum theory of organic processes is tied up with their solution.

Whereas the quantum structure described in Appendix C has for the present been accepted by physics as the ultimate layer, the skeptic philosopher may wonder whether this reduction is more than one step, the last at the moment, in a *regressus ad infinitum*. But, so warns the scientist, nothing is cheaper and on the whole more barren than to play with such possibilities in one's thoughts before new discoveries place one before a concrete situation that enforces a further Tieferlegung of the foundations.

Physical phenomena are spread out in the continuous extensive medium of space and time; it was this aspect which dominated to a considerable degree the epistemological thoughts about natural science that the main part of this book tried to collect in 1926. This was historically justified, and the accomplishments of general relativity, still very fresh at that time, lent additional emphasis to this point of view. In the last two decades, however, discontinuous and combinatorial structures underlying the natural phenomena have become of increasing significance. Here a deeper layer seems to come to light, for the description of which our ordinary language is woefully inadequate. The preceding Appendices bear witness to this changed outlook. However, we could not do much more than assemble relevant material; the philosophical penetration remains largely a task for the future.

REFERENCES

M. BORN, *Chemische Bindung und Quantenmechanik*, Ergebnisse der exakten Naturwissenschaften, vol. 10, Berlin, 1931.

LINUS PAULING, *The Nature of the Chemical Bond*, second ed., Cornell University Press, 1945.

W. G. PALMER, *Valency, Classical and Modern*, Cambridge, 1944.

Physics and Biology

1. One of the profoundest enigmas of nature is the contrast of dead and living matter. However one may characterize life phenomenologically: animate matter is obviously separated from inanimate by a deep chasm. Life dwells only in material systems that from a physico-chemical standpoint are to be considered as highly complex. In a descriptive way and without claiming completeness we enumerate some of the typical features of the living organism: its composition of cells, living units that are uniform in their more basic characteristics; wholeness as form (morphé, Gestalt) and as functional complex, with mutual adjustment of all cell differentiations to each other ("geprägte Form, die lebend sich entwickelt," Goethe); endowed through metabolism with the capacity of using alien matter as food and incorporating it into its own organization; development by assimilation of food, by growth and differentiation from relatively simple to more complicated states; in spite of inner lability, far-reaching though not unlimited capacity to maintain itself as this differentiated whole under changing external influences, in particular in the turmoil of the molecular heat motion, and to restore itself after disturbing encroachments; limitation of individual existence in time (birth and death); the capacity of propagation and of transmitting its specific constitution to its progeny. While dead matter is inert, the organism is a source of activity that bears the stamp of spontaneity ever more manifestly (with volitional action as its climax) the higher one climbs in the world of organisms. It is at the same time susceptible to stimuli (perceptions on the highest level) and endowed with the capacity of storing stimulative experiences (mneme). Life has unfolded into a vast multitude of species of typically diverse constitution, and the organisms are woven into a dense net of adaptations and relations to each other and their surroundings.

The last unit and its one basic property to which the essential characteristics of living matter seem to have been reduced by scientific analysis is the *gene* and its power of *self-duplication.* By this process a copy of the model gene is synthesized from the material available in the living cell. Incidentally, the gap between organic and inorganic matter has been bridged to a certain extent by the discovery of viruses. Viruses are submicroscopic entities that behave like dead inert matter unless placed in certain living cells. As parasites in these cells, however, they show the fundamental characteristics of life — self-duplication and mutation. On the other hand many viruses have the structure typical of inorganic matter; they are crystals. In size they range from the more complex protein molecules to the smaller bacteria. Chemically they consist of nucleo-protein, as the genes do. A virus is clearly something like a naked gene. The best studied virus, that of the tobacco mosaic disease, is a nucleo-protein of high molecular weight consisting of 95 per cent protein and 5 per cent nucleic acid; it crystallizes in long thin needles.

The elementary laws of matter that physics reveals and chemistry is ruled by are no doubt also binding on living matter. Hence such a profound change of physics as brought about by quantum theory must have its repercussions in biology. As long as progress from simple to more complicated configurations remains the methodologically sound way of science, biology will rest on physics, and not the other way around. The specific properties of living matter will have to be studied within the general laws valid for all matter; the viewpoint of holism that the theory of life comes first and that one descends from there by a sort of deprivation to inorganic matter must be rejected. It is therefore significant that certain simple and clearcut traits of wholeness, organization, acausality, are ascribed by quantum mechanics to the elementary constituents of all matter. A *rapprochement* between physics and biology has undoubtedly taken place in this regard. Structure and organization are not peculiar to living beings; physics is thoroughly familiar with this aspect and represents it by the symbolic apparatus of the theory that precedes all dynamical laws. The quantum physics of atomic processes will become relevant for biology wherever in the life cycle of an organism a moderate number of atoms exercises a steering effect upon the large scale happenings. (The radio tube is today the most familiar inorganic example of such a steering mechanism.)

On a broad empirical foundation, *genetics* furnishes the most convincing proof that organisms are controlled by processes of atomic range, where the acausality of quantum mechanics may make itself felt. At the turn of the century, when Planck introduced the action quantum into physics, de Vries discovered the jumplike mutations of the genetic constitution of Oenothera (the larger part of which, to be sure, are today recognized as structure rather than point mutations). For a physical understanding of mutations, their artificial generation by exposing chromosomes to X-rays has proved of momentous importance. The mere fact of such X-ray induced mutations proves that the genes are physical structures. When X-rays fall upon matter this or that photon relinquishes all or a large part of its energy to a fast secondary electron, and this in turn loses its energy in a number of steps by ionization (or excitation) of atoms. The average energy of

277

ionization amounts to about 30 electron volts. By ingenious methods H. J. Muller, N. W. Timoféeff-Ressowsky, and others have succeeded in establishing simple quantitative laws concerning the rate of induced mutations. These results indicate that the mutation is brought about by a single hit, not by the concerted action of several hits, and that this hit consists of an ionization, and is not, as one might have thought, a process directly released by the X-ray photon or absorbing the whole energy of the secondary electron.

These facts suggest the hypothesis that a gene is a (nucleo-protein) molecule of highly complicated structure, that a mutation consists in a chemical change of this molecule brought about by the effect of an ionization on the bonding electrons, and that thus allele genes are essentially isomeric molecules. The most elementary chemical changes which quantum physics can devise are localized two-step quantum jumps — first the molecule is lifted from an energy level 1 to a higher level 2, and from there it drops to a new stable state of energy level 3. The difference 2 minus 1 is the necessary activation energy U. The rate at which a specific quantum jump requiring the activation energy U occurs spontaneously at a given temperature depends essentially on U alone, but varies extremely strongly with U (according to an exponential law). At the temperature at present prevailing on the earth's surface such quantum jumps as correspond to values of U between say 1.4 and 1.7 would be occasionally occurring yet rare events. (For lower values of U the corresponding quantum jumps are so frequent that the statistical law of large numbers comes into power; they give rise to such ordinary chemical reactions as take place in the development of an organism.) Thus one is tempted to complete the picture by interpreting mutation as a rare quantum jump with an activation energy within the range just mentioned (Delbrück's model).[1] The observed absolute rate of mutations would be explained if a specific mutation requires that a hit occurs within a critical volume ('target') in the gene, the magnitude of which amounts to about 5–10 A cube (5–10 atomic distances cube). The physicist finds it, if not plausible at least acceptable, that a quantum jump at a specific point requiring an activation energy of about 1.5 is released by a hit of 30 electron volts within a sensitive volume of 5–10 A cube. The observed thermic variation of the spontaneous mutation rate (van't Hoff's factor) is in good quantitative agreement with the picture.

There are several methods for estimating size and molecular weight

---

[1] Cf. N. W. Timoféeff-Ressowsky, K. G. Zimmer, M. Delbrück, *Über die Natur der Genmutation und der Genstruktur*, (Nachr. Gött. Ges. Wissensch., Math.-physik. Kl., Fachg. VI, 1), 1935, pp. 189–245.

of a gene. Most of the radiation experiments are concerned with mutations called recessive lethals. A certain high percentage of these is due to an ionization depriving the gene of its reproductive power (while others are brought about by gross structure mutations). It is plausible to put the first kind of lethals in analogy to the inactivation of enzymes and viruses. For these latter processes, which are also due to single ionizations, one can determine the target size, either by means of the absolute dose of X-rays or by the relative efficiency of the various radiations. One finds target radii that are between one and five times as small as the radius of the enzyme molecule or the virus. Hence the target size of a gene for the totality of recessive lethal mutations 'of the first kind' ought not to be much smaller than the size of the gene. Another method for ascertaining the size of a gene or at least an upper bound for it is the following. The greatly enlarged chromosomes of the salivary glands of Drosophila show a cross striation by bands of characteristically different width and design, and the parallelism of the genetic and cytological findings vindicates the hypothesis that these bands correspond to genes or small groups of genes in the many parallel threads of which the giant chromosome consists. The several methods agree in making it likely that the molecular weight of genes is of the order of magnitude of one million (times the atomic weight of hydrogen). That is exactly what one would have expected, considering that the weights of the threadlike molecules of nucleic acids range from fifty thousand to several hundred thousand while the weight of the individual tobacco mosaic virus molecule reaches the figure of forty millions.

The investigations of the last ten years have not been favorable to the special hypothesis that a mutation is due to a quantum jump localized in and restricted to a few atoms. Several complications have come to light. To give one extreme example, W. M. Stanley found that a certain spontaneous mutation of the tobacco mosaic virus changes its chemical composition by adding about one-thousand molecules of histidine. One is thus forced to think of some mechanism by which the individual ionization releases a chain of (enzymatic?) reactions with the complex mutation as its end result. Be this as it may, the direction in which our model points is hardly deceptive; the gene is to be considered as a complex molecule and mutations are closely connected with quantum jumps. The latter can be brought about by single ionizations, and thus one may conclude with P. Jordan, that "the steering centers of life are not subject to macrophysical causality but lie in the zone of microphysical freedom." Incidentally viruses that can be isolated and observed by means of the electron microscope are in many respects better objects for the investigation of the physical foundations of the mutation process than the invisible genes in the chromosomes of cells.

The nucleus of a fertilized egg is supposed to furnish by its genetic constitution the complete determinants for the development of the organism. In earlier times one often found great difficulty in harmonizing this view — so closely related to the issue of 'preformation' versus 'epigenesis' — with the vast manifold of animals and plants, all their various courses of development and all their minute differentiations. However, the fantastically high number of possible combinations of atoms in a gene molecule (cf. the characteristic numbers for combinations of symbols in Appendix A) exceeds by far all that is needed for this purpose. It is thus not inconceivable that the miniature code contained in the gene molecules of the cell nucleus should precisely correspond with a highly complicated and specified plan of development and should somehow contain the means to put it into operation. In a famous experiment Driesch observed that the cut-off upper third of a Clavellina, its gill basket, reverts to an amorphous conglomerate of cells from which there develops a new complete Clavellina of reduced scale. He saw in this experiment a proof for the existence of an entelechy not expressible in terms of physical structure. Today we have a rather definite picture of the physical structure that can serve as such an entelechy. The question of selection among the combinatorial possibilities is something else; it points (i) towards the physicochemical problem of the stability of complex molecular compounds, (ii) to the mechanism by which the 'code' is translated into the development of an organism, and (iii) to the process of evolution.

The stability of a molecule stems from the chemical bonds between its atoms. As mentioned before, it was quantum physics that threw the first light upon the previously rather obscure nature of chemical bonds. A crystal (like diamond) is a regular pattern of atoms (C-atoms in the case of diamond) periodic in three independent spatial directions. Here the bonds extend between all atoms and thus the entire crystal is as it were a single molecule. The stability of solid bodies is that of crystals, and if the Delbrück model is basically correct, then the stability of genes rests on the same quantumtheoretic foundation. Yet while in a crystal the same building bricks are repeated periodically, each atom in the gene has its specific non-interchangeable place and role. Schrödinger therefore speaks of the gene as an aperiodic crystal and ascribes to it a higher degree of order and organization than to the periodic crystal. Whereas the macroscopic order and regularity of nature is based by statistical thermodynamics upon microscopic disorder, we encounter here in the crystals and the chromosomes of cell nuclei an order that is not overwhelmed by thermic disorder. In contrast to ordinary chemical reactions, the laws of which are obtained by averaging over an enormous number of molecular processes, mutations attack single genes. Inasmuch as (a) the order of the zygote is transmitted by self-duplication and mitosis to all somatic cells, and (b) the relatively minute speck of well-ordered atoms in the chromosomes of the cells controls the development of the living being, the dislocation of a few atoms in the mutant gene results in a well defined change in the macroscopic hereditary character of an organism. Before we have gained insight into the mechanism underlying the processes (a) and (b) we cannot claim to understand ontogenetic development.

While formal genetics has advanced by leaps and bounds during the last forty to fifty years, our knowledge in these fields is still very sketchy. As to the central problem of self-duplication, M. Delbrück has recently (1941) ventured to give a detailed but admittedly hypothetical picture of how amino-acids might conceivably be strung together in a pattern emulating a preexisting gene model by quantum-mechanical resonance at the site of the peptide links. Connection between gene and visible character, e.g. between the wing form of Drosophila called jaunty and its gene, is certainly the resultant of a chain of intermediary actions. It is therefore an important step ahead that recently attention has been concentrated on genetic control of enzymatic action; many experiences point to a close relation between genes and specific enzymes (cf. the recent work of G. W. Beadle and others on Neurospora). When a tiny speck of solid crystalline substance causes a saturated solution of the same substance to crystalize, we witness how a germ of order is capable of spreading order. Although we are as yet unable to pursue this physical process in theoretical detail there is no doubt that it lies within the scope of our known physical laws. Science will press on to analyze the manifold processes on which the order in living organisms depends in fundamentally the same way, i.e. on the ultimate basis of quantum physics with its primary statistics. But there may be a bifurcation in the following sense: as order is derived from disorder by means of the secondary statistics of thermodynamics, so may a parallel but different type of macro-law account for the production of large-scale order from small-scale order in an organism (Schrödinger).

2. With the mutations a clearly recognizable non-causal element penetrates into the behavior of organisms. Whereas my perceptions and actions are in general the resultants of innumerable individual atomic processes and thus fall under the rule of statistical regularity, it is a noticeable fact that, if favorable circumstances prevail, a few

photons (not more than 5 to 8) suffice to set off a visual perception of light. From here, from the quantum mutations in the gene molecule and the translation of a stimulus of a few photons into visual perception, it is still a long, long way to the full psychophysical reality with which man finds himself confronted, and to an integrated theoretical picture of it that would account for the facts of free insight and free will. "Although the door of human freedom is opened," says Eddington in *New Pathways in Science* (p. 87), "it is not flung wide open; only a chink of daylight appears. But it is no longer actually barred and efforts to prise it further open are encouraged." How far, we may ask, is the example of mutations representative, how far may organic processes be ascribed to the trigger action of small groups of atoms of unpredetermined behavior? The physicist P. Jordan has argued the point that to a considerable extent this is indeed the case, but has met with much opposition among biologists. Also Schrödinger warns that, without the almost complete precision and reliability of the macroscopic thermodynamical laws ruling the nervous and cerebral processes of the human body and its interactions with the surrounding world, perception and thought would be impossible. Niels Bohr, however, is inclined to widen the domain of uncertainty by adding a specific biological principle of indeterminacy (the precise content of which is still unknown) to Heisenberg's well-established quantum-mechanical principal of indeterminacy. He has pointed out in this connection that an observation of the state of the brain cells exact enough for a fairly definite prediction of the victim's behavior during the next few seconds may involve an encroachment of necessarily lethal effect — and thereby make the organism predictable indeed. Bohr maintains that in this way analysis of vital phenomena by physical concepts has its natural limits; just as one had to put up with complementarity as expressed by Heisenberg's principle of indeterminacy in order to explain the stability of atoms, so are further renouncements demanded of him who tries to account for the self-stabilization of living organisms.

Such theoretical acts as the judgment that $2 + 2$ makes 4 have served in the main text to bring out the salient point of the problem of freedom. Thought as thought would be abrogated if it were denied that in my judging thus the mental fact that $2 + 2$ actually makes 4 gains power over an individual psychic act, and not only over the psychic act but also over the movements of my lips that form the corresponding words pregnant with meaning, or over the movements of the hand that, perhaps in the context of a mathematical proof, writes down the marks '$2 + 2 = 4$' on paper. Punching a hole in the strict causality of nature does not therefore suffice; a representation within the theory must be found for vital, psychic, and spiritual factors that in some way direct and steer the atomic process. It is certainly important in that grasp of the totality of nature that precedes all theory not to lose sight of such traits. H. Spemann in his last work (1937) comes to the conclusion "that the processes of development . . . in their mutual association bear closer resemblance than to any other vital processes to those of which we have the most intimate knowledge, namely the psychic ones." Such voices as this — and it is not an isolated one — should be heeded. And yet, I believe that in a theory of reality the ideal factors which are here in question must be represented in basically the same way as the physical elementary particles and their forces, namely by a structure expressed in terms of symbols. I put no undue confidence in the suggestion made in the main text, that this purpose could be served by correlations between such atomic events as are treated as statistically independent by thermodynamics. Indeed the example of quantum mechanics has once more demonstrated how the possibilities with which our imagination plays before a problem is ripe for solution are always far surpassed by reality. Even so, the explanation of the chemical bond by Pauli's exclusion principle is perhaps a hint that the radical break with the classical scheme of statistical independence is an opening of the door as significant as the quantum mechanical complementarity.

Scientists would be wrong to ignore the fact that theoretical construction is not the only approach to the phenomena of life; another way, that of understanding from within (interpretation), is open to us. Woltereck, in a broadly executed *Philosophie der lebendigen Wirklichkeit*, has recently ventured to describe in some detail the "within" of organic life. Of myself, of my own acts of perception, thought, volition, feeling and doing, I have a direct knowledge entirely different from the theoretical knowledge that represents the 'parallel' cerebral processes in symbols. This inner awareness of myself is the basis for the understanding of my fellow-men whom I meet and acknowledge as beings of my own kind, with whom I communicate, sometimes so intimately as to share joy and sorrow with them. Even if I do not know of their consciousness in the same manner as of my own, nevertheless my 'interpretative' understanding of it is apprehension of indisputable adequacy. Its illuminating light is directed not only on my fellow men; it also reaches, though with ever increasing dimness and incertitude, deeply into the animal kingdom. Albert Schweitzer is right when he ridicules Kant's narrow opinion that man is capable of compassion, but not of sharing joy with the living creature, by the question, "Did he never see an ox coming home from the fields drink?"

It is idle to disparage this hold on nature 'from within' as anthropomorphic and elevate the objectivity of theoretical construction. Both roads run, as it were, in opposite directions: what is darkest for theory, man, is the most luminous for the understanding from within; and to the elementary inorganic processes, that are most easily approachable by theory, interpretation finds no access whatsoever. For objective theory the understanding from within can serve as a guide to important problems although it cannot provide their objective solution. A recent example is provided by investigations about the direction of the instinctive behavior of animals by 'appetences.'

It is tempting to stretch Bohr's idea of complementarity far enough to cover the relation of the two opposite modes of approach we are discussing here. But however one may weigh them against each other, one cannot get around the following significant and undeniable fact: the way of constructive theory, during the last three centuries, has proved to be a method that is capable of progressive development of seemingly unlimited width and depth; here each problem solved poses new ones for which the coordinated effort of thought and experiment can find precise and universally convincing solutions. In contrast the scope of the understanding from within appears practically fixed by human nature once for all, and may at most be widened a little by the refinement of language, especially of language in the mouth of the poets. Understanding, for the very reason that it is *concrete* and *full*, lacks the freedom of the 'hollow symbol.' A biology from within as advocated by Woltereck will, I am afraid, be without that never-ending impetus of problems that drives constructive biology on and on.

REFERENCES

W. M. STANLEY, Chemical properties of viruses, in *The Study of Man*, Bicentennial Conference, Univ. of Pennsylvania Press, 1941.

G. W. BEADLE, The Gene, *Proc. Am. Phil. Soc.*, *90* (1946), pp. 422–431.

H. J. MULLER, The Gene, *Proc. Roy. Soc.*, B*134* (1947), pp. 1–37.

N. W. TIMOFÉEFF-RESSOWSKY, K. G. ZIMMER, M. DELBRÜCK, *Über die Natur der Genmutation und der Genstruktur* (Nachr. Ges. Wiss. Göttingen, Math. physik. Kl., Fachgr. VI, *1*) 1935, pp. 189–245.

E. SCHRÖDINGER, *What is Life?*, Cambridge and New York, 1945.

D. E. LEA, *Actions of Radiations on Living Cells*, Cambridge and New York, 1946.

R. GOLDSCHMIDT, *Physiological Genetics*, New York, 1938.

M. DELBRÜCK, *Cold Spring Harb. Symp. on Quant. Biology*, 9 (1941), pp. 122–126.

NIELS BOHR, Light and Life, *Nature*, 131 (1933), pp. 421, 457.

—— Biology and Atomic Physics, *Rend. gener. celebr. Galvani*, Bologna, 1938.

P. JORDAN, *Naturwissenschaften*, *20* (1932), p. 815; *22* (1934), p. 485; *26* (1938), p. 537.

—— *Erkenntnis*, *4* (1934), p. 215.

—— Zur Quanten-Biologie, *Biol. Zbl.*, *59* (1939), pp. 1–39.

—— *Die Physik und das Geheimnis des organischen Lebens*, Braunschweig, 1941.

R. WOLTERECK, *Philosophie der lebendigen Wirklichkeit*. Bd. 1: *Grundzüge einer allgemeinen Biologie*, Stuttgart, 1931; Bd. 2: *Ontologie des Lebendigen*, Stuttgart, 1940.

The Main Features of the Physical World; Morphe and Evolution

----

The whale is a spouting fish with a horizontal tail.—H. MELVILLE, *Moby Dick*, Chap. XXXII.

Life must not cease. That comes before everything. It is silly to say you do not care. You do care. It is that care that will prompt your imagination; inflame your desires; make your will irresistible, and create out of nothing.—The Serpent to Eve, in G. B. SHAW, *Back to Methuselah*, Act I.

Farewell, farewell! but this I tell
To thee, thou Wedding-Guest!
He prayeth well, who loveth well
    Both man and bird and beast.
                COLERIDGE, *The Rime of the Ancient Mariner*.

Whence this creation has its origin,
Whether created whether uncreated,
He who looks down from heaven's highest seat,
He only knows — or does He know not either?
                Rig-Veda, Mandala X, Hymn 129.

Not content with an answer to the question 'How is it?' we wish to know 'How did it come to be so?' Man, wherever he awakens to ponder the riddles of existence, is prone to expect *evolution* to enlighten him about the essence of things. The idea of evolution plays a predominant role in mythology and the primitive philosophical thoughts of mankind. The Indian speculations about Brahma as the eternal self-existing being unfolding itself into the manifold world of material objects and individual souls continue the same line. So does the Neo-Platonists' doctrine of "emanation," while the early Ionian philosophers try to account for the genesis of the universe on a more physical basis. Important steps toward the modern conception of physical evolution are taken by Empedocles and then by Lucretius. This is not the place to review in detail the special forms which the idea of evolution assumed in Aristotle (with his contrast of potentiality and actuality), in Plato and the great philosophical systems of our era from Descartes on. Kant's name is associated with one of the first scientific genetical theories, that of the origin of the planetary system. More universal and speculative is Schelling's notion of the self-realization of nature in a succession of forms, points of arrest, as it

285

were, brought about by a limitation of her infinite productivity, or Hegel's immanent and timeless dialectical process, that only in the realm of the spirit gives rise to a truly historic development. With Buffon, Lamarck, Treviranus and Goethe we approach the modern conception of organic evolution.

The experience of science accumulated in her own history has led to the recognition that evolution is far from being the basic principle of world understanding; it is the end rather than the beginning of an analysis of nature. Explanation of a phenomenon is to be sought not in its origin but in its immanent law. Knowledge of the laws and of the inner constitution of things must be far advanced before one may hope to understand or hypothetically to reconstruct their genesis. For want of this knowledge the speculations on pedigrees and phylogeny let loose by Darwinism in the last decades of the nineteenth century were mostly premature. Even today, after all the new and great revelations of genetics, our knowledge of facts and laws does not by a long shot suffice to explain either ontogenetic development or phylogenetic evolution. Without such groundwork as Newton's gravitational law, hypotheses about the origin of the planets would have been futile. Only on the basis of the spectroscopic investigation of stars and modern atomic physics, and only after well-founded opinions about the spatial order of the stellar universe had been derived by analysis of vast observational material could the astronomers undertake to draw a picture, first of the inner constitution and then also of the temporal development, of stars. Cosmogony still remains a rather problematic enterprise.

In our survey of the formation of concepts and theories by science (Sections 20–21) we saw how causal analysis proper is preceded by ordering and classification. Perhaps more stress should have been laid on this preliminary stage, that still plays a major role in biology while it has become of subordinate importance in physics. The spectacle of the immense variety of plant and animal species displayed by nature has been an early and persistent stimulus for biology to develop to great perfection the art of morphological and taxonomic classification. The remarkable fact that the diverse species, notwithstanding their range of variation, mostly exhibit clearly recognizable typical differences, has facilitated the task. The typical may be elusive in terms of well-defined concepts, and yet we handle it with instinctive certitude, e.g. in recognizing persons. Nor is it easy to describe in general terms how the process of classification, step by step and ever more convincingly, succeeds in separating essential from unessential features. In an individual animal or plant the several organs are distinguished by form, structure, function. Plato finds no better way to illustrate his diaeretic process of continued division than through the biological simile of the dissection of a sacrificial animal. By form, structure, and function, comparative morphology determines the homology of organs. Whatever the logical foundations, in the end no zoologist doubts that whales belong among the mammals and not the fishes (see, however, Ishmael's and the Nantucket whalemen's contrary view in Melville's *Moby Dick*, Chap. XXXII) or that certain cranial bones of the higher groups of vertebrates are homologous to the gill arches of fish. The purely morphological classification of the forms of crystals observed in nature is an analogue in the field of physics; here we can go to the end of the road where description gives way to dynamical theory. Much labor had to be spent on ordering the material of spectroscopy before the quantum theory of spectra and atoms could come into existence.

Classification looks for forms of distinct stamp, for regularities of arrangement in space and of sequence in time, for a permanent structure as the "ruhende Pol in der Erscheinungen Flucht." Under the title of *morphé* we shall now briefly pass in review the primary structures encountered by science in its search for order and law.

2. Let us begin with the *absolute constants of nature*. Today's physical theories can claim to provide a radical understanding for two of them, the velocity of light, c, and Planck's action quantum, h. According to relativity theory there is but one arbitrary standard unit that enters into the measurement of distances in the four-dimensional world; the velocity of light thus measured as the ratio of two equal distances (one 'spatial,' one 'temporal'), turns out to be 1. The universal connection $U = h\nu$ established by quantum theory between energy U and frequency ν (and also between momentum and wave number) suggests measuring energy directly by frequency. Indeed what the basic operator L of the dynamical law [Appendix C, formula (2)] represents may be described either as frequency or as energy. Thereby the value of h also reduces to unity.

More difficulties are caused by the elementary electrical charge e. All known kinds of elementary particles — as far as they are not neutral, like the photon and the neutron — have, except for the sign, the same charge e. Thus we cannot but accept this value e as a constant of nature no less fundamental than c and h. Measured in terms of the 'natural' units that make c and h equal to unity, the electrical repulsion of two electrons obeying Coulomb's inverse square law with respect to their variable distance r assumes the value α/r^2. The constant factor α (the square of e) is a pure dimensionless number equaling approximately $\frac{1}{137}$. A complete theory ought to account for this value by mathematical reasons — just as geometry predicts the value $\pi = 3.1415 \ldots$ of the ratio between circumference and diameter of a circle. Whatever Eddington may have thought, no such theory is available today.

The first step would be to comprehend how it happens that the same particle with its definite charge and mass (the same 'entity' in the terminology of Appendix B) occurs in the world in a large number of copies. Classical physics derives the conservation of charge and mass from a tendency of perseverance, but permits bodies of arbitrary charge and mass to exist. This viewpoint is unsatisfactory as far as the fixed charges and masses of elementary particles are concerned. Their conservation must depend on *adjustment* rather than on *perseverance*. The direction of the axis of a rotating top (e.g. the position of Earth's axis) is indeed transferred from moment to moment by means of a tendency of perseverance or inertia — we have called it the inertial field — whereas the direction of a magnetic needle is determined by adjustment to the magnetic field. If conservation of a quantity depends on inertia then its initial value may be chosen arbitrarily; but since perturbation can never entirely be eliminated, deviations are apt to occur in the course of time. Adjustment however enforces a definite value that is independent of past history and hence reasserts itself after any disturbances and any lapse of time as soon as the old conditions are restored. The rigid rods and the clocks by which Einstein measures the fundamental quantity ds^2 of his metric theory of the gravitational field preserve their length and period in the last instance because charge e and mass m of the composing elementary particles are preserved. The systematic theory, however, proceeds in the opposite direction; it starts with a metric ground form and thus introduces a primitive field quantity to which the Compton wave length m^{-1} of the particle adjusts itself in a definite proportion. (Again we apply the natural units in which c and h equal unity.) The behavior of rods and clocks comes out as a remote consequence of the fully developed theory. (In certain hypothetical generalizations of Einstein's theory of gravitation, in Weyl's metrical and Eddington-Einstein's affine field theories, this field quantity appears in the disguise of the radius of curvature of the universe and is derived from more primitive field quantities; but that does not essentially alter the basic relationship just described.)

G. Mie and others have tried to modify Maxwell's equations of the electromagnetic field in such a way that they possess only *one* or at most a small number of static spherically symmetric solutions. Had one succeeded, then adjustment would have been explained in the framework of classical field physics. But so far this idea has led nowhere. Quantum theory on the other hand solves the riddle, at least to a certain extent, by the quantization of field equations, a process by which one passes indeed from one to an indeterminate number of equal particles. Thus equality is accounted for, yet the particular values of charge and mass remain as unexplained as before.

If anything in nature has the right to be considered a "simple and eternal mode of being," it is the electron, the best known among the elementary particles. Already the proton betrays signs of complex structure.

An electron, a proton, and a meson all have different masses. Thus the *mass* of elementary particles seems to be of a less primitive and universal nature than their charge. Charge is related to the electric field in the same manner as mass is related to the gravitational field. Measured in the natural units, the gravitational attraction of two electrons amounts to ϵ/r^2, where the pure number ϵ has a value of about 10^{-41}. This is still more mysterious than the factor α. Indeed a simple mathematical theory may lead to numbers like $\frac{1}{2}$ or 8π, but hardly to a non-dimensional number of the extravagant order of magnitude 10^{41}. Explaining the red shift of the spectral lines of spiral nebulae by means of the cosmological term in Einstein's equations, one arrives at a world radius of the order of magnitude 10^{27} cm. In a spatially closed world of this dimension a mass of 10^{55} grams, distributed uniformly throughout space, would be in static equilibrium. This amount of matter is in reasonable agreement with estimates of the density of the cosmic clouds as inferred from observation. Dividing by the mass of the electron (and thus by implication ignoring the other kinds of particles), one finds that the number N of particles present in the world amounts to about 10^{81}. Thus the mysterious numerical factor $\epsilon \approx 10^{-41}$ seems to be connected with this number N, which may well be accepted as accidental, by a relation like $\epsilon \approx 1/\sqrt{N}$. If this be taken seriously it would indicate that the gravitational attraction of two particles depends on the total mass of the universe! The idea is not as strange as it may first sound. Long ago E. Mach tried to interpret the inertial mass of a body as an inductive effect of the other masses of the universe. Einstein's theory of gravitation does not satisfy Mach's postulate, though historically the latter played a role in its conception. A theory that meets Mach's challenge remains a desideratum (would it be a statistical theory of gravitation, as the square root in the law $\epsilon \approx 1/\sqrt{N}$ seems to indicate?). For the moment we can say no more than that the construction of the world seems to be based on two pure numbers, α and ϵ, whose mystery we have not yet penetrated.

Atoms are compounds of elementary particles. A neutral atom consists of a highly stable nucleus around which a number Z of electrons revolve. The laws for the motion of this outer shell of electrons and for the related energy levels have been completely elucidated by quantum theory; the structure of the nucleus, on the other hand, is not yet understood to the same degree. An atom remains the same atom as long as its nucleus stays unaltered; it changes its state as an electron is raised to a higher or drops to a lower energy level. Because they are composed of protons and neutrons, nuclei of the same charge may differ in mass. The fact that the chemical characteristics of elements are essentially determined by the charge of their nuclei has long stood in the way of separating such isotopes. Transmutations of atoms into one another occur spontaneously (radioactive disintegration) or are brought about by artificial means; they require far greater energies than the changes of state. The several possible kinds of atoms are 'predestined' by the natural laws, from which they can be deduced by purely mathematical means. They are configurations as eternal and primordial as the elementary particles themselves; there is no trace of an evolution from simpler to more complicated atoms. The fact that the lines of hydrogen prevail in the spectrum of Sirius and iron lines in that of the Sun does not point to a development of chemical elements underlying the development of stars but is readily accounted for by the higher surface temperature of Sirius. What holds for atoms holds in principle also for *molecules;* their composition and their conditions of existence are fixed once for all by the universal laws of nature. But the probability of a jump from one constellation to another, for instance from one oxygen and two hydrogen molecules into two water molecules, depends on temperature; these probabilities determine the velocities of reactions. Thus molecules unstable at high temperatures may become much more stable at lower temperatures; and in this way the cooling off of the Earth may enable molecules not found at an early phase of Earth's history to exist at a later phase. Since genes are highly complex molecules, this viewpoint has some bearing on the origin and evolution of life on earth.

In the *crystals* we encounter impressive macroscopic structures that are obviously governed by simple harmonious laws. The symmetry of a crystal is exhibited not only by its external shape but by all its physical characteristics. Suppose that the crystalline substance fills the entire space. Its macroscopic symmetry finds its expression in a group g of rotations; only such orientations of the crystal in space are physically indistinguishable as are carried into each other by a rotation of this group. For example, light, which in general propagates with different velocities in different directions in the crystalline medium, will propagate with the same speed in any two directions that arise from each other by a rotation of the group g. So for all other physical

290

properties. For an isotropic medium the group g consists of all rotations, but for a crystal it is made up of a finite number of rotations only, sometimes even of nothing but the identity. Early in the history of crystallography the law of rational indices was derived from the arrangement of the plane surfaces of crystals. It led to the hypothesis of the lattice-like atomic structure of crystals. This hypothesis, which explains the law of rational indices, has now been definitely confirmed by the Laue interference patterns, that are essentially X-ray photographs of crystals. Thus we know that the atoms of a crystallized chemical element form a regular set of points S, i.e. a set lying in the same manner around each of its points. More precisely, in the group G of motions that carry all points of S into points of S there is always a motion carrying one point of S into an arbitrarily chosen other such point. Any motion in point space induces a rotation in vector space. In this manner the group of motions G induces the group g of rotations in vector space, $g = \{G\}$. G describes the hidden atomic morphé (Gestalt), g the manifest macroscopic spatial and physical morphé of the crystal. For g there are only 32 distinct possibilities while the possible G's form a continuous web.

It is possible to decompose the atomic symmetry as described by the groups G into a purely discontinuous and a purely continuous component. As this duality, discrete versus continuous, is of fundamental importance for all morphological investigations and is obviously closely connected with the distinction between 'nature' and 'orientation' that we tried to characterize on p. 87, and also with the general contrast between fixed internal constitution and variable external conditions, it seems worth while to describe the group-theoretic analysis of the structure of crystals in some detail.

{The translations contained in G form a group L that may be generated from three independent translations $\vec{e}_1$, $\vec{e}_2$, $\vec{e}_3$. Translations are vectors and thus L is the parallelepipedic lattice of vectors consisting of all the vectors $x_1\vec{e}_1 + x_2\vec{e}_2 + x_3\vec{e}_3$ with integral coordinates (x_1, x_2, x_3) relative to the basis $e = (\vec{e}_1, \vec{e}_2, \vec{e}_3)$. Transition from one lattice basis e to another e^* and the opposite transition $e^* \to e$ is effected by two reciprocal linear substitutions U, U^{-1} of the variables x_1, x_2, x_3 with integral coefficients (unimodular substitutions). The group g of rotations leaves the vector lattice L invariant. When expressed in terms of the 'lattice-adapted' affine coordinates it appears as a group $\mathfrak{g}$ of homogeneous linear substitutions with integral coefficients, and G appears as a group $\mathfrak{G}$ of non-homogeneous linear substitutions of the point coordinates x_1, x_2, x_3 that induces $\mathfrak{g}$ in the vector space, $\mathfrak{g} = \{\mathfrak{G}\}$. In view of the arbitrariness involved in the choice of the lattice basis e, two groups $\mathfrak{g}$ arising from each other by a unimodular transformation U are to be considered equivalent. In this sense there are exactly 70 unimodularly inequivalent possibilities for $\mathfrak{g}$ (while there were 32 'orthogonally inequivalent' possibilities for g), and, what is the decisive fact, also only a finite number, namely 230, of possibilities for $\mathfrak{G}$. But something must still be capable of continuous variation! Indeed, in stressing the lattice structure of the crystal we lost sight of the metric structure of space; that has to be corrected. The metric ground form of the space, i.e. the square of the length of an arbitrary vector (x_1, x_2, x_3) is a positive-definite quadratic form of its coordinates x_1, x_2, x_3 with respect to the lattice basis e, a form that is left invariant by all the substitutions of the group $\mathfrak{g}$. Now the positive quadratic forms left invariant by the substitutions of $\mathfrak{g}$ form a continuous pencil (of very elementary nature), and the metric ground form is one individual in this continuous manifold. Hence the symmetry of a crystal is finally described by one discontinuous feature, namely one among the 230 groups $\mathfrak{G}$, and one element to be picked from a continuous manifold, namely from the pencil of positive quadratic forms invariant under $\mathfrak{g} = \{\mathfrak{G}\}.\}$

The crystals actually occurring in nature display the possible types of symmetry in an abundance of different forms that are influenced by the prevailing external circumstances. Think of the marvelous decorative patterns of snow crystals, of twin formations and the like! The morphological laws are today understood in terms of atomic dynamics: if equal atoms exert forces upon each other that make possible a definite stable state of equilibrium for the atomic ensemble, then the atoms will of necessity arrange themselves in a regular system of points in our strict sense. The nature of the atoms composing the crystal determines, under given external conditions, their metric disposition, for which the purely morphological investigation had still left a continuous range of possibilities. The dynamics of the crystal lattice is also responsible for the crystal's physical behavior, in particular for the manner of its growth, and this in turn determines the peculiar shape it assumes under the influence of the environmental factors.

The visible characters of physical objects usually are the resultants of constitution and environment. Whether water, whose molecule has a definite chemical constitution, is solid, liquid, or vaporous depends on temperature. The examples of crystallography, chemistry, and genetics cause one to suspect that this duality is in some way bound up with the distinction between discrete and continuous. Here is one tentative suggestion. For a character like the symmetry group $\mathfrak{G}$ of a crystal that by its very nature (in agreement with an adopted theory) is capable of discrete values only, a specific one among these values is constitutive, whereas for a character with a continuous range, such as the character 'metric compatible with the given group $\mathfrak{G}$,' merely the range (here the pencil of all positive quadratic forms invariant under $\mathfrak{g} = \{\mathfrak{G}\}$) is constitutive. An individual stationary quantum state with its energy level is a good constitutional element (in spite of quantum jumps due to interaction); not so a wave state or, more generally, a statistical ensemble of wave states. Temperature is the environmental factor *kat' exochen*. I think that this whole problem is in need of epistemological clarification.

With greater abandon than anywhere else has Nature given free course to her *nisus formativus* in the realm of *living organisms*. Because of the colloidal state of organic matter, the forms are here less rigid than in the inorganic world, but they combine flexibility with an astonishing tenacity in maintaining their basic organization in the face of disturbances. The scheme of inner permanent constitution versus changeable external conditions, both of which factors influence appearance and development, has stood the test here as well as in the realm of molecules and crystals, and the science of genetics has given us a clear picture of the inner genetic constitution of organisms.

Finally, the *stellar world* shows us organization of matter on the largest scale. The fact most exigent of explanation is the conglomeration of matter into individual luminous stars whose density exceeds the average spatial density of matter about 10^{27} times. With relatively few exceptions all these stars have approximately the same mass, the variations being hardly greater than the differences in size among human beings of all age groups. Stars form large clusters separated from each other by wide interstices; one such group of enormous dimensions, of which our Sun is a member, is the galactic system. A certain part of matter is scattered through space in the form of rarefied gaseous nebulae and clouds. Classification of the heavenly objects is based above all on the nature of the light they emit. With the help of atomic physics the spectroscopical findings yield detailed information about the interior of the stars. One central question, that of the source from which the luminous stars restore the energy lost by radiation, no longer embarrasses us; we have witnessed man's success in harnessing the energy released by atomic transmutations for his own destructive ends. Our information on the interior of stars and the chemical analysis of meteorites seems to indicate that everywhere throughout the universe the various chemical elements are mixed in much the same proportion. This is an argument for the common origin of all stars.

The statement that the natural laws are at the bottom, not only of the more or less permanent structures occurring in nature, but also of all processes of temporal development, must be qualified by the remark that chance factors are never missing in a concrete development. Classical physics considers the initial state as accidental. Thus 'common origin' may serve to explain features that do not follow from the laws of nature alone. Statistical thermodynamics combined with quantum physics grants chance a much wider scope but shows at the same time how chance is by no means incompatible with 'almost' perfect macroscopic regularity of phenomena. Evolution is not the foundation but the keystone in the edifice of scientific knowledge. Cosmogony deals with the evolution of the universe, geology with that of the earth and its minerals, paleontology and phylogenetics with the evolution of living organisms.

3. As his external features betray a person's age, so are the spectral lines emitted by stars clues to their stage in life, and we have thus been enabled to write with some authenticity the 'life' of a typical star. James Jeans in our day put forward a cosmogonic theory based on observation and exact computations that traces the evolution from a slow rotating gas ball over a spiral nebula to a cluster of stars like the galaxy. A century earlier Laplace had advanced his hypothesis about the birth and development of the planetary system; the fact that all planets circle around the Sun in the same direction in nearly coinciding planes points very clearly to a common origin. Lemaître has recently ventured still further back in the history of the universe than did Jeans. The decisive factor in his cosmogony is the expansive force as expressed by the cosmological term in Einstein's equations of gravitation. Under the numerical conditions assumed by Lemaître, gravitational attraction almost balances the expansion, so that at a certain precarious phase of evolution minute local variations of density give rise to accumulative condensations. He surmises that the world has its origin in the radioactive disintegration of a single giant atom. There is certainly much that is hypothetical and preliminary in such cosmogonies; to mention but one point, deeper insight into the basic nature of gravitation will very likely result in radical modifications. But in view of all the achievements of astrophysics, it can hardly be doubted that the chosen approach is fundamentally right, that one has to appeal to atomic physics in order to explain the inner constitution of the stars and the evolution of the stellar system.

Among the three inferred evolutions mentioned above, that of the Earth is the least hypothetical. The empirical evidence by which the reconstruction of Earth's past history is supported is by far the strongest, and the physical interpretation of the relevant geological processes is nowhere beset by difficulties of a principal character.

A profound mystery, however, is the *evolution of life* on earth. The idea of organic evolution (conceived long before Darwin) was raised by Darwinism to the rank of the most dominant scientific issue of our times. Does it really deserve this position? The nineteenth century, so blindly addicted to the gospel of progress, welcomed with open arms a doctrine according to which the general trend of development is in the direction from simple to complex, from lower to higher forms. More important, however, than the question of man's ancestors in the animal kingdom, seems to me another fact revealed by biology, namely the deep inner affinity of all living beings. Building from cells, the basic structure of cells and the basic cellular processes, such as metabolism, chromosome duplication, cell-division, also the processes of meiosis and fertilization underlying procreation, all these are the same in man, animal, and plant. As far as the fundamental features of his organization are concerned, man is one with all other living creatures. In the conceit and pride of his singularity, he has always found it easier to believe with Genesis I:1, that he is created "in God's own image" and has dominance over the Earth, than to bow his head under this recognition of the deep community of all life. "Ehrfurcht vor dem Leben" in each and every form (Albert Schweitzer), "love and reverence to all things that God made and loveth" (Coleridge, *The Rime of the Ancient Mariner*, last gloss) is its ethical implication, St. Anthony's sermon to the fishes a moving religious expression of it.

We deem the atoms and molecules occurring in nature satisfactorily explained if we derive them as possible structures from the natural laws holding at all times. Why then our appeal to evolution when we come to the gene aggregates that as zygotes determine the characteristic forms, structures, and functions of plants and animals and their organs? Individual ontogenesis from a fertilized egg to a highly differentiated organism still has its analogies in inorganic nature, remote though they may be, e.g. the formation of the crest of a surf wave or the geological formation of a serrated mountain range. The question proper concerns phylogenesis. In attempting to answer it, let us first repeat a previous remark to the effect that the stability of molecules depends on temperature and other environmental conditions and that a molecule will in general be the more labile the more complex it is. Hence it is not surprising that the molecules characteristic of living matter, in particular the genes, are tied to a definite epoch of the history of the earth and that their emergence is closely connected with the earth's own evolution. This remark, however, is clearly insufficient to explain the mighty drama of organic evolution that led from tiny blobs of jelly-like material to the highest animals with their wonderfully adaptive sense organs. The decisive point is perhaps this: when one deals with complex molecules consisting of something like a million atoms, the manifold of possible atomic combinations is immensely larger than those actually occurring in nature. Such combinations as are capable of functioning as genes are extremely rare and can only be 'found' by a selective process that by probing many possibilities and using previously conquered positions as bases for further advance, slowly gropes its way from simple to more complicated structures. But this formulation of the problem does not give more than the vaguest hint for its solution.

The evidence for organic evolution is furnished by paleontology, embryology, comparative anatomy, and genetics. Paleontological evidence is the most direct, but sporadic. Embryology gives direct information only about ontogenesis. True, the fact for example, that the embryos of man and fish have similar gill slits, suggests the assumption of common ancestors — although a more cautious interpretation may be content with the inference that similar genotypes manifest themselves in phenotypes bearing a closer resemblance to each other in their earlier than in their later stages of development. T. H. Morgan has occasionally pointed out how the ontogenesis of antlers in deer mocks phylogenesis. In any case Haeckel's conception of ontogenesis as an abbreviated repetition of phylogenesis (biogenetic law) stands on very weak foundations. The affinity between species disclosed and evaluated by comparative anatomy need not be interpreted as consanguinity — as little as the systematics of chemical compounds found in the handbooks of inorganic and organic chemistry reflect an historical development. But an evolutionary interpretation is suggested when one realizes that large parts of the system display a 'tree-like' iterated ramification in one direction.

The evidence of genetics was not yet available to Darwin. His doctrine rests on the fact that a common hereditary equipment does not exclude chance variations or variations caused by changeable surroundings. Since the number of individuals from one generation to the next fluctuates but slightly in spite of the production of an abundance of offspring, a struggle for survival must take place among the variously endowed individuals. Assuming the variations to be inheritable, Darwin concludes that this natural selection will have a cumulative effect through the generations and act constantly to maintain and improve the adjustment of animals and plants to their surroundings and their way of life.[1] One of the main pillars of the

[1] The lengthening of the neck of a giraffe, trying to "get at the tender leaves high up on the tree" is the classical example widely discussed at the time.

theory, the heredity of phenotypical characters, including characters acquired by use during the individual development (Lamarck), was overthrown by Johannsen's experiments with pure lines. The modification of phenotype brought about by a change of environment disappears with the return to the original environment.

4. But here the modern science of genetics has provided a substitute; recombination and mutation have taken the place of Darwin's inheritable continuous chance variations. Mendel and his successors disclosed the combinatorial game that meiosis and syngamy play with genetic constitution. From a strain that is not completely uniform (is not a pure line) segregation and recombination will produce new inheritable variations. The sexual process in the life-cycle of organisms is of momentous aid in the emergence of ever new forms. The germ plasm of both parents is not blended; the constituent genes preserve their nature unmingled. In one regard this 'particulate inheritance' carried by genes serves the ends of evolution even better than blending inheritance as assumed by Darwin. Blending would drive a population which exhibits a large variability at the start rapidly and irretrievably toward greater and greater homogeneity unless new variations are constantly developed at a rate altogether incompatible with the frequency of observed mutations; the more extreme variants would quickly disappear by swamping.

Segregation and recombination account for the unceasing variety in nature, but they alone could not explain evolution if genes and gene structures were not allowed to mutate. It has previously been described (Appendix B) under what circumstances species (in a strict combinatorial sense) are conserved. Wherever these conditions are violated new species come into being. Among the structure mutations that alter the number of genes, the duplication of the entire chromosome outfit of a cell plays a particularly important role (polyploidy). Diploid zygotes will give rise to tetraploids, sometimes by hybridization (allopolyploidy); sometimes when under the influence of chemical agents, or a thermic shock, or for some other reason maturating cells fail to carry out meiosis (autopolyploidy). It may be expected, moreover, that genes not only change their mutant states but that one gene changes into another (just as atoms not only change their quantum states but are transmuted into each other), or new genes originate. A natural tetraploid contains four homologues of each kind of chromosomes, and yet one observes not infrequently that only two of its points seem to be occupied by the same gene — as if in the sequence of tetraploid generations one of the two equal gene pairs had changed into a different one. It is perhaps not entirely excluded that directed adaptive mutations occur, e.g. that in response to an

297

adaptive phenotypical chance alteration the rate of such mutations as would support that alteration tends to increase. The thesis of the non-inheritance of acquired characters is based on such solid and diversified evidence that one hesitates to tamper with it, and yet experiments with unicellular organisms propagating by asexual division show that it has its limits. Continued action of a chemical agent on one hundred generations of infusoria, after first attacking the cell plasm, then the macro-nucleus, ultimately brings about a change of inner constitution localized in the micro-nucleus; this change is inherited, to be sure, not through an indefinite time but with gradual slackening effect, over perhaps the next sixty generations.

Even when we know the origin of the constantly appearing inheritable variations, we still have to explain the extraordinary fact that the result is not a chaos of interfertile types but that a definite sorting out of variants into isolated races, species, and higher taxonomic categories takes place. Indeed the process of evolution has two aspects, since it involves the development of diversity as well as that of discontinuity in the living world (Dobzhanski). One will first have to investigate how newly created mutations fare in spreading through the successive generations of a population under definite assumptions about viability and mating. Mutants will differ in their adaptive value with respect to a given environment, and thus natural selection will do its work. The genes appear to be organized into genotypic systems such that the adaptive value attaches to the entire systems rather than to the constituent genes. The sorting out of races and biological species would be impossible without isolating factors, of which we mention only the two most important: infertility except between gametes of the same or very similar species makes for genetic isolation, difference in habitats for geographical isolation.

The ideas of progress and retrogression have no necessary connection with evolution, but if evolution is viewed under these aspects it is certainly not always progressive. The line from autotrophic forms that require only inorganic compounds and light (green algae and certain bacteria), over various steps of dependence on preformed specific organic molecules, to the ultimate in parasitic specialization that is reached in a virus, a naked gene synthesizing copies of itself from the living protoplasm of its host, this line may well be the picture of a retrogressive evolution that actually took place in nature.

Our present knowledge about the laws of inheritance is certainly not inconsistent with the doctrine of evolution, and it has scored some success in the interpretation of geographic and ecological speciation on a small scale. But — if a layman is entitled to an opinion on these matters — we are still far from a genetic explanation of evolution in the large and its most conspicuous features. The findings of paleontology seem to indicate that organic evolution has followed definite directions continuously over extended periods. Such 'orthogenetic' trends may be ascribable to the nature of protoplasm itself. One school holds that they are largely unadaptive and that "adaptive sequences in evolution are superposed on the great orthogenetic trends but are entirely independent of them" (R. F. Griggs); others, like H. F. Osborn, find that new structures often arise in response to a future functional need and therefore speak of secular adaptation. A classical example is provided by the cones on the teeth of various vertebrates that first arise as almost imperceptible prominences and at that stage of evolution are of very doubtful functional use, while later they develop into conspicuous features of obvious functional importance.

Consideration of evolution in the large will of necessity lead to the question of the origin of life. The evidence of genetics makes one incline to see in life the chance success of a play of creative accidents. Not some predictable macrophysical or macrochemical process that with a certain natural necessity came to pass at a certain stage of evolution and would repeat itself wherever the appropriate conditions prevailed seems responsible for the historic beginnings of life, but a molecular event of singular character occurring once by accident and then starting an avalanche by autocatalytic multiplication (P. Jordan). Jordan adduces as a strong argument for this opinion "the fact that all the more complicated molecules found in plants and animals, especially the protein molecules, are stereochemically different from their mirror images." Indeed had they an independent origin at many places and many times their levo- and dextro-varieties should show nearly the same abundance. Thus it looks as if there is some truth in the story of Adam and Eve, if not for the origin of mankind then for that of the most primitive forms of life.

Oparin has suggested that the first self-duplicating units ("protogenes") arose after a great variety of organic molecules had accumulated in a long pre-life evolutionary period during which they had a much better chance to survive than under present-day conditions, because of the absence of living organisms with their catalytic enzymes. The chain of reactions found necessary for a gene to reproduce may have been built up by mutations in retrograde steps, each step conferring a new selective advantage. In the remote past genes may have existed and reproduced in a free state. Whatever the merits of such hypotheses, when we try to imagine what mechanisms of search, presumably without direction and mneme, could have built up within geological times the enormously complicated structures of genes, all the complicated biochemical processes that condition the life of animals, and such marvels of adaptive differentiation as the human eye, then we cannot help realizing how speculative the whole field still is.

The temptation of an interpretation in terms of an overall plan of evolution is almost irresistible. One of the theses at which J. C. Willis arrives in his book *The Course of Evolution* (1940) asserts, "The process of evolution appears not to be a matter of natural selection of chance variations of adaptational value. Rather it is working upon some definite law that we do not yet comprehend. The law probably began its operations with the commencement of life, and it is carrying this on according to some definite plan."

Even an author so scornful of all teleology as Julian Huxley feels himself forced to admit, "The biological process culminating for the evolutionary moment in the dominance of Homo sapiens . . . could apparently have pursued no other general course than that which it has historically followed: or, if it be impossible to uphold such a sweeping and universal negative, we may at least say that among the actual inhabitants of the earth, past and present, no other lines could have been taken which would have produced speech and conceptual thought, the features that form the basis for man's biological dominance." (*Evolution*, p. 569.)

Whether or not the view is tenable that the organizing power of life establishes correlations between independent individual atomic processes, there is no doubt that wherever *thought* and the causative agent of *will* emerge, especially in man, that power is increasingly controlled by a purely spiritual world of images (knowledge, ideas). Is it conceivable that immaterial factors having the nature of images, ideas, 'building plans,' also intervene in the evolution of the living world as a whole? Some biologists answer in the affirmative and set out to describe these factors and their workings in closer detail. Henri Bergson has developed his philosophy of *évolution créatrice*, and it is essentially the same doctrine that, amidst all the fireworks of his wit, G. B. Shaw propounds in his play *Back to Methuselah*. The eternal life of pure thought freed from the bondage of matter is, according to this "metabiological Pentateuch," the ultimate goal of evolution. Thus speaks Lilith: "I am Lilith: I brought Life into the whirlpool of force, and compelled my enemy, Matter, to obey a living soul. But in enslaving Life's enemy I made Life's master; for that is the end of all slavery; and now I shall see the slave set free and the enemy reconciled, the whirlpool become all life and no matter." Scientists in general will be more cautious. As things stand now, the positing of transcendental creative agents possessing the nature of ideas, whether philosophically dangerous or desirable, is of no help in solving the actual concrete problems of biology. And it is a fact that (unless Lilith tells us) we know nothing of them, at least not in the same manner in which, by interpretative understanding from within, we know of the thoughts and impulses of ourselves and our fellow beings.

REFERENCES

P. Niggli, *Geometrische Kristallographie des Diskontinuums*, Leipzig, 1920.

F. Rinne, *Das feinbauliche Wesen der Materie*, Berlin, 1922.

A. Naef, *Idealistische Morphologie und Phylogenetik*, Jena, 1919.

D'Arcy Wentworth Thompson, *On Growth and Form*, Cambridge, 1917.

A. S. Eddington, *New Pathways in Science*, Cambridge, 1935.

—— *The Expanding Universe*, Cambridge, 1933.

—— *Fundamental Theory*, Cambridge, 1946.

H. P. Robertson, Relativistic Cosmology, *Rev. of Modern Physics*, 5 (1933), pp. 62–90.

J. H. Jeans, *Astronomy and Cosmogony*, Cambridge, 1928.

G. Lemaître, L'hypothèse de l'atome primitif, *Actes Soc. Helv. Sci. Nat.*, Fribourg, 1945, pp. 77–96.

H. S. Jennings, *Genetic Variations in Relation to Evolution*, Princeton, 1935.

R. A. Fisher, *The Genetical Theory of Natural Selection*, Oxford, 1930.

J. C. Willis, *The Course of Evolution* (by differentiation or divergent mutation rather than by selection), Cambridge, 1940.

Th. Dobzhanski, *Genetics and the Origin of Species*, second ed., New York, 1941.

A. I. Oparin, *The Origin of Life*, New York, 1938.

H. Bergson, *L'évolution créatrice*, Paris.

G. B. Shaw, *Back to Methuselah;* see also C. E. M. Joad, J. D. Bernal, W. R. Inge on Shaw's philosophy, Shaw the Scientist and Shaw as a Theologian, in: S. Winsten, *G. B. S. 90*, London, 1946.

Lecomte du Noüy, *Human Destiny*, New York, 1947.

M. Scheler, *Die Stellung des Menschen im Kosmos*, Darmstadt, 1930.

Charles Sherrington, *Man on his Nature*, Cambridge, 1940.

A. Portmann, *Biologische Fragmente zu einer Lehre vom Menschen*, Basel, 1944.

Index

aberration, 115

absolute constants of nature, 287 *et seq.;* — space and time, 99; — temperature, 140

abstraction, originary and mathematical, 11; definition by— 11

accuracy of measurements, 141; is a limiting idea, 143

ACKERMANN, W., 60

active future, 102

actuality and potentiality, 46; *see also* possible

additive quantities, 140

adjustment and perseverance, 288

affine geometry, 69, 123

affinity of all living beings, 295

aggregate, 239

aitemata, 19

ALEXANDER APHRODISIENSIS, 228

ALEXEJEFF, W., 274

all, 6, 14, 46, 51

ALVERDES, 214

ambiguity of truth, 153

AMPÈRE, 169

analogy, 161, 163

analysis of nature vs. description, 145

analytic geometry, 21; — propositions, 13, 18

ANAXAGORAS, 41

and, in classical logic, 5; in quantum logic, 257

ANGELUS POLITIANUS, 229

antisymmetry and exclusion principle, 262

apriori character of arithmetics, 63; of geometry, 132, 135; of the causal law, 192

arbitrariness of initial state, 191

ARCHIMEDES, 19, 39, 159

ARCHYTAS, 132

area, 28

ARISTOTLE, 54, 63, 96, 129, 132, 136, 149, 150, 158, 177, 178, 220, 285

arithmetical axioms, 33; — models of geometries, 32

arithmetization of theories, 227

ars combinatoria, 27, 62, 237 *et seq.*

ASTON, 186

atom, as center of force, 169; the modern conception, 184 *et seq., especially* 187; atomic theory, how it is built

up, 161; atomism and its history, 166; atomic energy levels, 247

Ausdehnungslehre, 68

automorphism, 72; —s of the physical world, 82

AVENARIUS, 122

axiomatic method, 18 *et seq.*

axioms, their consistency, 20; independence, 20; completeness, 24; categoricity, 25; special axioms: of arithmetic, 33, of geometry, 19, 69, of formalized mathematics, 57 *et seq.,* 221, of logic, 15, 57, 58, 221; axiom of choice, 232, of parallels, 19, of reducibility, 50

BACON, 132, 145

BAIN, A, 125, 129

Baralipton, Barbara, 13

basic relations, 6

BEADLE, G. W., 281

BECKER, O., 91, 129, 137

Being and Becoming, 46

BERGMANN, M., 212

BERGSON, 212, 300

BERKELEY, 112, 129

BERNAYS, P., 17, 60, 231

BERNOULLI, JACOB, 34, 195, 196

BERNOULLI, JOHANN, 44

BERTALANFFY, 214

BERZELIUS, 169, 212

binocular vision, 128

biogenetic law, 296

blank, 3

BOHR, NIELS, 119, 185, 187, 188, 206, 264, 282

BOLTZMANN, 203, 204, 251; Boltzmann's constant k, 208, 251

BOLYAI, 19, 67

BOLZANO, 48, 137

BORN, M., 173

BOSCOVICH, 169

Bose-Einstein statistics, 240

BRAHE, TYCHO, 157, 158

Brahma, 285

BRENTANO, F., 131

DE BROGLIE, 246

BROUWER, L. E. J., 50, 51, 52, 61, 65, 234

BRUNO, G., 43, 98, 108, 159

BÜCHNER, G., 112

BUFFON, 286

303